Road Vehicles Surroundings Supervision

Road Vehicles Surroundings Supervision

On-Board Sensors and Communications

Special Issue Editor

Felipe Jiménez

MDPI • Basel • Beijing • Wuhan • Barcelona • Belgrade

MDPI

Special Issue Editor
Felipe Jiménez
Universidad Politecnica de Madrid
Spain

Editorial Office
MDPI
St. Alban-Anlage 66
4052 Basel, Switzerland

This is a reprint of articles from the Special Issue published online in the open access journal *Applied Sciences* (ISSN 2076-3417) from 2017 to 2018 (available at: http://www.mdpi.com/journal/applsci/special_issues/Surroundings_Supervision)

For citation purposes, cite each article independently as indicated on the article page online and as indicated below:

LastName, A.A.; LastName, B.B.; LastName, C.C. Article Title. *Journal Name* **Year**, *Article Number*, Page Range.

ISBN 978-3-03897-568-7 (Pbk)
ISBN 978-3-03897-569-4 (PDF)

Contents

About the Special Issue Editor

Felipe Jiménez, has a PhD in Mechanical Engineering, is Full Professor at the Universidad Politecnica de Madrid and Head of the Intelligent Systems Unit of the University Institute for Automobile Research. His fields of interest include the automotive industry, vehicle safety, mechanical design, driver assistance systems, intelligent transport systems and connected and autonomous driving. He has been involved in several national and European research projects and has developed engineering studies for relevant companies. He is the author of several chapters, books and papers in pertinent scientific journals and has participated in several national and international congresses.

applied
sciences

MDPI

Editorial

Road Vehicles Surroundings Supervision: Onboard Sensors and Communications

Felipe Jiménez [ORCID]

University Institute for Automobile Research, Universidad Politécnica de Madrid (UPM), 28031 Madrid, Spain; felipe.jimenez@upm.es; Tel.: +34-91-336-53-17

Received: 25 June 2018; Accepted: 9 July 2018; Published: 11 July 2018

Abstract: This Special Issue covers some of the most relevant topics related to road vehicle surroundings supervision, providing an overview of technologies and algorithms that are currently under research and deployment. This supervision is essential for the new applications in current vehicles oriented to connected and autonomous driving. The first part deals with specific technologies or solutions, including onboard sensors, communications, driver supervision, and traffic analysis, and the second one presents applications or architectures for autonomous vehicles (or parts of them).

Keywords: vehicle surroundings supervision; vehicle surroundings reconstruction vehicle positioning; autonomous driving; connected vehicles; sensors; communications

---

1. Introduction

New assistance systems, cooperative services and autonomous driving of road vehicles imply an accurate knowledge of vehicle surroundings. This knowledge can come from different sources, such as onboard sensors, sensors in the infrastructure, and communications.

Among onboard sensors, short- and long-range sensors can be distinguished. In the first case, ultrasonic, infrared, and capacitive sensors can be cited. Among the second group, laser scanners and computer vision technologies appear to provide the best performance, although there are many others that can complement the information using data fusion processes. In any case, the goal is to have a representation of vehicle surroundings that is as complete as possible. Sensor fusion algorithms are a common solution to overcome the sensors' individual limitations.

Vehicle positioning is also essential. For this purpose, satellite positioning is commonly used, but when it is not sufficiently reliable or the signal is lost, the same technologies as those used for the recognition of the surroundings can provide an acceptable solution. In this regard, the SLAM (Simultaneous Localization and Mapping) problem should be noted, which tries to build or update the map of an environment that is not known a priori, and position the vehicle on that map simultaneously. This technique, widely used and proven in robotics, has also been implemented in the vehicular field for the perception of the environment in real time, for support and accuracy improvements in autonomous global navigation satellite systems (GNSS) navigation, and for generation of digital maps or calculation of trajectories followed when no GPS (Global Positioning System) signal is received.

Vehicle-to-vehicle (V2V) communications and vehicle-to-infrastructure (V2I) communications allow the vehicle to have greater knowledge of surroundings and to obtain information that is far away from the onboard sensors. These communications provide additional data that could be used for driver information purposes or for decision modules in autonomous driving, for example. In this sense, connected and cooperative driving appears as a catalyst of autonomous driving, because it enables real deployment under complex driving situations.

2. The Present Issue

This Special Issue consists of 11 papers covering some of the most relevant topics related to road vehicle surroundings supervision, providing an overview of technologies and algorithms that are currently under research and deployment.

These papers could be divided in two main groups. The first deals with specific technologies or solutions [1–7] and the second presents applications or architectures for complete autonomous vehicles (or parts of them) [8–11]. Furthermore, the first group presents a quite wide vision of research fields that could be involved such as onboard sensors [1–3], communications [4], driver supervision [5], and traffic analysis [6,7].

Onboard perception systems provide knowledge of the surroundings of the vehicle, and some algorithms have been proposed to detect road boundaries and lane lines. This information could be used to locate the vehicle in the lane. However, most proposed algorithms are quite partial and do not take advantage of a complete knowledge of the road section. An integrated approach to the two tasks is proposed in [1] that provides a higher level of robustness of results for road boundary detection and lane line detection. Furthermore, the algorithm is not restricted to certain scenarios such as the detection of curbs; it could be also used in off-road tracks.

The next paper [2] also considers that global navigation satellite systems (GNSS), to achieve the necessary performance, must be combined with other technologies into a common onboard sensor set that allows the cost to be kept low and which features the GNSS unit, odometry, and inertial sensors. However, odometers do not behave properly when friction conditions make the tires slide. The authors introduce a hybridization approach that takes into consideration the sliding situations by means of a multiple model particle filter (MMPF). Tests with real datasets show the goodness of the proposal.

Considering that monitoring systems for intelligent vehicles that employ remote sensors, such as cameras and radar, require the tracking of moving objects, adaptive tracking techniques are commonly used for this purpose. To this end, a gain design strategy to compose an optimal α-β-η-θ filter is proposed [3].

Vehicular communications, both between separate vehicles and between vehicles and infrastructure can be seen as an extension of the knowledge a vehicle can obtain from the surroundings. VANETs (Vehicular Ad-hoc Networks) are an emerging offshoot of MANETs (Mobile Ad-hoc Networks) with highly mobile nodes. They are envisioned to play a vital role in providing safety communications and commercial applications to the on-road public. Establishing an optimal route for vehicles to send packets to their respective destinations in VANETs is challenging because of the quick speed of vehicles, dynamic nature of the network, and intermittent connectivity among nodes. A novel position-based routing technique called Dynamic Multiple Junction Selection based Routing (DMJSR) is proposed for the city environment [4].

Even in intelligent vehicles and sometimes for the design of assistance systems, driver's intention classification and identification is identified as the key technology. To study driver's steering intention under different typical operating conditions, five driving school coaches of different ages and genders were selected as the test drivers for a real vehicle test [5].

Furthermore, the knowledge of traffic flow and its modelling is relevant information that could be taken into account for decision-making in many intelligent systems. A localized space–time autoregressive (LSTAR) model is proposed and a new parameter estimation method is formulated based on the Localized Space–Time ARIMA -autoregressive integrated moving average- (LSTARIMA) model to reduce computational complexity for real-time traffic prediction purposes [6]. A roundabout traffic model based on cellular automata for computer simulation that takes into account various sizes of roundabouts, as well as various types and maximum speeds of vehicles, is presented [7].

The four remaining papers that are included deal with autonomous vehicles. An adaptive global fast sliding mode control (AGFSMC) for Steer-by-wire system vehicles with unknown steering parameters is proposed [8]. Due to the robust nature of the proposed scheme, it can not only handle the tire–road variation, but also intelligently adapts to the different driving conditions and ensures that the tracking error and the sliding surface converge asymptotically to zero in a finite time.

Appl. Sci. **2018**, *8*, 1125

A Driving Decision-making Mechanism (DDM) is formulated [9] in order to take into account the road conditions and their coupled effects on driving decisions. The results demonstrate the significant improvement in the performance of DDM with added road conditions. They also show that road conditions have the greatest influence on driving decisions at low traffic density. Among them, the most influential is road visibility, followed by adhesion coefficient, road curvature, and road slope; at high traffic density, they have almost no influence on driving decisions.

Finally, two papers include autonomous vehicle architectures. An open and modular architecture is proposed [10], capable of easily integrating a wide variety of sensors and actuators which can be used for testing algorithms and control strategies and including a reliable and complete navigation application for a commercial vehicle.

The last experimental platform [11] consists of a platform for research on the automatic control of an articulated bus, and the paper also focuses on the development of a human–machine interface to ease progress in control system evaluation.

Funding: This research received no external funding.

Acknowledgments: First of all, I would like to thank all researchers who submitted articles to this Special Issue for their excellent contributions. I am also grateful to all reviewers who helped in the evaluation of the manuscripts and made very valuable suggestions to improve the quality of contributions. I am also grateful to the Applied Sciences Editorial Office staff who worked thoroughly to maintain the rigorous peer review schedule and timely publication.

Conflicts of Interest: The author declares no conflict of interest.

References

1. Jiménez, F.; Clavijo, M.; Castellanos, F.; Álvarez, C. Accurate and Detailed Transversal Road Section Characteristics Extraction Using Laser Scanner. *Appl. Sci.* **2018**, *8*, 724. [CrossRef]
2. Toledo-Moreo, R.; Colodro-Conde, C.; Toledo-Moreo, J. A Multiple-Model Particle Filter Fusion Algorithm for GNSS/DR Slide Error Detection and Compensation. *Appl. Sci.* **2018**, *8*, 445. [CrossRef]
3. Saho, K.; Masugi, M. Performance Analysis and Design Strategy for a Second-Order, Fixed-Gain, Position-Velocity-Measured (α-β-η-θ) Tracking Filter. *Appl. Sci.* **2017**, *7*, 758. [CrossRef]
4. Abbasi, I.A.; Khan, A.S.; Ali, S. Dynamic Multiple Junction Selection Based Routing Protocol for VANETs in City Environment. *Appl. Sci.* **2018**, *8*, 687. [CrossRef]
5. Hua, Y.; Jiang, H.; Tian, H.; Xu, X.; Chen, L. A Comparative Study of Clustering Analysis Method for Driver's Steering Intention Classification and Identification under Different Typical Conditions. *Appl. Sci.* **2017**, *7*, 1014. [CrossRef]
6. Chen, J.; Li, D.; Zhang, G.; Zhang, X. Localized Space-Time Autoregressive Parameters Estimation for Traffic Flow Prediction in Urban Road Networks. *Appl. Sci.* **2018**, *8*, 277. [CrossRef]
7. Małecki, K.; Wątróbski, J. Cellular Automaton to Study the Impact of Changes in Traffic Rules in a Roundabout: A Preliminary Approach. *Appl. Sci.* **2017**, *7*, 742. [CrossRef]
8. Iqbal, J.; Zuhaib, K.M.; Han, C.; Khan, A.M.; Ali, M.A. Adaptive Global Fast Sliding Mode Control for Steer-by-Wire System Road Vehicles. *Appl. Sci.* **2017**, *7*, 738. [CrossRef]
9. Zhang, J.; Liao, Y.; Wang, S.; Han, J. Study on Driving Decision-Making Mechanism of Autonomous Vehicle Based on an Optimized Support Vector Machine Regression. *Appl. Sci.* **2018**, *8*, 13. [CrossRef]
10. Borraz, R.; Navarro, P.J.; Fernández, C.; Alcover, P.M. Cloud Incubator Car: A Reliable Platform for Autonomous Driving. *Appl. Sci.* **2018**, *8*, 303. [CrossRef]
11. Montes, H.; Salinas, C.; Fernández, R.; Armada, M. An Experimental Platform for Autonomous Bus Development. *Appl. Sci.* **2017**, *7*, 1131. [CrossRef]

applied
sciences

MDPI

Article

Accurate and Detailed Transversal Road Section Characteristics Extraction Using Laser Scanner

Felipe Jiménez * [iD]**, Miguel Clavijo** [iD]**, Fernando Castellanos and Carlos Álvarez**

University Institute for Automobile Research (INSIA), Universidad Politécnica de Madrid (UPM);
28031 Madrid, Spain; miguel.clavijo@upm.es (M.C.); f.castellanos@alumnos.upm.es (F.C.);
c.alomas@alumnos.upm.es (C.Á.)
* Correspondence: felipe.jimenez@upm.es; Tel.: +34-91-336-53-17

Received: 15 February 2018; Accepted: 20 April 2018; Published: 5 May 2018

Abstract: Road vehicle lateral positioning is a key aspect of many assistance applications and autonomous driving. However, conventional GNSS-based positioning systems and fusion with inertial systems are not able to achieve these levels of accuracy under real traffic conditions. Onboard perception systems provide knowledge of the surroundings of the vehicle, and some algorithms have been proposed to detect road boundaries and lane lines. This information could be used to locate the vehicle in the lane. However, most proposed algorithms are quite partial and do not take advantage of a complete knowledge of the road section. This paper proposes an integrated approach to the two tasks that provides a higher level of robustness of results: road boundaries detection and lane lines detection. Furthermore, the algorithm is not restricted to certain scenarios such as the detection of curbs; it could be also used in off-road tracks. The functions have been tested in real environments and their capabilities for autonomous driving have been verified. The algorithm is ready to be merged with digital map information; this development would improve results accuracy.

Keywords: positioning; road boundary; curb; lane line; laser scanner; algorithm; accuracy

---

1. Introduction

New driver assistance systems (ADAS) require a more precise and detailed knowledge of the vehicle environment and its positioning [1]. These requirements are even greater when autonomous driving functions are introduced where, in many cases, knowledge of "where in the lane" is required [2]. However, conventional GNSS (Global Navigation Satellite System)-based positioning systems are not able to achieve these levels of accuracy under real traffic conditions, even when receivers that accept DGPS (Differential Global Positioning Systems) are available, since centimeter accuracy levels cannot be guaranteed along a whole route [3], much less so, in urban areas [4].

A solution that has been used for many years is the fusion of GNSS information with inertial systems (e.g., [5,6]). These inertial sensors have the disadvantage of drift along time, that causes a cumulative error that makes its use unviable over very large distances unless corrections are introduced by means of another positioning source. However, in areas where GNSS signal loss or deterioration is short, this fusion provides adequate results.

However, problems reappear in situations when the GNSS signal is zero, or bad over long distances, such as in tunnels or in urban areas of narrow streets. In such cases, even navigation systems that do not require high levels of precision do not work reliably [7]. Particularly critical examples are deviations inside tunnels.

Another solution for positioning that has been explored is the use of smartphones [8–10]. However, accuracy is far below what is required for many assistance systems, and autonomous driving and coverage problems could be common in certain scenarios.

The incorporation of sensors in vehicles for the perception of the environment for safety functions and autonomous driving (such as LiDAR—e.g., [11–13], computer vision—e.g., [14,15], radar—e.g., [16], or sensor fusion of some of them—e.g., [17,18]) has opened another field to improve positioning. Thus, different methods of visual odometry and SLAM (Simultaneous Localization and Mapping), based on computer vision and laser scanners, have emerged in recent years (e.g., [19–22]).

In this way, the sensors used for obstacle detection are also used as primary or secondary sensors for positioning and road mapping [23,24]. Thus, lane keeping systems are based, generally, on computer vision systems, and base their correct operation on the perception of the lines that delimit the lanes. However, the detection of lane lines presents deficiencies in complex scenarios with dense traffic, where other vehicles cover these marks, or due to the limitations of the sensors themselves, such as the malfunctioning under lighting changes in the case of computer vision [25], or for errors in the detection of badly maintained lines, areas with complex patterns on the road (merging lanes, exits, intersections, change in the number of lanes, or confusion with areas on the roadway due to maintenance operations). In addition, visual odometry methods also introduce cumulative errors in the calculation of trajectories, and SLAM algorithms require significant computational calculations when high precisions are required [26].

In this paper, a 3D laser scanner is used for transversal road characteristics detection, taking advantage of the use of this equipment for other purposes such as safety critical assistance systems or for autonomous driving. In this sense, there are solutions that seek to partially solve this problem. Such is the case of [27,28], that proposes a method for the detection of the curvature of the road with a 3D laser scanner, and introduces the use of the robust regression method—named least trimmed squares—to deal with occluding scenes. The curb detector is also used as an input to a Monte Carlo localization algorithm, which is capable of estimating the position of the vehicle without an accurate GPS sensor. Also, in [29] an approach for laser-based curb detection is presented. Other approaches to achieving higher levels of robustness in the detection of road lines are raised in [30]. Similarly, with similar equipment, [31] proposes algorithms for the detection of road boundaries.

The authors of [32] propose a system that is based on a formulation of stereo with homography as a maximum a posteriori (MAP) problem in a Markov random field (MRF). This solution, that uses computer vision, provides quite robust results even in complex scenarios. In [33], the free road surface ahead of the ego-vehicle using an onboard camera is detected. The main contribution is the use of a shadow-invariant feature space combined with a model-based classifier. The model is built online to improve the adaptability of the algorithm to the current lighting, and to the presence of other vehicles in the scene.

In other cases, the fusion of laser scanners and computer vision is used, as in [34], where the algorithm is based on clouds of 3D points and is evaluated using a 3D information from a pair of stereo cameras, and also from the laser scanner. To obtain a dense point cloud, the scanner cloud has been increased using Iterative Closest Point (ICP) with the previous scans.

Algorithms proposed in this paper are oriented for the positioning of the vehicle in the road with centimetric precision, even when the signals obtained from GNSS or inertial systems are not appropriate. The solution is based on the use of 3D laser scanner, and the algorithms pursue the detection of the boundaries of the road and the lanes, identifying the lane through which it circulates, and the lateral position (and, therefore, the identification of lane change maneuvers).

Unlike previous approaches, the algorithm proposed in this paper tries to increase the robustness of the results with a low computational cost (much faster than real time), using only laser scanner data and a digital map when available, even in complex scenarios, and to go through the detection of road boundaries and, subsequently, the detection of lanes in an integrated way, taking advantage of both results to increase global reliability and to improve partial approaches. The algorithm also accept the input of digital map information in order to improve results, but this additional information is not strictly necessary. Furthermore, the algorithms should work under different scenarios, rural and urban, considering that the road boundaries and the configurations of the lane lines could vary.

2. Algorithm

In order to locate the vehicle laterally on the road it is necessary firstly to know the transversal characteristics of the road. To do this, the algorithm shown in Figure 1 is proposed. The algorithm begins by obtaining the raw data from the laser scanner. From that set of points, the function of road boundaries detection that offers as output the width of the roadway is implemented. Two methods are considered in parallel. The same set of points serves as input data for the lane lines detection function, where the lane through which the ego-vehicle moves is distinguished from the rest, since it is expected that a greater density of points will be available for fitting. This second function offers lane widths as output. We proceed to evaluating whether the width obtained for each lane is included in a preset range, with which the shoulder can be differentiated from the other lanes, as well as checking if a line has been lost due to the lack of points of it; this implies the generation of a new estimated line between those already calculated. In the cases of both the lanes and the road boundaries calculation, when the functions do not offer satisfactory results, the information of the previous scan is used. Finally, the possibility of having a digital map is contemplated, which can be used to corroborate the number of lanes obtained (and use this feedback for the selection in the lane identification function) and classify them, distinguishing the lanes of the main roadway from the additional lanes as the merging or exit ones.

In the following subsections, the main functions of lane detection are described in detail.

Finally, after the road characteristics extraction process, it is simple to determine the lane through which the vehicle moves, its lateral position in, and if it should perform a lane change maneuvers.

2.1. Road Boundaries Detection Functions

The determination of the limits of the road is of interest for two main reasons. On the one hand, this calculation allows delimiting the area of interest for subsequent operations, such as the detection of obstacles [35], for example. On the other hand, in some roads, especially in urban areas, external lanes are not delimited by lines on one side, but their boundary is the curb that separates the road from the sidewalk.

Initially, two methods are proposed that make use of the geometric characteristics that suppose the end of the roadway.

○ Method I: Study of the variations in the detection of each layer of the laser scanner.

The intersection of each layer of the laser scanner with the plane of the road causes a roughly circular or elliptical section, depending on the relative inclinations, and may become other conics such as hyperbolas or parabolas, although these special cases are not relevant for this paper. The presence of the sidewalk at a height which is greater than the roadway involves cuts at different heights, so that, at the position of the curb, gradients occur at the Y (lateral) and Z (vertical) coordinates, which are significantly greater than those detected on the roadway (Figure 2).

In this way, laser scanner spots are studied in both dimensions, and the points where both gradient levels are higher than preset thresholds are detected. These thresholds have been adjusted using practical data from several scenarios, firstly using ideally simulated environments with typical dimensions, and then with practical data. Subsequently, the DBSCAN clustering function is applied in order to group all the points belonging to the curb and eliminate potential false detections, for example, because of vertical elements in the vicinity of the road. This process provides a set of points that distinguish the roadway from the surroundings because of the abrupt changes detected in curves detected by each layer of the laser scanner.

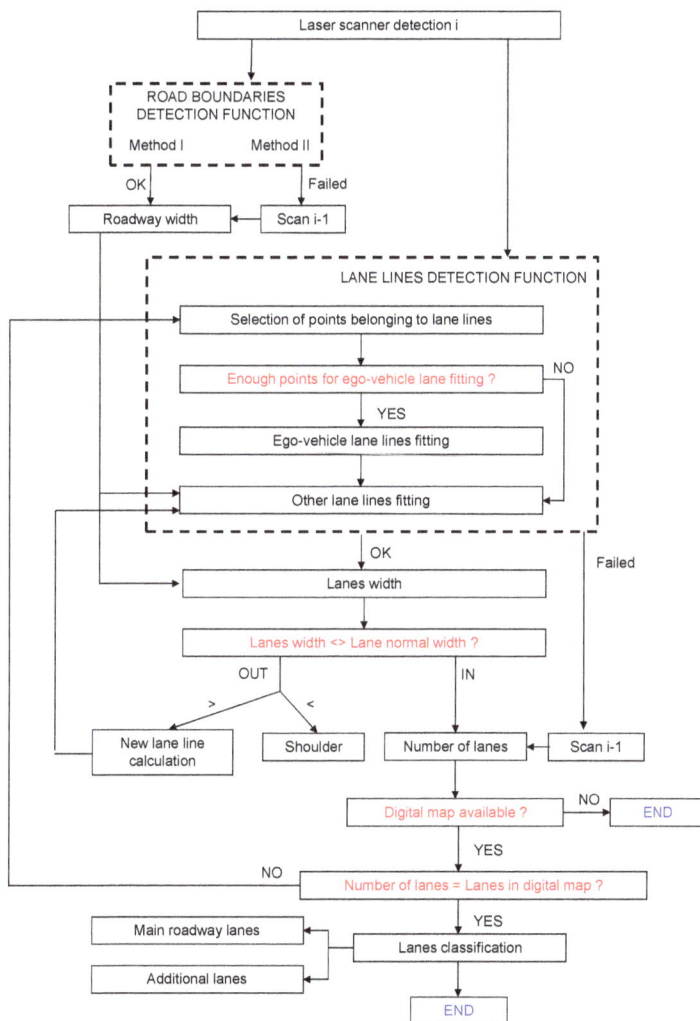

Figure 1. Flowchart for the extraction of the transversal characteristics of the road.

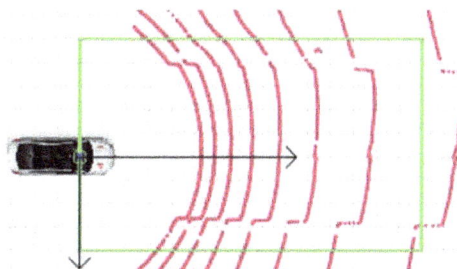

Figure 2. Illustration for Method I for the detection of roadway boundaries.

Then, in order to eliminate other false points that do not belong to the road, we proceed to adjust the points set to a plane using the M-estimator algorithm SAmple Consensus (MSAC), a variant of the RANdom algorithm SAmple Consensus (RAMSAC).

○ Method II: Study of the separation between intersecting sections of consecutive laser scanner layers

This method is based on [27]. Considering the intersections of the layers of the laser scanner with the ground surface, the radius of the circumferences (assuming perpendicularity between the vertical axis of the laser scanner and the ground) depends on the height, so considering that the roadway is not in the same vertical dimension as the adjacent zones, a gradient in the radius of said intersection curves can be expected. It should be noted that both methods are based on the same concept (heights differences), but they provide quite different results, as practical tests demonstrate.

Figure 3 shows an example of the pavement-sidewalk transition and how the radius difference Δr_i of the detection circles between two consecutive layers varies. Although the curb could be identified as the zones of greater gradients, it has been shown that there are several cases where this solution induces errors in the detection of them, because of the presence of other obstacles near the curbs or the path boundaries; this fact inhibits the identification of those boundaries (there are multiple candidates). The solution proposed in the method is to try to identify areas with constant radius differences within a predefined tolerance, which allows determining the roadway area.

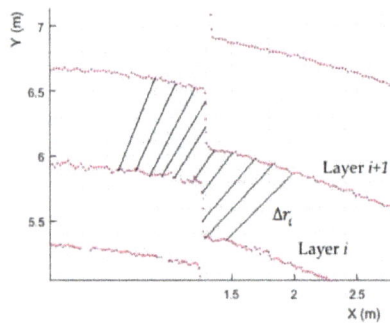

Figure 3. Example of evolution of radius difference between 2 consecutive laser scanner layers in the roadway-sidewalk transition.

Since it is possible to accept points on the road that do not correspond to it, two additional filters are applied on the set of points detected: a height filter with which the points of each layer that differ significantly from the average height of the rest are discarded from points identified as roadway (potentially), and DBSCAN clustering algorithm.

2.2. Lane Lines Detection Function

From this first detection of the roadway, we proceed to the determination of the number of lanes, the lane through which the vehicle moves, and the lateral position of the vehicle in the lane. This function can be divided into the following steps:

Step 1: On the area definition belonging to the roadway, it is possible to fit mathematical curves at each of the intersections of the layers of the laser scanner with the ground. The intersections of laser scanner layers and the ground form conics (in general, ellipses, parabolas and hyperbolas). Considering the common positioning of the scanner on vehicles, the intersection uses ellipses, so a least-square fitting method is applied to obtain the conical parameters. The points that fall away from this curve cannot be considered to belong to the roadway, but to obstacles on it, which allows them to be removed from the points set (Figure 4).

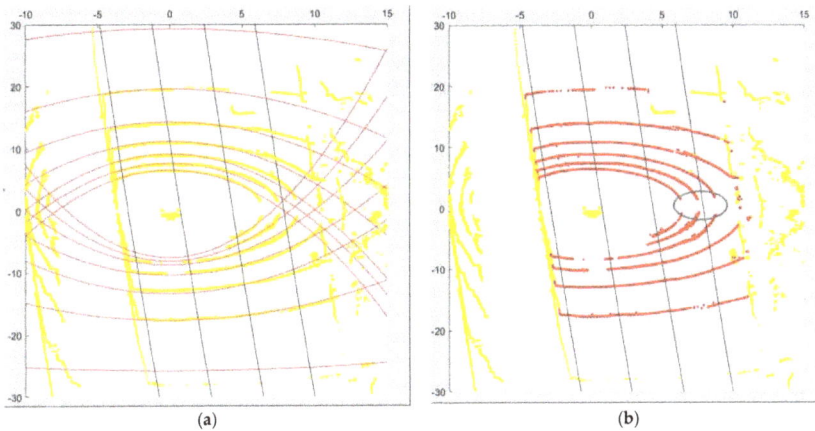

Figure 4. Example of curve fitting to points of intersection of the layers of the laser scanner and the ground. (**a**) Total set of points (yellow); (**b**) Filtered points (red).

Step 2: Then, on the filtered point set, those points that show a greater reflectivity are extracted, having checked experimentally the remarkable contrast in this variable between the asphalt and the lines that delimit the lanes.

Step 3: On these points, geometric considerations are applied in order to eliminate false signals due to irregularities on the asphalt with high reflectivity that do not correspond to the road markings of lane delimitation. In this way, it is considered that, for the same layer, the dispersion of valid contiguous points must be small in the longitudinal direction, and large in transverse, if they are detected in front or behind the vehicle, and vice versa if they are located on the sides, due to the shape of the intersection curve of the layer with the ground. Figure 5 shows the output of steps 2 and 3 on a scanner frame. In most cases, a filtering process for this last step is not necessary because step 1 provides quite a good set of selected points, so few false points are considered after Step 2.

Step 4: The points clustering belonging to the same lane delimitation is done using knowledge of the relative orientation of the road with the vehicle (extracted from the previous function of road boundaries detection), and regions of interest are established around each set of points that, by proximity criteria, are considered to correspond to the same line. These regions of interest maintain the longitudinal direction of the points set, and extend a distance such that, considering the regulations about horizontal road markings, they must intersect if they correspond to the same line (Figure 6a).

Step 5: When points that may belong to the delimitation of lanes are detected but there is not enough data to establish the orientation of the area of interest, we proceed to generating new areas with the average orientation of those areas in which the density of nearby points allowed for the calculation (green areas in Figure 6b). It must be taken into account that, with the knowledge of the expected lane width, it is possible to determine if those isolated points can be part of a lane line or not. Additionally, the lane where the ego-vehicle is usually provides more information, although this is not necessarily true, due to possible occlusions of obstacles on the roadway.

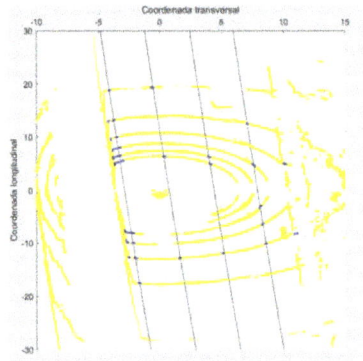

Figure 5. Definition of points belonging to lane lines.

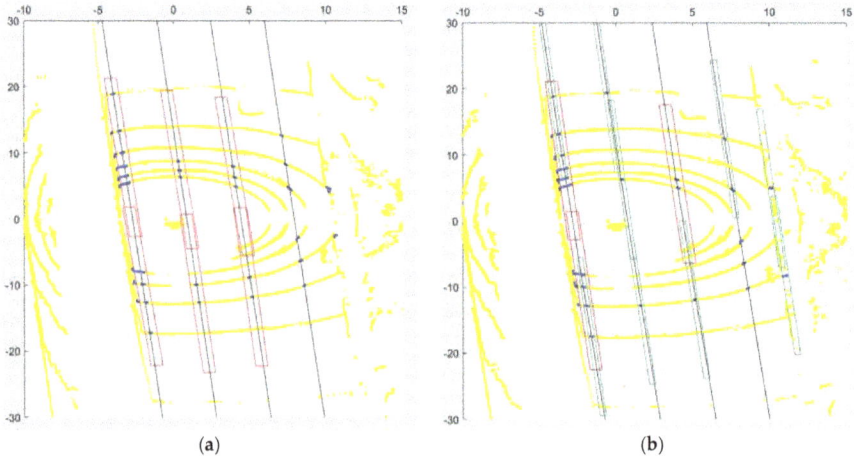

(a)

(b)

Figure 6. Definition of regions of interest for grouping line sections. (**a**) Case of having enough information of each line section (areas marked in red); (**b**) Case of not having enough information on each of the sections (areas amrked in green).

Step 6: Then, we proceed to adjust the sets of points corresponding to lane delimitation lines. To do this, we proceed to a fitting process using quadratic polynomials, since they are considered to approximate reality [30]. Additionally, conditions of parallelism between the lines are considered, so equality conditions are imposed on linear and quadratic terms of the polynomials, as shown in Equation (1) for the curve equation representing lane line i (only coefficient a_i varies from one line to another, but b and c are equal for all of them).

$$p_i(x) = a_i + bx + cx^2 \qquad (1)$$

The least-squares fitting method is applied to obtain the $N + 2$ different coefficients, where N is the number of lanes lines. Equation (2) shows the calculation of the quadratic error committed in the curve fitting of lane line i, considering that N_i points have been selected in previous Steps for this line and their coordinates are (x_{ij}, y_{ij}) with j ranging from 1 to N_i.

$$E_i = \sum_{j=1}^{N_i} (a_1 + bx_{ij} + cx_{iji}^2 - y_{ij})^2 \tag{2}$$

Finally, the total error is calculated. It has been verified that the introduction of weighting factors that give greater relevance to the curves with a greater number of points increases the robustness of the overall adjustment. In particular, these coefficients are considered equal to the number of points detected for each line, so the equation that should be minimized is written as follows, considering that N lines have been distinguished:

$$E_T = \sum_{i=1}^{N} N_i E_i \tag{3}$$

Minimization of Equation (3) provides values of polynomials coefficients a_i, b and c.

3. Results

In order to evaluate the results of the developed algorithm, road tests have been conducted in highway and urban areas. In them, a 16-layer Velodyne VLP-16 laser scanner was installed on top of a passenger car. Additionally, the trajectory was recorded by means of a Trimble R4 DGPS receiver, and the scenario was captured by a camera.

Firstly, the accuracy and robustness of the specific functions is assessed without the information from previous scans and information from digital map. Afterwards, the complete algorithm is considered to analyze whether misdetections of isolated functions could be solved in the proposed iterative loops.

In the first place, the differences of both methods for roadway delimitation are analyzed. Figure 7 shows an example of the use of the second method. Table 1 shows the comparison of the detection of curbs.

Detections at distances less or greater than 10 m are distinguished. This fact is relevant since the first layers acquired with the laser scanner provide a higher resolution and the first method presented above is especially sensitive to this fact. In this case, only in 28% of the frames is it possible to detect the curb more than 10 m in advance. In 67%, the detection is performed correctly, but at distances of less than 10 m, which greatly limits its applicability since the anticipation for any action or automatic control is very small. On the contrary, method II improves these results in a significant way, reaching a reliability index in the detection with a range greater than 10 m of 94% of the frames.

It should be noted that the first method is based on information from each laser layer, so the results are quite sensitive to any deviations that could appear in one layer. Other obstacles near the boundaries could lead to incorrect detections because they could be considered as the road boundaries following the method criteria. The second method relies on the information of two consecutive layers, so the impact of obstacles is reduced if it is only detected by one of the layers.

These results are obtained in real scenarios with traffic around the vehicle which causes occlusions of the curbs. Despite this, the degree of robustness detected for Method II is very high, and there are not a significant number of consecutive frames with erroneous detection, which would be quite limiting for autonomous driving applications. Furthermore, results take into account neither the information from previous scans nor a digital map.

Table 1. Comparison of roadway detection methods (analysis over 875 frames).

	Method I	Method II
Correct detection with a range greater than 10 m	28%	94%
Correct detection with a range lower than 10 m	67%	2%
Incorrect detection	5%	4%

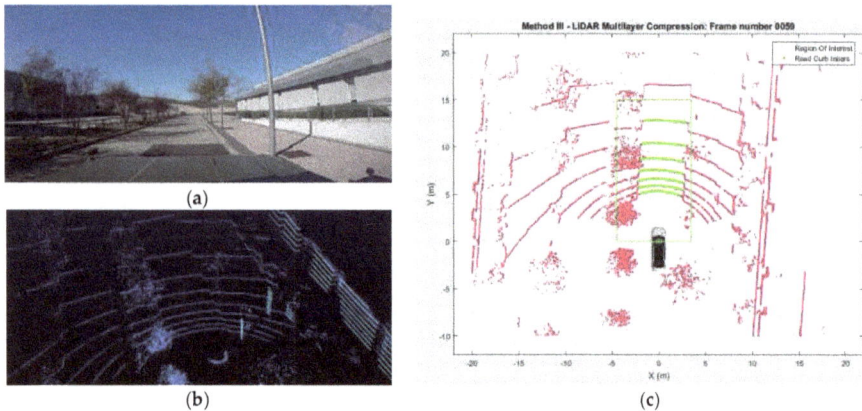

Figure 7. Roadway detection using Method II. (**a**) Scenario; (**b**) Tridimensional detection of the laser scanner; (**c**) Function results (points belonging to the passable path inside the region of interest marked in green).

Additionally, it should be noted that the roadway detection function (Method II) has also been tested in off-road areas in order to differentiate passable areas from those that may not be. With a small computational cost, it is observed that satisfactory results are obtained, as can be seen in Figure 8, where the two possible paths are seen as passable, and they differ from the rest of the paths defined as non-passable.

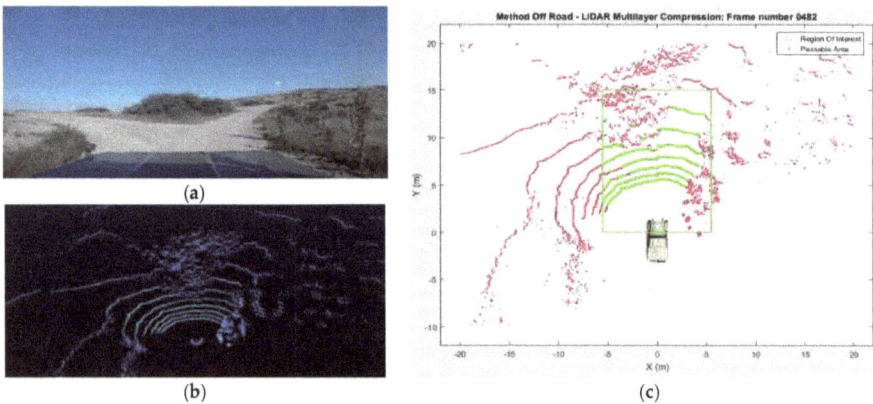

Figure 8. Detection of passable areas in an off-road environment. (**a**) Scenario; (**b**) Three-dimensional detection of the laser scanner; (**c**) Function results (points belonging to the passable path inside the region of interest marked in green).

Subsequently, the reliability of the function for determining the number of lanes has been checked. In this sense, it is necessary to distinguish the cases in which there are no singularities, such as merging lanes, exits or changes in the number of lanes, with scenarios in which one of those situations does occur. Table 2 shows the reliability of the function described in Section 3 in both scenarios. As can be seen, in areas without singularities, robustness is quite high even when the information from previous scan is not used (78% correct), while it is reduced in the case of road sections with singularities. Note that 47% accuracy is guaranteed, although this value is considered a lower threshold since it does not include those special cases where, for example, an additional lane appears as a merging or exit lane. In these cases, in the absence of a reference of a digital map, the function is not able to differentiate them from the lanes of the main roadway, so a conservative approach has been chosen not to include them in the correct detections, since they require additional information.

Table 2. Reliability in lanes detection.

	Scenario without Singularities (200 Frames)	Scenario with Singularities (700 Frames)
Correct detection	78%	>47% *
Calculation not performed	8%	18%
Incorrect detection	14%	<35% *

* Every case that involves a merging or exit lane detection is not considered as correct, because it is not possible to determine its nature, and it could be confused as an additional lane.

In this case, as in the case of the road boundaries detection, there are not many consecutive frames with incorrect detection, and these are mainly due to the occlusion of the lane lines by other vehicles. Furthermore, it should be noted that the reason for many incorrect detections or non-performed calculations is the presence of obstacles. But, as these perturbations do not remain for long periods of time, estimations of road boundaries and lane lines could be done using previous information.

The study of the most frequent situations of erroneous or unrealized calculations shows the following conclusions:

○ The shoulder can be identified as an additional lane, but has no influence on the detection of the other lanes after detection of the roadway boundaries in the first phase of the function (Figure 9a)

○ Due to the presence of obstacles on the road (mainly, other vehicles), occlusions of the lane lines occur and, in extreme cases, complete information is lost relative to any of the lines (Figure 9b,c).

  ○ In this case, the loss of information in the lane through which the vehicle moves is more critical, since the initial estimates of the function are based on it.

  ○ The loss of information from the farthest lanes is less relevant, and causes errors in lane counting but not in the position of the vehicle on the road.

  ○ The loss of information from an intermediate lane is usually solved approximately by the function when estimating standard lane widths in the absence of information.

○ The presence of additional lanes can be reliably detected, but, in the absence of other information, they are not distinguished with respect to the main lanes of the roadway, as indicated above (Figure 9d)

In the two final cases, it should be noted that the identification of the boundaries of the roadway allows us to obtain approximate estimates of the information of the undetected lanes.

Finally, it should be noted that previous results from both functions show only their performance in an isolated mode; interactions between them or use of previous scans as Figure 1 states have not been considered in results of Tables 1 and 2. Even in such circumstances, the road characteristics extraction partial functions provide quite accurate results in many cases. When using the information of previous scans, transversal road characteristics estimation is possible throughout the test. Furthermore, when using the information from a digital map, the lane lines and the number of lanes have been always properly identified. It should be noted that this number, and the lane widths, are parameters that imply interactions in the algorithm; however, some incorrect estimations could lead to convergence problems. These problems have not arisen, and this fact could be taken as a complete success rate (100%) in the tests performed in on-road scenarios.

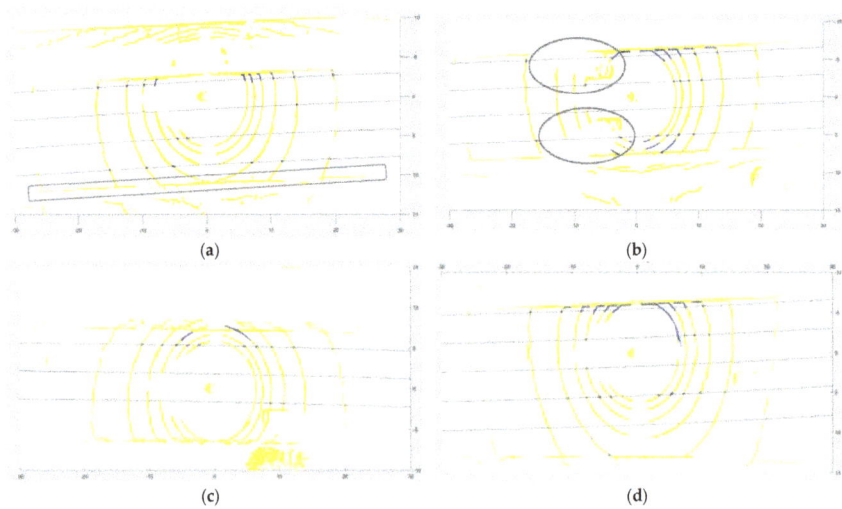

Figure 9. Examples of situations of erroneous or uncertain detections. (**a**) Road shoulder detection; (**b**) Occlusions; (**c**) Loss of information of any of the lane lines; (**d**) Merging lane detection.

4. Conclusions

In this paper, an algorithm has been presented that is able to characterize the transversal road section using 3D laser scanner, so that the boundaries of the road and the lines of separation of the lanes are identified, positioning the vehicle laterally in it, performing these functions in a coordinated manner to increase the reliability of the results. This positioning improves the accuracy that could be obtained with conventional means of GNSS or inertial systems when the loss or deterioration of satellite positioning extends over long stretches.

The tests carried out in real scenarios show the reliability of the detection of the road boundaries and the difficulties in estimating the number of lanes, especially in sections of the road where singularities, such as merging lanes, exits, crossings, etc. occur. In such situations, the reliability of the function falls notably, although it is estimated that final rates of correct detection could increase significantly with knowledge of the environment, for example, through the information of a digital map; therefore, the algorithms should contrast the information captured by the laser scanner with known information, instead of estimating information without any reference. However, results only show the operating mode without that external help of a digital map.

Additionally, the detected errors—although they are abundant in complex scenarios—do not appear in a continuous way, so reliable detections are mixed with other erroneous ones, but the first ones allow establishing a reference in positioning.

It should be noted that the proper functioning of the boundaries detection function in off road scenarios shows its robustness and versatility, which behaves satisfactorily in an environment for which it was initially not foreseen, since its calibration is oriented towards the transition between road and sidewalk. This function also correctly identifies the boundaries of the roadway with or without curbs, so it is not limited to the case of urban areas with these elements.

Despite the accurate results, the proposed approach could only be used for positioning in a practical way (from an economic point of view, and considering current state-of-the-art technology and costs) if the sensor is used for other purposes. This fact is not really a limitation when dealing with autonomous driving applications, because of the common current use of LiDAR technology, but this fact is not strictly the same when dealing with only assistance systems.

Finally, integration of computer vision could be used to enhance the extracted information, because of the capacity of managing other kinds of image properties (such as colors, contrast, etc.), and limitations of this technology could be mitigated by sensor fusion with the current laser scanners.

Author Contributions: F.J. conceived the complete algorithm, the applications for autonomous driving and supervised the results. M.C. performed the tests and helped in the algorithm approach. F.C. implemented and tested the lane lines detection function. C.Á. implemented and tested the roadway boundaries detection functions.

Acknowledgments: This work has received the support of the Spanish Ministerio de Economía y Competitividad MINECO (CAV project TRA2016-78886-C3-3-R), the Spanish Ministrio de Defensa (REMOTE DRIVE project 2015/SP03390102/00000035-10032/15/0029/00) and Madrid Region Excellence Network SEGVAUTO-TRIES.

Conflicts of Interest: The authors declare no conflict of interest.

References

1. Jiménez, F.; Naranjo, J.E.; García, F.; Zato, J.G.; Armingol, J.M.; de la Escalera, A.; Aparicio, F. Limitations of positioning systems for developing digital maps and locating vehicles according to the specifications of future driver assistance systems. *IET Intell. Transp. Syst.* **2011**, *5*, 60–69. [CrossRef]
2. Jiménez, F. Improvements in road geometry measurement using inertial measurement systems in datalog vehicles. *Measurement* **2011**, *44*, 102–112. [CrossRef]
3. Naranjo, J.E.; Jiménez, F.; Aparicio, F.; Zato, J. GPS and inertial systems for high precision positioning on motorways. *J. Navig.* **2009**, *62*, 351–363. [CrossRef]
4. Jiménez, F.; Monzón, S.; Naranjo, J.E. Definition of an Enhanced Map-Matching Algorithm for Urban Environments with Poor GNSS Signal Quality. *Sensors* **2016**, *16*, 193. [CrossRef] [PubMed]
5. Sasiadek, J.Z.; Hartana, P. GPS/INS Sensor Fusion for Accurate Positioning and Navigation Based on Kalman Filtering. *IFAC Proc. Vol.* **2004**, *37*, 115–120. [CrossRef]
6. Melendez-Pastor, C.; Ruiz-Gonzalez, R.; Gomez-Gil, J. A data fusion system of GNSS data and on-vehicle Sensors data for improving car positioning precision in urban environments. *Expert Syst. Appl.* **2017**, *80*, 28–38. [CrossRef]
7. Jiménez, F. The Role of Digital Road Maps in the Future. In *Highways: Construction, Management and Maintenance*; Nova Publishers: Hauppauge, NY, USA, 2010; ISBN 978-1617288623.
8. Zandbergen, P.A. Accuracy of iPhone locations: A comparison of assisted GPS, WiFi and cellular positioning. *Trans. GIS* **2009**, *13*, 5–25. [CrossRef]
9. Engelbrecht, J.; Booysen, M.J.; van Rooyen, G.J.; Bruwer, F.J. Survey of smartphone-based sensing in vehicles for intelligent transportation system applications. *IET Intell. Transp. Syst.* **2015**, *9*, 924–935. [CrossRef]
10. Wahlström, J.; Skog, I.; Händel, P. Smartphone-based Vehicle Telematics—A Ten-Year Anniversary. *IEEE Trans. Intell. Transp. Syst.* **2017**, *18*, 2802–2825. [CrossRef]
11. Asvadi, A.; Premebida, C.; Peixoto, P.; Nunes, U. 3D Lidar-based static and moving obstacle detection in driving environments: An approach based on voxels and multi-region ground planes. *Robot. Auton. Syst.* **2016**, *83*, 99–311. [CrossRef]
12. Guo, J.; Tsai, M.; Han, J. Automatic reconstruction of road surface features by using terrestrial mobile lidar. *Autom. Constr.* **2015**, *58*, 165–175. [CrossRef]
13. García, F.; Jiménez, F.; Naranjo, J.E.; Zato, J.G.; Aparicio, F.; Armingol, J.M.; de la Escalera, A. Environment perception based on LIDAR sensors for real road applications. *Robotica* **2011**, *30*, 185–193. [CrossRef]
14. Caraffi, C.; Cattani, S.; Grisleri, P. Off-road path and obstacle detection using decision networks and stereo vision. *IEEE Trans. Intell. Transp. Syst.* **2007**, *8*, 607–618. [CrossRef]
15. Hernández-Aceituno, J.; Acosta, L.; Piñeiro, J.D. Pedestrian Detection in Crowded Environments through Bayesian Prediction of Sequential Probability Matrices. *J. Sens.* **2016**, *2016*, 4697260. [CrossRef]
16. Tokoro, S.; Kuroda, K.; Nagao, T.; Kawasaki, T.; Yamamoto, T. Pre-crash sensor for pre-crash safety. In Proceedings of the International Conference on Enhanced Safety of Vehicles, Gothenburg, Sweden, 8–11 June 2015.
17. Zhang, X.; Zhang, A.; Meng, X. Automatic fusion of hyperspectral images and laser scans using feature points. *J. Sens.* **2015**, *2015*, 415361. [CrossRef]

18. Chavez-Garcia, R.O.; Aycard, O. Multiple Sensor Fusion and Classification for Moving Object Detection and Tracking. *IEEE Trans. Intell. Transp. Syst.* **2016**, *17*, 525–534. [CrossRef]
19. Demim, F.; Nemra, A.; Louadj, K. Robust SVSF-SLAM for Unmanned Vehicle in Unknown Environment. *IFAC-PapersOnLine* **2016**, *49*, 386–394. [CrossRef]
20. Pierzchała, M.; Giguère, P.; Astrup, R. Mapping forests using an unmanned ground vehicle with 3D LiDAR and graph-SLAM. *Comput. Electron. Agric.* **2018**, *145*, 217–225. [CrossRef]
21. Nicolai, A.; Skeele, R.; Eriksen, C.; Hollinger, G.A. *Deep Learning for Laser Based Odometry Estimation*; Automation and the Oregon Metal Initiative; University of Oregon: Eugene, OR, USA, 2016.
22. Zhang, J.; Singh, S. LOAM: Lidar Odometry and Mapping in Real-time. In Proceedings of the Robotics: Science and Systems Conference, Berkeley, CA, USA, 12–16 July 2014.
23. Castro, M.; Lopez-Cuervo, S.; Paréns-González, M.; de Santos-Berbel, C. LIDAR-based roadway and roadside modelling for sight distance studies. *Surv. Rev.* **2016**, *48*, 309–315. [CrossRef]
24. Holgado-Barco, A.; Riveiro, B.; González-Aguilera, D.; Arias, P. Automatic Inventory of Road Cross-Sections from Mobile Laser Scanning System. *Comput.-Aided Civ. Infrastruct. Eng.* **2017**, *32*, 3–17. [CrossRef]
25. Fuerstenberg, K.C.; Baraud, P.; Caporaletti, G.; Citelli, S.; Eitan, Z.; Lages, U.; Lavergne, C. Development of a pre-crash sensorial system—The CHAMELEON project. In Proceedings of the Joint VDI/VW Congress Vehicle Concepts for the 2nd Century of Automotive Technology, Wolfsburg, Germany, November 2001.
26. De la Escalera, A.; Izquierdo, E.; Martín, D.; Musleh, B.; García, F.; Armingol, J.M. Stereo visual odometry in urban environments based on detecting ground features. *Robot. Auton. Syst.* **2016**, *80*, 1–10. [CrossRef]
27. Hata, A.; Osorio, F.S.; Wolf, D. Curb Detection and Vehicle Localization in Urban Environments. In Proceedings of the IEEE Intelligent Vehicles Symposium, Dearborn, MI, USA, 8–11 June 2014.
28. Hata, A.; Wolf, D.F. Road Marking Detection Using LiDAR Reflective Intensity Data and its Application to Vehicle Localization. In Proceedings of the IEEE 17th International Conference on Intelligent Transportation Systems (ITSC), Qingdao, China, 8–11 October 2014.
29. Zhao, G.; Yuan, J. Curb Detection and Tracking Using 3D-Lidar Scanner. In Proceedings of the 19th IEEE International Conference on Image Processing (ICIP), Orlando, FL, USA, 30 September–3 October 2012.
30. Jimenez, F.; Clavijo, M.; Naranjo, J.E.; Gómez, O. Improving the lane reference detection for autonomous road vehicle control. *J. Sens.* **2016**, *2016*, 9497524. [CrossRef]
31. Zhang, W. Lidar-based road and road-edge detection. In Proceedings of the Intelligent Vehicles Symposium, 2010 IEEE Intelligent Vehicles Symposium, San Diego, CA, USA, 21–24 June 2010.
32. Guo, C.; Mita, S.; McAllester, D. Robust road detection and tracking in challenging scenarios based on markov random fields with unsupervised learning. *IEEE Trans. Intell. Transp. Syst.* **2012**, *13*, 1338–1354. [CrossRef]
33. Alvarez, J.; Lopez, A. Road detection based on illuminant invariance. *IEEE Trans. Intell. Transp. Syst.* **2011**, *12*, 184–193. [CrossRef]
34. Fernandez, C.; Llorca, D.F.; Stiller, C.; Sotelo, M.A. Curvature-based Curb Detection Method in Urban Environments using Stereo and Laser. In Proceedings of the 2015 IEEE Intelligent Vehicles Symposium (IV), Seoul, Korea, 28 June–1 July 2015.
35. Jiménez, F.; Naranjo, J.E. Improving the obstacle detection and identification algorithms of a laserscanner-based collision avoidance system. *Transp. Res. Part C Emerg. Technol.* **2011**, *19*, 658–672. [CrossRef]

applied
sciences

MDPI

Article

A Multiple-Model Particle Filter Fusion Algorithm for GNSS/DR Slide Error Detection and Compensation

Rafael Toledo-Moreo [1,*] , Carlos Colodro-Conde [2] and Javier Toledo-Moreo [1]

[1] Department of Electronics and Computer Technology, Universidad Politécnica de Cartagena,
    30202 Murcia, Spain; javier.toledo@upct.es
[2] Instituto de Astrofísica de Canarias, La Laguna, 38205 Tenerife, Spain; ccolodro@iac.es
* Correspondence: rafael.toledo@upct.es; Tel.: +34-968-325948

Received: 6 February 2018; Accepted: 13 March 2018; Published: 15 March 2018

Abstract: Continuous accurate positioning is a key element for the deployment of many advanced driver assistance systems (ADAS) and autonomous vehicle navigation. To achieve the necessary performance, global navigation satellite systems (GNSS) must be combined with other technologies. A common onboard sensor-set that allows keeping the cost low, features the GNSS unit, odometry, and inertial sensors, such as a gyro. Odometry and inertial sensors compensate for GNSS flaws in many situations and, in normal conditions, their errors can be easily characterized, thus making the whole solution not only more accurate but also with more integrity. However, odometers do not behave properly when friction conditions make the tires slide. If not properly considered, the positioning perception will not be sound. This article introduces a hybridization approach that takes into consideration the sliding situations by means of a multiple model particle filter (MMPF). Tests with real datasets show the goodness of the proposal.

Keywords: positioning; navigation; data fusion; odometry; particle filter; multiple-model filter

---

1. Introduction

Today's most outstanding technology for road vehicle positioning is global navigation satellite systems (GNSS). Among all GNSS, the US GPS (global positioning system) is the most popular. GNSS systems provide absolute positioning based on the information gathered at the GNSS receiver from the GNSS satellite constellation. Because of this, GNSS is subject to relevant problems such as lack of satellite visibility, multipath problems due to the reflection of the satellite signals on surfaces around the receiver's antenna, jamming, or spoofing [1]. For this reason, it is recommended that the positioning system features some other sensors that, together with the GNSS, provide an overall better performance. The most common configuration is the GNSS, plus the odometry of the vehicle and inertial sensors for attitude determination (a gyro). The values of travelled distance and the heading provided by the odometry and the gyro respectively, work together to provide a dead reckoning (DR) position that supports the GNSS in case of no coverage, and add redundant values to improve the reliability of the whole system [2]. The extra cost of this sensor set is relatively low, as vehicles have already embedded odometry and gyroscopes.

The main concept of DR is that the estimation of the vehicle position is based on the previous one, as spatial increments are calculated from the sensor values. Consequently, the necessary integration process leads to drifts when no absolute updates are obtained. Odometry errors may be due to mechanical or electrical (quantization) issues; tire wear, which changes the wheel diameter over time; and slippages and slides. Among these, the latter is the one that accounts for larger errors in the provision of instant positioning. However, GNSS/DR fusion algorithms do not usually account for slide errors. Therefore, when it happens, the solution is not only more inaccurate, but also inconsistent, since the errors considered by the algorithm are underestimated. This may lead to the malfunction of applications that are based on the confidence of the position solution. For instance, in lane-level applications, the confidence on the system on being on a specific lane at a given instant is crucial: the knowledge of the goodness of the position is as important as the position itself to decide whether a certain action must be taken. This is the concept of position integrity [3].

This article presents a multiple-model particle filter (MMPF) based solution that features two different models running in parallel, one that accounts for slides errors, and another one that does not, choosing at any time which model represents better the vehicle behavior. The simple concept behind this approach lays on the fact that particles will follow different behaviors depending on their assumed model and probability density function, and only those that are a better match with the vehicle motion will remain strong (with relevant weights), serving as the germ for new offspring particles.

The rest of the paper is organized as follows. Section 2 analyzes the most relevant related works. Later, the model for the slide error is introduced in Section 3. Sections 4 and 5 present the formulations of both the MMPF implementation and the consistency checks, respectively. Section 6 shows and discusses the results. Finally, Section 7 concludes the paper.

2. Related Work Materials and Methods

The literature of the field has a good number of articles featuring GNSS/DR solutions [2,4–8]. Despite the fact that some other technologies may be used to support the estimates along the longitudinal axis of the vehicle, such as accelerometers [4], the use of the odometry is the most widely spread approach [1]. With respect to the estimates of the heading, some authors employ electronic compasses, but many others discard them for lack of consistency and poor performance [9]. To compensate for the errors in the estimates of travelled distance, some authors use visual odometry, like in [10,11]. An interesting approach based on stereo vision for navigation in natural environments can be found in [12]. The robotics community is also quite active in this field, like the work presented in [13] based on cellular automata.

Regarding positioning filters, the most popular solution for fusing GNSS and DR is the extended Kalman filter (EKF), which is a variation of the Kalman filter for non-linear systems [14]. Its success is based on the relative good performance at a low computational cost and simplicity [14]. However, the nature of the vehicle motion is clearly non-linear, and in these cases, some other approaches provide better performances [15]. In these conditions, particle filters (PF), as genetic-type Monte Carlo (MC) methods, use weighted samples to generate approximations of the probability density function. This flexibility (when compared to Kalman), serves well to solve the positioning filtering problem [16,17].

On another note, multiple model (MM) filters have been used with success before by the first author of this article in the problem of positioning error estimation. In [13,18], different interactive multiple model versions of the extended Kalman filter (IMM-EKF) showed good performance for either improving the positioning solution, or maneuver prediction, respectively.

Multiple model particle filters have been applied in the field of tracking of ground, 3D targets and [19–23]. However, there are not many articles with MMPF applied to road vehicle position. The authors of [24] used it for detection and estimation of multipath effects on GPS measurements.

3. Error Model of the Odometry

As it has been aforementioned, odometry suffers from different kind of errors. The one introduced by tire wear has a very low frequency and may be compensated by calibration of the step value. It may be represented as a calibration error with the form $\omega(k)\delta r$, where $\omega(k)$ is the angular rate of the wheels at instant k and δr the difference between the real wheel radius, and the nominal value r.

Errors due to mechanical and electrical noises can be grouped together in a common term with a Gaussian characteristic and zero mean, $n(k) \sim N(0, \sigma_{odo})$. We have tuned the value of σ_{odo} to 0.26 m, corresponding with the odometry step in our vehicle setup.

A wheel slide appears when the wheel rotates slower than the value calculated according to the current vehicle speed. Wheel slip is the opposite case, when the wheels rotate faster than they would do in normal friction conditions.

Although both effects may partially compensate one another in a long term, the estimation of the vehicle pose at a given time suffers from slips and slides. In addition, slides appear more often than slips in car navigation. This is the reason why we focus on slides in this paper. Nevertheless, analogous considerations apply for slippages.

The characterization of the odometry error may be completed now to represent slides in the sensor model

$$\varepsilon_{dt}^{slide}(k) = \delta^{slide}(k) + \omega(k)\delta r + n(k) \tag{1}$$

where $\delta^{slide}(k) \geq 0$ stands for the sliding error, that may be also developed as

$$\delta^{slide}(k) = (1 - scv) \times c(k) \tag{2}$$

being $c(k) \sim N(0, \sigma_{odo})$, and scv (slide compensation value) varying from 0 (maximum compensation) till 1 (minimum).

4. Multiple Model Particle Filter Based Method

The principle of the particle filter is to represent by means of a set of N samples $\{X^i(k)\}^N_{\{i=1\}}$ and their corresponding weights $\{w^i(k)\}^N_{\{i=1\}}$ the probability density function $p(X(k)/Y_{\{1:k\}})$ at instant k of the state vector $X(k)$, given past observations, following the expression

$$p(X(k)/Y_{1:k}) \simeq \sum_{i=1}^{N} X^i(k) \cdot w^i(k) \tag{3}$$

where $Y_{\{1:k\}}$ stands for the observations collected from the initialization till instant k. Each sample $X^i(k)$ can be described as a Dirac delta expression $\delta^i(k)(X(k))$ in the form

$$\delta^i(k)(X(k)) = 0 \; if \; X(k) \neq X^i(k) \tag{4}$$

$$\delta^i(k)(X(k)) = 1 \; if \; X(k) = X^i(k) \tag{5}$$

The process of particle filtering can be summarized as follows:

- Initialization: Generation of N particles, or samples of the state vector, $X^i(0)$, with equal weights $1/N$. The proposed state vector is $[x(k) \; y(k) \; \psi(k)]$, representing east, north and heading (from north to east) at the center of the rear axle of the vehicle.
- Prediction: Estimation of $X^i(k + 1)$ following the prediction model. We use a classical 2D kinematical model for a vehicle on a plane. The measurements of the odometry and the gyroscope work as inputs to the filter. The travelled distance measured by the odometer, $ds(k)$, is estimated when a gyroscope value, $\dot{\psi}(k)$, is processed. The equation for pose prediction is:

$$
\begin{aligned}
x(k+1) &= x(k) + ds(k)\sin c(\dot{\psi}/2)\cos(\psi(k) + \dot{\psi}(k)T/2) - \dot{\psi}(k)T(Dx\sin\psi(k) + Dy\cos\psi(k)) \\
y(k+1) &= y(k) + ds(k)\sin c(\dot{\psi}/2)\sin(\psi(k) + \omega(k)T/2) - \dot{\psi}(k)T(Dx\cos\psi(k) - Dy\sin\psi(k)) \\
\psi(k+1) &= \psi(k) + \dot{\psi}(k)T
\end{aligned}
\tag{6}
$$

being T the sampling period, Dx, Dy the coordinate of the GPS antenna in the body frame, and sinc the cardinal sinus.

- Measurement update: Update of the weights of the particles with the observations $Y(k)$. In our case, the observation vector is $[x_{GPS}, y_{GPS}]$, standing for east and north values coming from the GPS. The update is done following the expression

$$w^i(k) = w^i(k-1) \times e\{-0.5\,(Y(k)-hX^i(k))'R^{-1}(Y(k)-hX^i(k))\} \tag{7}$$

where $h(X^i(k))$ is the observation function that relates at instant k the state $X^i(k)$ and observations $Y(k)$ (in our case, simply the second order identity matrix), and R the covariance matrix of observations.
- Normalization of the weights: $w^i(k) = w^i(k) / \sum_{i=1}^{N} w^i(k)$.
- Resampling: To prevent high concentration of probability mass in only a few particles, (leading to the convergence of a single $w^i(k)$ to 1), particles are resampled if

$$\frac{1}{\sum_{i=1}^{N} w^i(k)^2} < 0.5N \tag{8}$$

- End of cycle: Making $k = k + 1$, and iterating to step 2.

In our approach, a fixed number of particles is assigned to the each model.

The addition of the weights of the particles of each model after the observation will indicate the probability that each model m represents properly the vehicle behavior, $\mu^m(k)$. This probability is

$$\mu^m(k) = \frac{\sum_i^{Nm} w(k)_i^m}{\sum_i^{Nm} w(k)_i^m + \sum_i^{Nm} w(k)_i^{\neq m}} \tag{9}$$

where $w(k)_i^m$ stands for the weight at instant k of the particle i that represents model m, Nm is the number of particles of each model, that will be equal for all of them, and $w(k)_i^{\neq m}$ stands for the weight at instant k of a particle i that is not driven by model m.

Our solution is based on two different models, named the normal condition model (NCM) and the slide condition model (SCM). Both models will actually share the state vector and prediction equation, but will have different assumptions for the odometry model with and without slides, as described in (1).

Finally, the selected output will be such of the model with higher cumulative weight (more probable). A weighted mixed output from both models, as the one used in [9], would be also possible, but it was found unsuitable for the problem under consideration, as will be shown later in the tests.

5. Filter Consistency

As important as the filter itself, is the capability to evaluate whether or not it performs well. In order to check the performance consistency of the filter under consideration, several methods have been considered, detailed next. Later on the paper, in the sections dedicated to results, the goodness of the proposal will be evaluated based on these methods.

5.1. Filter Covariance

The estimation of the covariance of the filter state at instant k is denoted as matrix $P^m(k)$, where m stands for the choice of odometry model. Its value can be used to compute the positioning reliability and can be calculated by

$$P^m(k) = w(k)_i^m \times \left(x(k)_i^m - \overline{x}(k)^m\right) \times \left(x(k)_i^m - \overline{x}(k)^m\right) \tag{10}$$

where $\overline{x}(k)^m$ is the weighted mean value at instant k of the N particles i that belong to model m, calculated as

$$\overline{x}(k)^m = w(k)_i^m \times x(k)_i^m \quad i = 1 \cdots N \tag{11}$$

For a quick check, one can plot ellipsis of reliability around a certain position estimate based on its covariance and compare them to the errors to realize whether the errors envelopes are consistent.

5.2. Time Average Autocorrelation

The time average autocorrelation test is based on the ergodicity of the innovation sequence. If the number of time samples, K, is enough, this statistic is normally distributed and its variance is $\frac{1}{K}$. Its calculation for innovations one step apart can be done using

$$\bar{\rho}(1) = \sum_{k=1}^{K} v(k)' v(k+1) \times \left[\sum_{k=1}^{K} v(k)' v(k) \times \sum_{k=1}^{K} v(k+1)' v(k+1) \right]^{\frac{1}{2}} \tag{12}$$

where $v(k)$ stands for the innovation at instant k, that can be estimated as the difference between the predicted position and the observation.

This test can be executed in real time, and therefore, in single-run trials [14].

5.3. Time Average Normalized Innovation Square

The value of NIS at instant k can be calculated as

$$\epsilon_v(k) = v(k)' \, S(k)^{-1} v(k) \tag{13}$$

being $S(k)$ the innovation covariance matrix, that can be calculated with

$$S(k) = H \, P_i(k) \, H + R \tag{14}$$

where R is the matrix of observation covariance, H is the matrix that relates the state vector space with the observation vector space, being in our case

$$H = \begin{bmatrix} 1 & 0 & 0 \\ 0 & 1 & 0 \end{bmatrix} \tag{15}$$

and $P_i(k)$ the estimated covariance of the particle filter at instant k, that can be obtained as

$$P_i(k) = w(k)_j^i \times \left(x(k)_j^i - \bar{x}(k)^i \right) \times \left(x(k)_j^i - \bar{x}(k)^i \right) \tag{16}$$

where $\bar{x}(k)^i$ is the weighted mean value at instant k of the N particles j that belong to model i, calculated as

$$\bar{x}(k)^i = w(k)_j^i \times x(k)_j^i \quad j = 1 \ldots N \tag{17}$$

The time-average normalized innovation squared can be calculated as

$$\bar{\epsilon}_v = \frac{1}{K} \sum_{k=1}^{K} \epsilon_v(k) \tag{18}$$

If the innovation is white, zero mean, and $S(k)$ represents its covariance, then $K \bar{\epsilon}_v$ follows a χ^2 distribution.

6. Tests

6.1. Test Setup

The sensor setup included:

- EGNOS capable GPS receiver by Trimble, L1.

- Dual-frequency RTK receiver by Ashtech, for ground reference.
- FOG (Fiber Optic Gyroscope) by KVH.
- Vehicle odometer with 26.15 cm resolution coupled to the rear wheels axle.

6.2. Results and Discussion

For proving the validity of our proposal, we collected measurements of several acceleration and deceleration conditions, more prone to sliding errors in the odometry along the longitudinal axis. The main objective of the tests is to analyze the capability of the filter to represent the vehicle motion at any time, for what we will use the filter consistency techniques introduced in previous sections. Since both Kalman and particle filters using single models show very similar results, only the results of the particle filter will be discussed.

Figure 1 shows the distribution of particles around the solution in an experiment where the vehicle drives from north to south. Cyan particles are estimated by the SCM while green ones are calculated by the NCM. Even though both solutions are within the possible confidence scenarios, both of them are not the same good. While in the upper cloud, the GPS-EGNOS value matches better the solution that assumes normal conditions, after the slide, the satellite solution fits better the filter output that assumed slides.

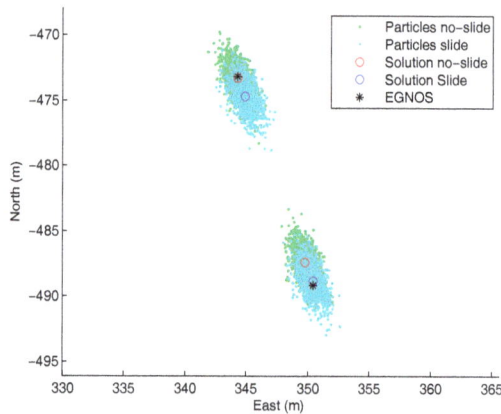

Figure 1. Cloud of particles during an experiment with a slide.

Figure 2 shows how model probabilities change at those instants, precisely during the slides. The MMPF manages to identify the sliding situations, and switches to the SCM model, returning to the NCM only when the slide has disappeared. For the same test, Table 1 provides the consistency results for all model combinations by using the time average correlation. Low values represent better consistency. For a properly tuned filter, the variance should be $\sigma^2 = \frac{1}{R}$. In our experiment, $\sigma = 0.0083$, and the 95% probability region results $(-0.01633, 0.01633)$. Both NCM and SCM fall out of this threshold, only satisfied by the MMPF.

Table 1. Time average correlation results for NCMPF, SCMPF, and MMPF.

Model	Value
NCMPF	0.5762
SCMPF	0.9110
MMPF	0.0140

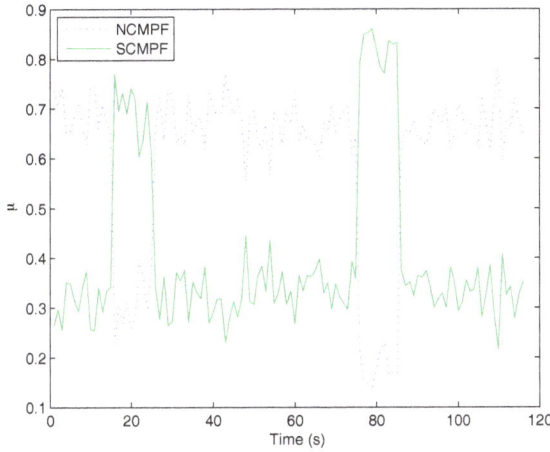

Figure 2. Probability values of both models during the test.

Similar conclusions can be achieved by using the time average NIS indicator (Figure 3). Even if the MMPF encounters some trouble to model the slide around instant 85, it is still the best solution. Additionally, switching models based on each model probability appears to be sounder than a weighted solution.

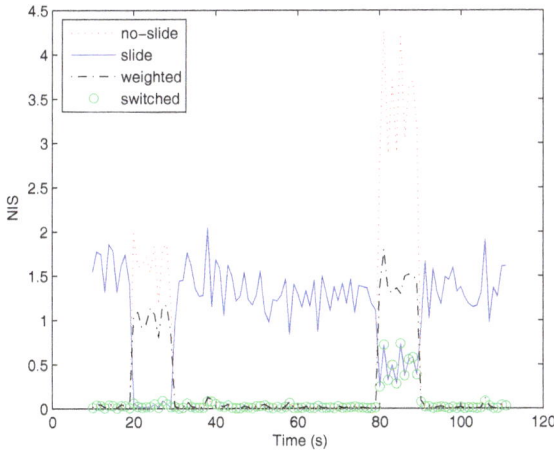

Figure 3. Time average NIS value.

7. Conclusions

The article focused on the problem of inaccurate positioning and error estimation when GNSS/DR positioning is subject to slides. To do so, the positioning system features two different models, one that accounts for the wheel slides (SCM), and one that does not (NCM). Each model is driven by a particle filter, thus allowing for a more flexible PDF (when compared to the most common extended Kalman filter). The system is capable of switching between both SCM and NCM by means of a multiple model implementation.

Different consistency checks for the filters under consideration are run, showing the goodness of the MMPF with real datasets when compared to the single model implementations. The final positioning system is able to characterize better the errors at any time. On the one hand, particles with stronger weights will survive, and only these are used in the resampling stage. On the other hand, the model whose errors better represent the innovation of the positioning loop is selected.

Acknowledgments: This research has been carried out by researchers of the Group of Diseño Electrónico y Técnicas de Tratamiento de señal/Universidad Politécnica de Cartagena, awarded as an excellence researching group in the frame of the Spanish Plan de Ciencia y Tecnología de la Región de Murcia, Fundación Seneca (19882/GERM/15).

Author Contributions: R.T.-M. conceived and designed the experiments. C.C.-C. and F.J.T.-M. performed the experiments; all authors analyzed the data and wrote the paper.

Conflicts of Interest: The authors declare no conflict of interest. The founding sponsors had no role in the design of the study; in the collection, analyses, or interpretation of data; in the writing of the manuscript, or in the decision to publish the results.

References

1. Bétaille, D.; Bonnifait, P.; Blanco Delgado, N.; Cosmen Schortmann, J.; Engdahl, J.; Gikas, V.; Gilliéron, P.Y.; Hodon, M.; Feng, S.; Machaj, J.; et al. *SaPPART White Paper. Better Use of Global Navigation Satellite Systems for Safer and Greener Transport*; IFSTTAR Techniques et Méthodes; IFSTTAR: Paris, France, 2015.
2. Toledo, R.; Zamora, M.A.; Ubeda, B.; Gomez-Skarmeta, A.F. An Integrity Navigation System based on GNSS/INS for Remote Services Implementation in Terrestrial Vehicles. In Proceedings of the 7th International IEEE Conference on Intelligent Transportation Systems, Washington, DC, USA, 3–6 October 2004; pp. 477–480.
3. Toledo-Moreo, R.; Bétaille, D.; Peyret, F. Lane-Level Integrity Provision for Navigation and Map Matching With GNSS, Dead Reckoning, and Enhanced Maps. *IEEE Trans. Intell. Transp. Syst.* **2010**, *11*, 100–112. [CrossRef]
4. Sukkarieh, S.; Durant-Whyte, H.; Nebot, E. A High Integrity IMU/GPS Navigation Loop for Autonomous Land Vehicle Applications. *IEEE Trans. Robot. Autom.* **1999**, *15*, 572–578. [CrossRef]
5. Wilson, C.; Wang, J. *Safety at the Wheel. Improving KGPS/INS Performance and Reliability*; GPS World: Duluth, MN, USA, 2003; pp. 16–26.
6. Berdjag, D.; Pomorski, D. DGPS/INS data fusion for land navigation. In Proceedings of the Fusion 2004, Stockholm, Sweden, 28 June–1 July 2004.
7. Zhang, P.; Gu, J.; Milios, E.E.; Huynh, P. Navigation with IMU/GPS/digital compass with unscented Kalman filter. In Proceedings of the IEEE International Conference on Mechatronics and Automation, Niagara Falls, ON, Canada, 29 July–1 August 2005; pp. 1497–1502.
8. Liu, B.; Adam, M.; Ibañez-Guzman, J. Multi-aided Inertial Navigation for Ground Vehicles in Outdoor Uneven Environments. In Proceedings of the 2005 IEEE International Conference on Robotics and Automation, Barcelona, Spain, 18–22 April 2005.
9. Toledo-Moreo, R.; Zamora-Izquierdo, M.; Ubeda, B.; Skarmeta, M.A. High-Integrity IMM-EKF-Based Road Vehicle. *IEEE Trans. Intell. Transp. Syst.* **2007**, *8*, 491–511. [CrossRef]
10. Takeyama, K.; Machida, T.; Kojima, Y.; Kubo, N. Improvement of Dead Reckoning in Urban Areas through Integration of Low-Cost Multisensors. *IEEE Trans. Intell. Veh.* **2017**, *2*, 278–287. [CrossRef]
11. Alonso, I.P.; Llorca, D.F.; Gavilan, M.; Pardo, S.Á.; Garcia-Garrido, M.A.; Vlacic, L.; Sotelo, M.Á. Accurate Global Localization Using Visual Odometry and Digital Maps on Urban Environments. *IEEE Trans. Intell. Transp. Syst.* **2012**, *13*, 1535–1545. [CrossRef]
12. Kunii, Y.; Kovacs, G.; Hoshi, N. Mobile robot navigation in natural environments using robust object tracking. In Proceedings of the IEEE 26th International Symposium on Industrial Electronics (ISIE), Edinburgh, UK, 19–21 June 2017; pp. 1747–1752.
13. Chaves, G.D.L.; Martins, L.G.A.; de Oliveira, G.M.B. An improved model based on cellular automata for on-line navigation. In Proceedings of the 2017 Latin American Robotics Symposium (LARS) and 2017 Brazilian Symposium on Robotics (SBR), Curitiba, Brazil, 8–11 November 2017.
14. Bar-Shalom, Y. *Estimation and Tracking: Principles*; Artech House: Nonvood, MA, USA, 1993.
15. Skog, I.; Handel, P. In-Car Positioning and Navigation Technologies—A Survey. *IEEE Trans. Intell. Transp. Syst.* **2009**, *10*, 4–21. [CrossRef]
16. Doucet, A.; de Freitas, N.; Gordon, N. *Sequential Monte Carlo*; Springer: New York, NY, USA, 2001.

17. Djuric, P.M.; Kotecha, J.H.; Zhang, J.; Huang, Y.; Ghirmai, T.; Bugallo, M.F.; Miguez, J. Particle filtering. *IEEE Signal Proceess. Mag.* **2003**, *20*, 19–38. [CrossRef]
18. Toledo-Moreo, R.; Zamora-Izquierdo, M. IMM-Based Lane-Change Prediction in Highways with Low-Cost GPS/INS. *IEEE Trans. Intell. Transp. Syst.* **2009**, *10*, 180–185. [CrossRef]
19. Ekman, M.; Sviestins, E. Multiple Model Algorithm Based on Particle Filters for Ground Target Tracking. In Proceedings of the 2007 10th International Conference on Information Fusion, Quebec, QC, Canada, 9–12 July 2007; pp. 1–8.
20. Hong, L.; Cui, N.; Bakich, M.; Layne, J.R. Multirate interacting multiple model particle filter for terrain-based ground target tracking. *IEE Proc. Control Theory Appl.* **2006**, *53*, 721–731. [CrossRef]
21. Guo, R.; Qin, Z.; Li, X.; Chen, J. IMMUPF Method for Ground Target Tracking. In Proceedings of the IEEE International Conference on Systems, Man and Cybernetics, Montreal, QC, Canada, 7–10 October 2007; pp. 96–101.
22. Foo, P.H.; Ng, G.W. Combining IMM Method with Particle Filters for 3D Maneuvering Target Tracking. In Proceedings of the 2007 10th International Conference on Information Fusion, Quebec, QC, Canada, 9–12 July 2007; pp. 1–8.
23. Wang, J.; Zhao, D.; Gao, W.; Shan, S. Interacting Multiple Model Particle Filter to Adaptive Visual Tracking. In Proceedings of the Image and Graphics (ICIG'04), Hong Kong, China, 18–20 December 2004; pp. 568–571.
24. Giremus, A.; Tourneret, J.-Y.; Calmettes, V. A Particle Filtering Approach for Joint Detection/Estimation of Multipath Effects on GPS Measurements. *IEEE Trans. Signal Process.* **2007**, *55*, 1275–1285. [CrossRef]

applied
sciences

MDPI

Article

Performance Analysis and Design Strategy for a Second-Order, Fixed-Gain, Position-Velocity-Measured (α-β-η-θ) Tracking Filter

Kenshi Saho [1,*] and Masao Masugi [2]

[1] Department of Intelligent Systems Design Engineering, Toyama Prefectural University,
     Imizu 939-0398, Japan
[2] Department of Electronic and Computer Engineering, Ritsumeikan University, Kusatsu 525-8577, Japan;
     masug@fc.ritsumei.ac.jp
*   Correspondence: saho@pu-toyama.ac.jp; Tel.: +81-766-56-7500

Academic Editor: Felipe Jimenez
Received: 29 June 2017; Accepted: 21 July 2017; Published: 26 July 2017

Featured Application: Design and evaluation of monitoring systems in intelligent vehicles, robots, and so on.

Abstract: We present a strategy for designing an α-β-η-θ filter, a fixed-gain moving-object tracking filter using position and velocity measurements. First, performance indices and stability conditions for the filter are analytically derived. Then, an optimal gain design strategy using these results is proposed and its relationship to the position-velocity-measured (PVM) Kalman filter is shown. Numerical analyses demonstrate the effectiveness of the proposed strategy, as well as a performance improvement over the traditional position-only-measured α-β filter. Moreover, we apply an α-β-η-θ filter designed using this strategy to ultra-wideband Doppler radar tracking in numerical simulations. We verify that the proposed strategy can easily design the gains for an α-β-η-θ filter based on the performance of the ultra-wideband Doppler radar and a rough approximation of the target's acceleration. Moreover, its effectiveness in predicting the steady state performance in designing the position-velocity-measured Kalman filter is also demonstrated.

Keywords: α-β-η-θ filter; tracking filter design; velocity measurements; Kalman filter; α-β filter; UWB Doppler radar

---

1. Introduction

Monitoring systems for robots and intelligent vehicles that employ remote sensors, such as cameras and radar, require the tracking of moving objects. Adaptive tracking techniques such as Kalman and extended Kalman filters [1–5] and particle filters [6,7] are commonly used for this purpose because of their accuracy. An alternative option is fixed-gain tracking filters, which have also been studied and used extensively for two reasons [8–12]:

- Simple implementation and low computational overhead: Optimal gain calculation is not required in the fixed-gain filters. Thus, the number of matrix operations is small compared with the Kalman filter and its variants [11].
- Applicability to the analytical evaluation of the Kalman filter: Fixed-gain filters are also useful for analytical evaluations of the Kalman filter because they can be characterized as steady state Kalman filters [12].

For these reasons, fixed gain filters are still being widely used in applications that strongly require real-time capability and simple implementation, such as tracking in ultrasonography in medicine [13], motor position control [14], human fall detection [15] and vehicular radar [16]. Additionally, the analysis of the Kalman filters assuming the steady state (fixed gain) is conducted to predict their tracking performance in the filter design process. The simplest second-order fixed-gain tracking filter is known as an α-β filter, which have been deployed in various tracking systems [12,17–22]. The design of the α-β filter has been discussed based on an efficient design parameter known as the tracking index [10,12,22].

However, α-β filters only consider position measurements and hence cannot make full use of modern sensors that can also measure velocity, such as ultra-wideband (UWB) Doppler radars, which have recently come into use [23–25]. In the near-field, these radars can achieve accurate sensing of moving objects, such as humans and cars. In [23,24], position and velocity estimates for pedestrians were achieved with centimeter and cm/s accuracy, respectively (see [25] for hardware implementation). Moreover, sensor fusion based on the Internet of Things technology also enables the simultaneous measurement of position and velocity possible (e.g., sensor data fusion based on the communication between radars and speedometers embedded in targets). Consequently, tracking filters for such systems have become an important area of research [26–29]. However, almost all conventional position-velocity-measured (PVM) trackers have been based on Kalman or particle filtering, whereas fixed-gain PVM filters have not seen wide use. This is because the computational performance is sufficient to drive PVM Kalman/particle filters in many applications. Additionally, the empirical design of these filters without steady state analyses can realize tolerable tracking performance. However, reiterating, fixed gain filter techniques are still important for various applications and analytical performance evaluation of tracking systems to find better parameter settings.

To address the above problem, we have proposed third-order fixed-gain (α-β-γ) PVM filters and have verified their performance [11]. However, a simpler second-order tracker such as an α-β filter is often required when the number of components and/or the size of hardware is quite limited and the complexity of target motion is predicted to be relatively small (i.e., a constant velocity model assuming a second-order tracker is sufficient). To this end, we also have investigated the fundamental properties of a PVM α-β filter [30]. However, this filter assumes an unrealistic assumption; specifically, correlated errors of position and velocity do not exist in the filtering process. As a realistic second-order fixed gain filter, Sudano [31] proposed a fixed-gain, position-velocity-measured, second-order tracking filter, described as an α-β-η-θ filter. This filter corresponds to the α-β filter in position-only-measured tracking problem, and the relationship between α-β-η-θ and the PVM Kalman filters is similar to that between the α-β and Kalman filters. Therefore, we believe that although the α-β-η-θ filter is underutilized at present, this filter will be widely used like the α-β filter after the spread of PVM systems such as the UWB Doppler radar. Thus, clarifying the analytical properties and design strategy of the α-β-η-θ filter is important for tracking technology in the near future. Although Sudano investigated the relationship of this filter with the PVM Kalman filter using a random-acceleration model based on the tracking indices, he did not discuss their performance or any design strategies. In addition, although Crouse [8] described a general solution for optimal fixed-gain trackers with steady state Kalman gains, he too did not discuss filter performance or a design strategy.

In this paper, a gain design strategy to compose an optimal α-β-η-θ filter is proposed, and efficient performance indices are derived. The strategy is based on the method presented in [4] for Kalman filters, which optimizes an analytical performance index for the tracking filter. The proposed strategy provides the easy design of filter gains and accurate tracking of the α-β-η-θ filter compared with the conventional empirical design methods. Furthermore, another important objective of this paper is to demonstrate, using numerical simulations, the effectiveness of the α-β-η-θ filter obtained with the proposed strategy for a realistic UWB Doppler radar application. In this application, we show its effectiveness in the design of the PVM Kalman filter that has better steady state performance.

The remainder of this paper is organized as follows: Section 2 defines the tracking problem dealt with in this paper. Section 3 reviews the α-β-η-θ filter of Sudano [31]. Its definition, relationship to conventional filters and design problems are described. Section 4 analytically derives the filter's performance indices and stability conditions for an appropriate gain design. Section 5 proposes our design strategy and explores its relationship to the Kalman filter. Section 6 analyzes the performance of α-β-η-θ filters designed with this strategy and compares it with the conventional filters. Section 7 presents a numerical application to realistic UWB Doppler radar tracking, and Section 8 offers concluding remarks.

2. Definitions of Problem and Symbols

This paper mainly considers the one-dimensional second-order moving object tracking filter assuming that only position and velocity measurements are considered. For the one-dimensional problem, only tracking along the x-axis is considered. Note that for the performance evaluation, assuming the realistic situation presented in Section 7, an actual two-dimensional tracking in the x-y plane is considered, and the one-dimensional tracking filter being considered is implemented for each axis in this simulation.

The inputs of the filter are measured target position x_o and velocity v_o. This assumes that the observed data are both position and velocity. Note that many conventional studies on the tracking system adopt a position-only analysis, whereas assuming position/velocity measurements is one of the features of our study. The errors in x_o and v_o conform to white Gaussian noise, and their correlations are not considered for simplicity. We assume that the variance of the position measurement errors B_x and that of the velocity measurement errors B_v are known.

The outputs are predicted target position x_p and velocity v_p and smoothed (estimated) target position x_s and velocity v_s. The focus of this paper is the optimization of the steady state accuracy in x_p. To achieve this, the inputs x_o and v_o are filtered by some gains. This study assumes that these gains are unknown parameters that we must design. Thus, the purpose of this paper is the design of the tracking filter gains that minimizes the errors in the predicted target position.

The detailed definitions and explanations of the tracking filters and their design methodology that we focus on are presented in the rest of this paper. Table 1 lists the symbols used in this paper. Furthermore, each symbol is defined at its first appearance.

Table 1. List of symbols.

Variables	Description	Unit
T	Sampling interval	(s)
k	Discrete sampling index	Dimensionless
$()_{,k}$	Parameter at index k	
x_p	Predicted position	(m)
v_p	Predicted velocity	(m/s)
x_s	Smoothed (estimated) position	(m)
v_s	Smoothed (estimated) velocity	(m/s)
x_o	Observed (measured) position	(m)
v_o	Observed (measured) velocity	(m/s)
α	Filter gain for x_s with respect to x_o	Dimensionless
β	Filter gain for v_s with respect to x_o	Dimensionless
η	Filter gain for x_s with respect to v_o	Dimensionless
θ	Filter gain for v_s with respect to v_o	Dimensionless
$\tilde{()}$	Forecasts	
$\hat{()}$	Estimates	
$()^T$	Transpose of matrix	
$()^{-1}$	Inversion of matrix	
x	State vector of target composed of position and velocity	
z	Measurement vector	
F	Transition matrix from k to $k+1$	
P	Error covariance matrix with respect to x	
Q	Covariance matrix of process noise	
K	Kalman gain matrix	
B	Covariance matrix of measurement noise	

Table 1. *Cont.*

Variables	Description	Unit
B_x	Error variance of x_o	(m^2)
B_v	Error variance of v_o	(m^2/s^2)
Q_{ra}	Process noise matrix in random-acceleration (RA) model	
q	Variance of random-acceleration (RA) process noise	(m^2/s^4)
R_{xv}	Ratio of B_x to $T^2 B_v$	Dimensionless
$E()$	Mean with respect to k	
σ_P^2	Smoothing performance index	(m^2)
e_{fin}	Tracking performance index	(m)
a_c	Acceleration assumed in the derivation of e_{fin}	(m/s^2)
ϵ_{rms}	Root-mean-square (RMS) index	(m)
μ	Evaluating function in the proposed gain design strategy	Dimensionless
a_D	Design parameter for the proposed strategy	Dimensionless
Q_{gen}	Arbitrary process noise matrix	
a	(1,1) element of Q_{gen}	(m^2)
b	(1,2) (or (2,1)) element of Q_{gen}	(m^2/s)
c	(2,2) element of Q_{gen}	(m^2/s^2)
ϵ	RMS prediction error of Monte Carlo simulations	(m)

3. The α-β-η-θ Filter

3.1. Definition

The α-β-η-θ filter proposed by Sudano [31] is a second-order fixed-gain PVM tracker. It can be considered to be an extension of the α-β filter, which uses position measurements only. The α-β-η-θ filter iterates prediction and smoothing (update) processes. The prediction process is conducted under the assumption that the target's velocity is constant over the sampling interval and yields a position prediction:

$$x_{p,k} = x_{s,k-1} + T v_{s,k-1}, \tag{1}$$

$$v_{p,k} = v_{s,k-1}, \tag{2}$$

where $x_{s,k}$ is the smoothed target position at time kT, T is the sampling interval, $x_{p,k}$ is the predicted position, $v_{s,k}$ is the smoothed velocity and $v_{p,k}$ is the predicted velocity. The smoothing process is defined as in [31]:

$$x_{s,k} = x_{p,k} + \alpha(x_{o,k} - x_{p,k}) + T\eta(v_{o,k} - v_{p,k}), \tag{3}$$

$$v_{s,k} = v_{p,k} + (\beta/T)(x_{o,k} - x_{p,k}) + \theta(v_{o,k} - v_{p,k}), \tag{4}$$

where $x_{o,k}$ is the measured position, $v_{o,k}$ the measured velocity and α, β, η and θ are fixed filter gains that we must design.

3.2. The α-β Filter

The α-β filter is well known and popular in tracking because of its simplicity and utility in real-time applications. Its prediction steps are the same as for the α-β-η-θ filter. The smoothing process is defined as in [10]:

$$x_{s,k} = x_{p,k} + \alpha(x_{o,k} - x_{p,k}), \tag{5}$$

$$v_{s,k} = v_{p,k} + (\beta/T)(x_{o,k} - x_{p,k}). \tag{6}$$

When $\eta = \theta = 0$, the α-β-η-θ filter is identical to the α-β filter. Thus, the difference between the two lies in whether measured velocities are used. Sudano verified the better performance of the α-β-η-θ filter compared with the α-β filter using velocity measurements [31].

The α-β filter is widely used in the position-only-measured tracking systems, and its performance has been sufficiently analyzed [12]. Various useful relationships between gains α and β and the design strategy based on a design parameter known as a tracking index have been applied in its design [10]. The α-β filter is derived from the Kalman filter equations in the limit $k \to \infty$. Thus, it is useful in the steady state performance analysis of the Kalman filter tracking. Similar properties of the α-β filter are expected for the α-β-η-θ filter, and clarifying these is useful for the PVM tracker design.

3.3. Relationship to Kalman Filters

Kalman filters are optimal tracking filters and are based on the adaptive calculation of a gain matrix. The α-β-η-θ filter (and the α-β filter) is equivalent to steady state Kalman filters [31]. Thus, we derive the optimal gains for the motion model under consideration from the Kalman filter equations [4]:

$$\tilde{x}_k = F\hat{x}_{k-1}, \tag{7}$$
$$\tilde{P}_k = F\hat{P}_{k-1}F^{\mathrm{T}} + Q, \tag{8}$$
$$K_k = \tilde{P}_k H^{\mathrm{T}}(H\tilde{P}_k H^{\mathrm{T}} + B)^{-1}, \tag{9}$$
$$\hat{x}_k = \tilde{x}_k + K_k(z_k - H\tilde{x}_k), \tag{10}$$
$$\hat{P}_k = \tilde{P}_k - K_k H\tilde{P}_k, \tag{11}$$

where x is a state vector, forecasts and estimates are denoted by tildes and hats, respectively, superscript "T" and "-1" denote transpose and inversion, z is a measurement vector, F is the transition matrix, P_k is the error covariance matrix at time kT, Q is the covariance matrix for process noise, K_k is the optimal gain (Kalman gain) at time kT and B is the covariance matrix for the measurement noise.

The α-β-η-θ filter is obtained by substituting into Equations (7) and (10) vectors $\tilde{x}_k = (x_{p,k} \; v_{p,k})^{\mathrm{T}}$, $\hat{x}_k = (x_{s,k} \; v_{s,k})^{\mathrm{T}}$, $z_k = (x_{o,k} \; v_{o,k})^{\mathrm{T}}$, and matrices:

$$F = \begin{pmatrix} 1 & T \\ 0 & 1 \end{pmatrix}, \tag{12}$$

$$H = \begin{pmatrix} 1 & 0 \\ 0 & 1 \end{pmatrix}, \tag{13}$$

$$K_k = \begin{pmatrix} \alpha & T\eta \\ \beta/T & \theta \end{pmatrix}, \tag{14}$$

$$B = \begin{pmatrix} B_x & 0 \\ 0 & B_v \end{pmatrix} \tag{15}$$

(see [31]), where B_x and B_v are the error variances of $x_{o,k}$ and $v_{o,k}$, respectively. With $k \to \infty$ for K_k, the appropriate gains of the α-β-η-θ filter are calculated as the Kalman filter predicts the state with the minimum error for the assumed target model.

3.4. Optimal Filter for a Random-Acceleration Model and Its Problems

The optimal α-β-η-θ filter has been derived as the steady state Kalman filter under a general random-acceleration (RA) model [31]. In this model, it is assumed that the process noise consists of random accelerations with Q expressed as in [10]:

$$Q_{\text{ra}} = \begin{pmatrix} T^4/4 & T^3/2 \\ T^3/2 & T^2 \end{pmatrix} q, \tag{16}$$

where q is the variance of the process noise. By calculating the limit of the K_k using Equation (16), we have the optimal gains of the α-β-η-θ filter presented in [31]. For example, the relationship between the optimal β and η is:

$$\eta = R_{\text{xv}}\beta, \tag{17}$$

where:

$$R_{\text{xv}} \equiv B_{\text{x}}/T^2 B_{\text{v}}, \tag{18}$$

corresponds to the ratio of the measurement accuracies in position and velocity. The other gains are expressed using tracking indices (see Equations (24)–(27) of [31]).

However, this filter is not optimal for other models, such as the frequently-used random-velocity model [9] and the diagonal Q, which does not include correlations in process noise [1,2]. Other process noise can be incorporated using arbitrary process noise; see [4]. The performance of this α-β-η-θ filter was evaluated in [31] only in terms of several simple numerical calculations, and strategies for designing tracking indices were not discussed. These problems must be solved to establish a design strategy and to properly evaluate the filter's performance.

4. Derivation of Performance Indices and Stability Conditions

To evaluate the performance of the tracking filter, steady state errors for the reduction of measurement noise and the tracking of accelerating targets are used [11,12]. These indices are more effective in evaluating steady state tracking accuracy than the error covariance matrix in the Kalman filter equations, as discussed by Ekstrand (see Section 9.8 of [12]). Moreover, a comprehensive performance index for measurement-error smoothing and tracking of an accelerating target based on these indices is presented in [4]. This comprehensive index is used for our proposed gain design strategy. Consequently, this section derives these performance indices of the α-β-η-θ filter. Stability conditions are also derived for practical filter designs.

4.1. Smoothing Performance Index

An important function of a tracking filter is the reduction of random errors caused by measurement noise. One such performance index is the steady state error for a target undergoing the same motion as in the motion model, but taking into account sensor noise. We assume that $x_{\text{o},k}$ contains noise with variance B_{x}, that $v_{\text{o},k}$ contains noise with variance B_{v} and that the target moves with constant acceleration. The variance of the predicted target position in a steady state is calculated from:

$$\sigma_{\text{p}}^2 = \lim_{k \to \infty} E[(x_{\text{p},k} - x_{\text{ts},k})^2], \tag{19}$$

(see [11,12]), where $x_{\text{ts},k}$ is the true target position, used to evaluate the smoothing performance and $E[\ ]$ denotes the mean. The quantity σ_{p}^2 is called the smoothing performance index.

The smoothing performance index for the α-β-η-θ filter is derived as:

$$\sigma_{\text{p}}^2(\alpha, \beta, \eta, \theta) = \frac{g_2(\alpha, \beta, \eta, \theta)B_{\text{x}} + g_3(\alpha, \beta, \eta, \theta)T^2 B_{\text{v}}}{g_1(\alpha, \beta, \eta, \theta)}, \tag{20}$$

where:

$$g_1(\alpha, \beta, \eta, \theta) = (\beta\eta - \alpha\theta - \beta)(\alpha\theta - \beta\eta - \alpha - \theta)(4 - 2\alpha - \beta - 2\theta + \alpha\theta - \beta\eta), \tag{21}$$

$$\begin{aligned}
g_2(\alpha, \beta, \eta, \theta) &= \alpha^3\theta(\theta - 2)(\theta - 1) + \alpha^2\beta(2 - 2\eta - 2\theta + 6\eta\theta - 3\eta\theta^2) \\
&+ \alpha^2\theta^2(2 - \theta) + 2\alpha\beta\theta(\eta - 2)(\theta - 2) \\
&+ \alpha\beta^2(1 + 2\eta - \theta - 3\eta^2 + 3\eta^2\theta) + \beta^3\eta(1 - \eta)^2 \\
&+ \beta^2\{\eta(2 - \eta)(\theta - 2) - \theta + 2\},
\end{aligned} \tag{22}$$

$$\begin{aligned}
g_3(\alpha, \beta, \eta, \theta) &= \alpha\theta(2\eta^2 + 2\eta\theta + \theta^2 - \theta) \\
&+ \beta\eta(2\eta + 2\theta - 2\eta^2 - \theta^2 - 2\eta\theta) + \theta^2(2 - \theta).
\end{aligned} \tag{23}$$

The derivation of $\sigma_p^2(\alpha, \beta, \eta, \theta)$ is given in Appendix A. Note that when $\eta = \theta = 0$,

$$\sigma_p^2(\alpha, \beta, 0, 0) = \frac{2\alpha^2 + 2\beta + \alpha\beta}{\alpha(4 - 2\alpha - \beta)} B_x. \tag{24}$$

This is the smoothing performance index of the conventional α-β filter [12].

4.2. Tracking Performance Index

The filter is required to track complicated motions. In second-order trackers, when tracking a target moving with constant acceleration, steady state bias error occurs as a result of the difference between the motion model and the actual target motion. This provides an index of the tracking performance for an accelerating target. When the true target position $x_{tt,k} = a_c(kT)^2/2$ (a_c denotes a constant acceleration) and measurement errors are not considered, the steady state predicted error is expressed as [11]:

$$e_{fin} = \lim_{k \to \infty}(x_{tt,k} - x_{p,k}). \tag{25}$$

which is called the tracking performance index.

For the α-β-η-θ filter, the tracking performance index becomes:

$$e_{fin}(\alpha, \beta, \eta, \theta) = \lim_{z \to 1}(1 - z^{-1})E_p(z) = \frac{2 - 2\eta - \theta}{2(\alpha\theta - \beta\eta + \beta)}a_c T^2. \tag{26}$$

The derivation of $e_{fin}(\alpha, \beta, \eta, \theta)$ is given in Appendix B. Note that when $\eta = \theta = 0$,

$$e_{fin}(\alpha, \beta, 0, 0) = a_c T^2/\beta, \tag{27}$$

which is the tracking performance index of the conventional α-β filter [12].

4.3. RMS Index

The smaller the tracking and smoothing performance indices are, the better a tracking filter is. However, there are trade-offs between these indices. To consider these trade-offs and practical performance evaluations, a comprehensive performance index in smoothing and tracking was proposed and its effectiveness for the Kalman filter verified in [4]. This index corresponds to the root-mean-square (RMS) prediction error for a constant-acceleration target (considering sensor noise) and is calculated as:

$$\epsilon_{rms} = \sqrt{\sigma_p^2 + e_{fin}^2}. \tag{28}$$

We refer to this as the RMS index.

4.4. Stability Condition

To apply the α-β-η-θ filter to real systems, it must be stable. Hence, stability conditions are now derived. As shown in Equations (A19) and (A20) of Appendix B, the characteristic polynomial of the α-β-η-θ filter is $f_{\alpha\beta\eta\theta}(z) = z^2 + (\alpha + \beta + \theta - 2)z + \alpha\theta - \eta\beta - \alpha - \theta + 1$. Applying Jury's stability test [32] to $f_{\alpha\beta\eta\theta}(z)$, we obtain stability conditions:

$$(1-\eta)\beta+\alpha\theta > 0 \quad \text{and} \quad 4-2\alpha-\beta-2\theta+\alpha\theta-\eta\beta > 0 \quad \text{and} \quad |\alpha\theta-\eta\beta-\alpha-\theta+1| < 1. \tag{29}$$

5. Optimal Gain Design Strategy

5.1. Optimal Gain Design Using the RMS Index

Conventional gain design strategies based on tracking indices [10,22] have the following difficulties:

- The selection of an appropriate model (e.g., RA, random-velocity) is not considered. Thus, this selection is conducted empirically [4].
- There are no general rules for the determination of a tracking index [12]. Sudano did not discuss how to set the tracking indices for the α-β-η-θ filter [31].

To resolve these problems, we have adapted the strategy based on optimizing the RMS index presented in [4] to the α-β-η-θ filter. Given Equations (20) and (26) and the normalization of the RMS index in Equation (28) and substituting Equation (17), we define the index for the gain design as:

$$
\begin{aligned}
\mu(\alpha, \beta, \theta, R_{xv}, a_{D}) &\equiv \epsilon_{rms}^2 / B_x \\
&= \frac{g_2(\alpha, \beta, \theta, R_{xv}) + g_3(\alpha, \beta, \theta, R_{xv})/R_{xv}}{g_1(\alpha, \beta, \theta, R_{xv})} + a_{D}^2 \left(\frac{2 - 2R_{xv}\beta - \theta}{2(\alpha\theta - R_{xv}\beta^2 + \beta)} \right)^2,
\end{aligned} \tag{30}
$$

where:

$$a_{D}^2 \equiv a_c^2 T^4 / B_x \tag{31}$$

is the important dimensionless parameter for the proposed strategy because μ depends on a_D. Although μ also depends on R_{xv}, it is determined from known measurements of the noise parameters (B_x, B_v) and sampling interval T as indicated in Equation (18). Thus, the optimal gains are determined by a_D, and its appropriate presetting is essential for the proposed strategy. Note that Equation (17) is always satisfied for α-β-η-θ filters derived from the Kalman filter (see Appendix C). The optimal gains are calculated by solving the minimization problem:

$$\arg\min_{\alpha,\beta,\theta} \ \mu(\alpha, \beta, \theta, R_{xv}, a_{D})$$

sub. to stability conditions Equation (29) are satisfied. $\tag{32}$

With Equation (32), optimal gains are determined for each R_{xv} and a_D. R_{xv} can be set from the performance of the sensors. Therefore, the main design parameter for the proposed strategy is a_D. As given in Equation (31), a_D is determined by the target's acceleration. Hence, presetting a_D appropriately is important in practical applications, and the value should be a typical value of the target acceleration (e.g., mean or maximum). The choice of a_D for the UWB Doppler radar application is discussed in Section 7.

5.2. Procedure and Notes of the Proposed Strategy

Our proposed strategy can be summarized as follows:

1. Set R_{xv} from the sensor performance.
2. Design a_D based on the approximate target acceleration.

3. Determine α, β and θ by solving Equation (32).
4. Determine η with Equation (17).

With respect to the proposed strategy, note that:

- Equation (32) can be solved by simple gradient descent with several initial values [33]. This is because the range of parameter searching is not so wide due to the stability conditions.
- This design process is conducted only once before using the filter. Although the computational costs of the above optimization process are not small, this does not affect the simple tracking process of the α-β-η-θ filters.

5.3. Relationship with Steady State PVM Kalman Filters

As described in Section 1, one of the reasons for considering the α-β and α-β-η-θ filters in practical use is the analytical performance predictions of the Kalman filter (and its variants). However, the proposed gain design strategy does not use the Kalman filter, and the relationship between the designed α-β-η-θ and Kalman filters is therefore unclear. This section clarifies this relationship analytically. Indeed, the proposed strategy corresponds to an optimization of the elements of the covariance matrix of the process noise with respect to the RMS index. To prove this, we now derive the relationship between steady state Kalman gains and the arbitrary covariance matrix of process noise. The covariance matrix is expressed as [4]:

$$Q_{\text{gen}} = \begin{pmatrix} a & b \\ b & c \end{pmatrix},$$ (33)

where $a > 0$, $b > 0$, $c > 0$ and the dimensions of a, b and c are m^2, m^2/s and m^2/s^2, respectively. For example, substituting $(a, b, c) = (qT^4/4, qT^3/2, qT^2)$ into Equation (33) gives the Q_{ra} of (16); substituting $(a, b, c) = (q_v T^2, q_v T, q_v)$ (q_v is the variance of the velocity noise) yields the random-velocity model [9]; and $b = 0$ leads to a diagonal Q, which is also a well-used setting in real applications [1,2].

The relationship between steady state Kalman gains and Q_{gen} is derived as:

$$
\begin{aligned}
a = {} & \frac{T^2 B_v}{1 - R_{xv}\beta^2 - (1-\theta)\alpha - \theta} \{ (\beta^2\alpha + 2\beta^3 + \beta^2 + (\theta-1)\theta\beta)R_{xv}^2 \\
& + \alpha^2((1-\theta) + 2\alpha\beta(1-\theta) + \beta^2\theta + \beta(-\theta^2 + 3\theta - 2))R_{xv} \\
& + \alpha\theta(1-\theta) + \theta(\theta-1) \},
\end{aligned}
$$ (34)

$$
b = \frac{\beta^3 R_{xv}^2 + (\alpha(\beta(1-\theta) - \theta^2 + \theta) + \beta^2\theta + \beta\theta + \theta^2 - \theta)R_{xv}}{1 - R_{xv}\beta^2 - (1-\theta)\alpha - \theta} TB_v,
$$ (35)

$$
c = \frac{(\alpha\beta\theta + \beta^2(\theta+1) - \beta\theta)R_{xv} + \alpha\theta(-\beta - \theta) + \beta\theta + \theta^2}{1 - R_{xv}\beta^2 - (1-\theta)\alpha - \theta} B_v.
$$ (36)

The derivation of these is given in Appendix C.

Equations (34)–(36) transform the optimal gains of the α-β-η-θ filter as elements of the covariance matrix of the process noise of the PVM Kalman filter and hence are useful in their design. In substituting these optimal designed gains into Equations (34)–(36), we obtain a Kalman filter having the same steady state performance. Similarly, solving Equations (34)–(36) with respect to (α, β, θ) and using a set Q_{gen}, we find the steady state gains of the designed PVM Kalman filter and its performance using Equations (20), (26) and (28). Moreover, optimization of the proposed strategy of Equation (32) with respect to (α, β, θ) is equivalent to optimization of the RMS index with respect to (a, b, c) using Equations (34)–(36).

The tracking filters and their design strategies that this paper considers are summarized in Table 2. In the following subsections, the performance of the α-β-η-θ filter realized with the proposed strategy is investigated and is compared with that of the conventional α-β filter and RA model-based filter design.

Table 2. Summary of the conventional and proposed tracking filters and design strategies.

Tracking Filter	Input	Design Strategy	Preset Parameter	$k \to \infty$
α-β filter	Position	Based on RA model [12] Proposed strategy	q or tracking index [10] a_D of Equation (31)	Kalman filter
α-β-η-θ filter	Position and velocity	Based on RA model [27] Proposed strategy	q or tracking index [10] a_D of Equation (31)	Position-velocity-measured (PVM) Kalman filter

6. Steady State Performance Analysis

This section presents theoretical performance analyses of the α-β-η-θ filter using the proposed design strategy. We compare the RMS index calculated using Equations (20), (26) and (28) for the following filters:

- Proposed filter: the α-β-η-θ filter with the proposed strategy.
- RA filter: the α-β-η-θ filter with the RA model using optimal q (from Equation (16)) with respect to the RMS index.
- Best α-β filter: the conventional α-β filter obtained with the proposed strategy, assuming $\eta = \theta = 0$.

Comparison with the RA filter indicates the effectiveness of the proposed strategy (i.e., considering Q_{gen}), and comparison with the best α-β filter illustrates the effectiveness of the velocity measurements. Note that the analysis in this section investigates the steady state performance, and this also corresponds to the steady state Kalman filter analysis. We assume that B_x and T are normalized to one.

6.1. Relationship between Performance and a_D

Figure 1 shows the relationship between the design parameter a_D and the RMS index ϵ_{rms} for $R_{xv} = 1$ and 10. From Figure 1a, the proposed filter realizes the best performance for relatively large a_D. This result verifies that the proposed strategy determines gains corresponding to a better covariance matrix of process noise than the RA filter. The proposed filter also achieves better performance compared with the best α-β filter, including small a_D, and even for $R_{xv} = 1$, which means that the measurement accuracy of the position and velocity is the same. Furthermore, when the velocity measurement accuracy is high, the proposed filter achieves greater accuracy than the best α-β filter (Figure 1b). In addition, although the difference between the RA and proposed filters is small for $R_{xv} = 10$, the proposed filter also achieves the best performance.

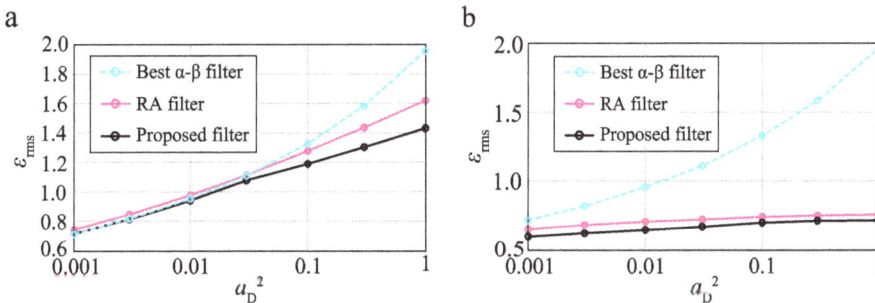

Figure 1. Relationship between a_D and ϵ_{rms} for (a) $R_{xv} = 1$ and (b) $R_{xv} = 10$; RA: random-acceleration.

6.2. Relationship between Performance and R_{xv}

Figure 2 shows the relationship between R_{xv} and ϵ_{rms} for $a_D^2 = 0.01$ and 0.1, and both cases exhibit the same trend. For both proposed and RA filters, better performance is achieved with better velocity measurement accuracy. The performance of the proposed filter is better than that of the best α-β filter including relatively small R_{xv} (the velocity measurement accuracy is low). In contrast, the performance of the RA filter is worse than that of the best α-β filter for small R_{xv}, because the covariance matrix of the RA filter is limited to Equation (16). Moreover, by comparing Figure 2a,b, we see the superior effectiveness of the proposed filter for relatively large a_D. These results indicate that the proposed filter is effective when the velocity measurement accuracy and/or target acceleration is relatively high.

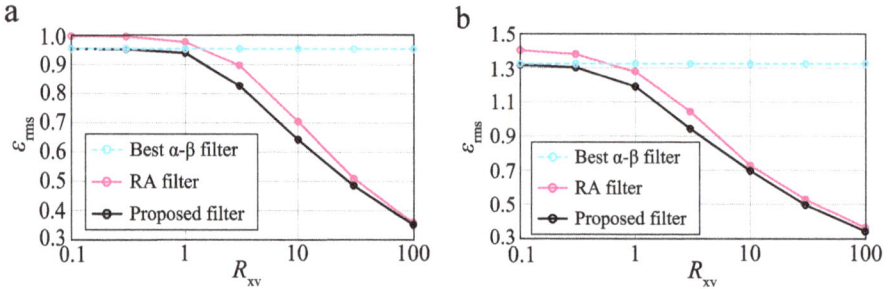

Figure 2. Relationship between R_{xv} and ϵ_{rms} for (a) $a_D^2 = 0.01$ and (b) $a_D^2 = 0.1$.

7. Application to UWB Doppler Radar Simulation

This section provides examples of the α-β-η-θ filter designed with the proposed strategy for a realistic application to UWB Doppler radar tracking. Numerical simulations show the effectiveness of the proposed strategy. We consider two scenarios:

- Medium maneuvering target assuming simple near-field sensing.
- High maneuvering target assuming the target executes an abrupt motion.

7.1. Tracking of Medium Maneuvering Target

7.1.1. Simulation Setup

First, we show the application examples for a maneuvering target that assumes near-field radar remote sensing for surveillance and robot monitoring systems. We simulate the UWB Doppler radar tracking [23–25] of a maneuvering target and compare the tracking errors of the filters assumed in the previous section and the PVM Kalman filter. For one-dimensional tracking assumed in the previous sections, the steady state performances of the PVM Kalman and the proposed filters are the same. However, for two-dimensional tracking assumed in this section, a difference in their tracking accuracy occurs because the Kalman filter considers correlations between the $x - y$ positions and velocities. Therefore, a comparison between the α-β-η-θ filters and the PVM Kalman filter is also necessary.

Figure 3 shows the simulation scenario and the true acceleration of a target. The true target position is $(x_{t,k}, y_{t,k}) = (0.5 + 0.3kT \sin(2\pi kT/10), 1.5 + 0.1(kT)^{1.2} \cos(2\pi kT/12))$. Two-dimensional tracking in the x-y plane of the point target is assumed. We consider two Doppler radars located at $(x, y) = (0.5 \text{ m}, 0)$ and $(1.0 \text{ m}, 0)$. The sampling interval T is 100 ms, and the observation time is 4 s. The transmitted signal is a UWB pulse with a center frequency of 26.4 GHz and a bandwidth of 500 MHz. The received radar signals are calculated using ray tracing with the addition of Gaussian white noise. The radars measure position using ranging results [24] and measure velocity using the Doppler shift with the method presented in [25]. We determine a standard derivation for this noise to set $B_x = 0.030^2 \text{ m}^2$ and $B_v = 0.10^2 \text{ m}^2/\text{s}^2$. These values are the averages along the two axes (x and y)

and are set based on the experimental results in [23,24]. Thus, the R_{xv} of the assumed radar system is 9.0. For simplicity, we use this value for both axes. In addition, to evaluate the performance for smaller R_{xv}, the case $R_{xv} = 1.0$ is generated by adding Gaussian white noise to the measured velocity data. B_v in this instance is 0.30^2 m^2/s^2.

The α-β and α-β-η-θ filters are implemented for each axis, for which we use the same gain. The implementation of the PVM Kalman filter is the same as in [27], and its covariance matrix for process noise is calculated from the optimal α-β-η-θ filter gains designed with the proposed strategy using Equations (34)–(36). The initial values of the state vectors and the error covariance matrix of the PVM Kalman filter are all zero. Using the RMS prediction error calculated from 1000 Monte Carlo simulations, the performance is defined as:

$$\epsilon_k = \sqrt{\frac{1}{1000} \sum_{m=1}^{1000} \{(x_{t,k} - x_{p,m,k})^2 + (y_{t,k} - y_{p,m,k})^2\}}, \tag{37}$$

where $x_{p,m,k}$ and $y_{p,m,k}$ are the predicted positions in the m-th Monte Carlo simulation.

7.1.2. Filter Design

The gain design for the UWB Doppler radar using the proposed strategy is presented here. We design an appropriate a_D and evaluate the resultant performance. We presume a rough approximate prediction of accelerations. For instance, when the maximum acceleration of the target in Figure 3b is approximately predicted as $a_c = 0.6$ m/s^2, a_D^2 is then 0.04 from Equation (31). Using this a_D^2, the above radar settings of $R_{xv} = 9.0$ (T = 100 ms, $B_x = 0.03^2$ m^2 and $B_v = 0.1^2$ m^2/s^2) and solving Equation (32), we have the optimal gains $\alpha = 0.315$, $\beta = 0.00801$, $\eta = 0.0721$ and $\theta = 1.15$. Optimal gains for other settings are similarly determined.

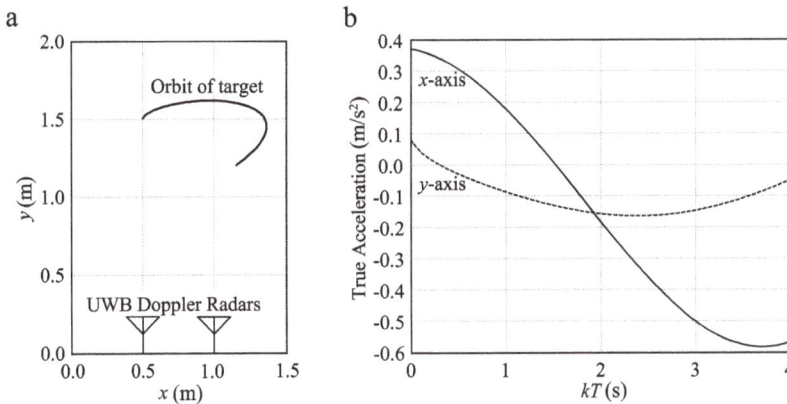

Figure 3. UWB Doppler radar simulation scenario: (**a**) radar positions and true orbit; (**b**) true acceleration; UWB: ultra-wideband.

7.1.3. Evaluation Results

Figure 4 shows the simulation results for $a_D^2 = 0.04$ ((**a**) $R_{xv} = 9.0$, (**b**) $R_{xv} = 1.0$). Clearly, the filters using velocity measurements achieve greater accuracy than the best α-β filter in both cases. For $R_{xv} = 9.0$, the mean steady state prediction RMS errors ($E[\epsilon_k]$ in $2 < kT < 4$) of the RA, PVM Kalman and proposed filters are 2.48, 2.30 and 2.33 cm, respectively. These results indicate that the proposed filter achieves better accuracy than the RA filter even for realistic situations. Although the PVM Kalman filter is slightly better than the proposed filter, because it considers the correlated noise in the x- and y-axes, their errors are almost the same. This shows that the steady state performance of the PVM filter is close to that of the α-β-η-θ filter, and the α-β-η-θ filter analysis is effective in performance predictions of the PVM Kalman filters. The relationship of these filters is similar to the traditional α-β and position-only-measured Kalman filters. The computational load of the α-β-η-θ filter is considerably smaller than that of the PVM Kalman filter; the mean calculation times for each k of the α-β-η-θ and PVM Kalman filters are 14.0 and 38.1 µs using an Intel Core i7-4600U CPU@2.10 GHz 2.70 GHz processor and Scilab 5.5.0. This is because the α-β-η-θ filter does not require the adaptive calculations of the gains and error covariance matrices. In addition, although the difference in steady state accuracy between the proposed and best α-β filters is small (Figure 4b), the proposed filter achieves better accuracy even for R_{xv}=1. The mean steady state prediction RMS errors of the RA, PVM Kalman and proposed filters are 3.84, 3.61 and 3.68 cm, respectively, and these results lead to the same conclusions as those from Figure 4a. These results are matched to the analysis results presented in the previous section.

We next investigate the performance of the proposed filter for different values of a_c to assess appropriate settings. Table 3 shows the steady state RMS error ($E[\epsilon_k]$ in $2 < kT < 4$) for different values of a_c for $R_{xv} = 9.0$; $a_c = 0.4$ m/s^2 is assumed as a rough approximation for the mean acceleration of the target, and $a_c = 0.1$ and 1 m/s^2 are considered as the order of the acceleration. The performance of the proposed filter deteriorates for $a_c = 0.1$ and 1 m/s^2. This is because the difference between the true target accelerations and these very rough approximate accelerations is too large. This means that the proposed strategy requires some minimum degree of accuracy in target acceleration prediction to perform adequately. However, the $a_c = 0.4$ and 0.6 m/s^2 cases have almost the same accuracy, implying that strict values of the target acceleration are in practice not required with UWB Doppler radars. As a method to obtain an approximated acceleration, communications of the tracking systems and the accelerometers embedded in targets can be considered. Many sensing targets have acceleration sensors, e.g., the robots and vehicles have inertial sensors, and humans have accelerometers embedded in smart phones. In the near future, the Internet-of-Things technology will make data communications between robots, smart phones and radar possible. Thus, we can obtain approximated acceleration based on this novel technology.

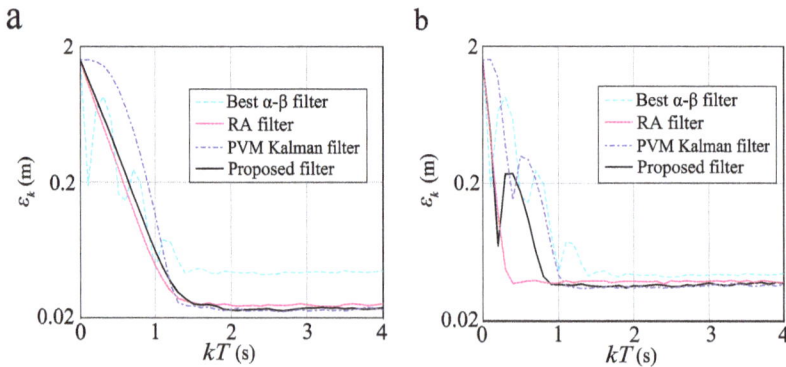

Figure 4. Simulation results for (**a**) $R_{xv} = 9$ and (**b**) $R_{xv} = 1$; PVM: position-velocity-measured.

Table 3. Steady state RMS prediction error of the proposed filter for various a_c ($R_{xv} = 9$).

a_c (m/s^2)	a_d^2	Mean Steady State RMS Error (cm)
0.1	0.00111	2.82
0.4	0.0178	2.32
0.6	0.040	2.33
1.0	0.111	2.57

7.2. Tracking of High-Maneuvering Target

Finally, an application to high-maneuvering targets with relatively high accelerations is presented to clearly show the effectiveness of the proposed filter compared with the RA filter. Given the true target position of $(x_{t,k}, y_{t,k}) = (k^2 T^2, 20 + (kT)^{1.5} \cos(2\pi kT/10))$, the true acceleration is plotted in Figure 5a. Compared with the previous section, a high acceleration is assumed. We set $B_x = 0.3^2$, $R_{xy} = 1.0$, $a_c = 3 \text{ m/s}^2$ and $a_D = 1.0$; other settings are the same as in the previous section.

Figure 5b shows the simulation results for the high-maneuvering target. The difference in the RMS prediction error between the RA and proposed filters becomes large compared with the moderate-maneuvering target assumed in the previous section. $E[\epsilon_k]$ in $2 < kT < 4$ s of the best α-β, RA, PVM Kalman and proposed filters are 0.589, 0.459, 0.190 and 0.201 m, respectively. Whereas the PVM Kalman filter is slightly better than the proposed filter for the same reason as for a moderate-maneuvering target, the proposed filter achieves greater accuracy than the RA filter. This is because the RA model cannot track the abrupt motion of the high-maneuvering target because of limitations in expressing process noise. In contrast, the proposed filter can set gains corresponding to the appropriate process noises to accurately track the high-maneuvering target. Our theoretical analyses presented in Figure 1a show that the performance difference between the assumed filters becomes large when a_D is relatively large. The above simulation results are consistent with these analyses. In addition, with respect to the steady state accuracy of the PVM Kalman and the proposed filter, the same result with the moderate-maneuvering target is obtained. Thus, the applicability to performance predictions of the PVM Kalman filters using the analysis of the α-β-η-θ filter is clearly indicated with these simulation results.

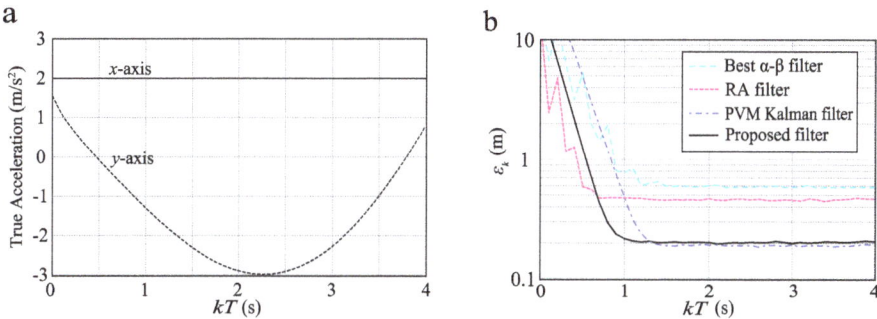

Figure 5. Simulation assuming a high-maneuvering target: (**a**) true acceleration; (**b**) results.

8. Conclusions

We have proposed a gain design strategy for α-β-η-θ filters and applied it to UWB Doppler radar simulations. Stability conditions and an efficient performance index (the RMS index) for the α-β-η-θ filter were analytically derived, and a design strategy was presented based on minimization of the RMS index. We clarified the design parameters of the proposed strategy and examined their relationship with those of the PVM Kalman filter. Numerical analyses using the derived performance index showed that the designed α-β-η-θ filter achieved better performance than the traditional α-β filter, as well as an α-β-η-θ filter designed using the general RA model. Finally, a numerical simulation assuming a realistic UWB Doppler radar application verified the effectiveness of an α-β-η-θ filter designed using the proposed strategy and the validity of the theoretical analyses. This simulation showed that the proposed strategy can be applied to realistic tracking systems by presetting several simple parameters, specifically an approximate acceleration and the radar measurement accuracy. Moreover, the possibility of the application to performance predictions for the PVM Kalman filter design was also indicated.

Acknowledgments: This work was supported in part by the JSPS KAKENHI Grant Number 16K16093.

Author Contributions: Kenshi Sahodeveloped the proposed design strategy and derived the fundamental properties presented in Section 4. Kenshi Saho and Masao Masugi conducted the numerical analyses and simulations. Kenshi Saho wrote the paper. Masao Masugi read and approved the final manuscript.

Conflicts of Interest: The authors declare no conflict of interest.

Abbreviations

The following abbreviations are used in this manuscript:

UWB	Ultra-wideband
PVM	Position-velocity-measured
RA	Random-acceleration
RMS	Root-mean-square

Appendix A. Derivation of Equation (20)

The σ_p^2 of the α-β-η-θ filter is now derived from Equations (1)–(4); $x_{\text{ts},k}$ represents the motion of the constant-velocity target:

$$x_{\text{ts},k} = x_{\text{ts},k-1} + T v_{\text{ts},k-1}, \tag{A1}$$

where $v_{\text{ts},k}$ is the true velocity. With Equations (1) and (A1), the predicted error is calculated as:

$$
\begin{aligned}
\sigma_p^2 &= E[(x_{p,k} - x_{\text{ts},k})^2] \\
&= E[(x_{s,k-1} - x_{\text{ts},k-1})^2] + 2TE[(x_{s,k-1} - x_{\text{ts},k-1})(v_{s,k-1} - v_{\text{ts},k-1})] \\
&\quad + T^2 E[(v_{s,k-1} - v_{\text{ts},k-1})^2].
\end{aligned}
\tag{A2}
$$

Thus, it is necessary to derive the error variance and covariance in the smoothing process to obtain σ_p^2.

The quantities $x_{\text{ts},k}$ and $v_{\text{ts},k}$ can be expressed as:

$$x_{\text{ts},k} = (1 - \alpha)x_{\text{ts},k} + \alpha x_{\text{ts},k} + T\eta(v_{t,k} - v_{t,k}), \tag{A3}$$

$$v_{\text{ts},k} = (1 - \theta)v_{\text{ts},k} + \theta v_{\text{ts},k} + (\beta/T)(x_{t,k} - x_{t,k}). \tag{A4}$$

Using Equations (3), (4), (A3) and (A4), we have:

$$
\begin{aligned}
x_{s,k} - x_{\text{ts},k} = &(1-\alpha)(x_{p,k} - x_{\text{ts},k}) + \alpha(x_{o,k} - x_{\text{ts},k}) \\
&+ T\eta((v_{o,k} - v_{t,k}) - (v_{p,k} - v_{t,k})),
\end{aligned}
\tag{A5}
$$

$$
\begin{aligned}
v_{s,k} - v_{\text{ts},k} = &(1-\theta)(v_{p,k} - v_{\text{ts},k}) + \theta(v_{o,k} - v_{\text{ts},k}) \\
&+ (\beta/T)((x_{o,k} - x_{t,k}) - (x_{p,k} - x_{t,k})).
\end{aligned}
\tag{A6}
$$

From Equations (1), (2) and (A1) and defining $\Delta x_{s,k} \equiv x_{s,k} - x_{\text{ts},k}$, $\Delta v_{s,k} \equiv v_{s,k} - v_{\text{ts},k}$, $\Delta x_{o,k} \equiv x_{o,k} - x_{\text{ts},k}$ and $\Delta v_{o,k} \equiv v_{o,k} - v_{\text{ts},k}$, we calculate:

$$\Delta x_{s,k} = (1-\alpha)(\Delta x_{s,k-1} + T\Delta v_{s,k-1}) + \alpha\Delta x_{o,k} + T\eta(\Delta v_{o,k} - \Delta v_{s,k-1}), \tag{A7}$$

$$\Delta v_{s,k} = (1-\theta)\Delta v_{s,k-1} + \theta\Delta v_{o,k} + (\beta/T)(\Delta x_{o,k} - (\Delta x_{s,k-1} + T\Delta v_{s,k-1})). \tag{A8}$$

Thus, the variances of the errors in the smoothed positions can be calculated using Equation (A7) as:

$$
\begin{aligned}
E[\Delta x_{s,k}^2] =& (1-\alpha)^2 (E[\Delta x_{s,k-1}^2] + T^2 E[\Delta v_{s,k-1}^2] + 2TE[\Delta x_{s,k-1}\Delta v_{s,k-1}]) \\
& + T^2\eta^2 (E[\Delta v_{o,k}^2] - 2E[\Delta v_{s,k-1}\Delta v_{o,k}] + E[\Delta v_{s,k-1}^2]) \\
& + 2\alpha(1-\alpha)(E[\Delta x_{s,k-1}\Delta x_{o,k}] + TE[\Delta v_{s,k-1}\Delta x_{o,k}]) + 2T\alpha\eta \\
& \cdot (E[\Delta x_{o,k}\Delta v_{o,k}] - E[\Delta x_{o,k}\Delta v_{s,k-1}]) + 2T\eta(1-\alpha) \\
& - E[\Delta x_{s,k-1}\Delta v_{s,k-1}] + TE[\Delta v_{s,k-1}\Delta v_{o,k}] - TE[\Delta v_{s,k-1}^2]).
\end{aligned}
\tag{A9}
$$

Because we have assumed $k \to \infty$, the variances and covariances of errors do not depend on k. Consequently, we can define the variances and covariances of the smoothing process as:

$$
\sigma_{sx}^2 \equiv E[\Delta x_{s,k}^2] = E[\Delta x_{s,k-1}^2], \sigma_{sv}^2 \equiv E[\Delta v_{s,k}^2] = E[\Delta v_{s,k-1}^2],
$$
$$
\sigma_{sxv}^2 \equiv E[\Delta x_{s,k}\Delta v_{s,k}] = E[(\Delta x_{s,k-1}\Delta v_{s,k-1})].
\tag{A10}
$$

In addition, the following relations are satisfied because the smoothed parameters are a linear combination of the measured parameters:

$$
E[\Delta x_{s,k}\Delta x_{o,k}] = E[\Delta x_{s,k}\Delta v_{o,k}] = E[\Delta v_{s,k}\Delta x_{o,k}] = E[\Delta v_{s,k}\Delta v_{o,k}] = 0.
\tag{A11}
$$

Substituting Equations (A10) into (A9) and simplifying by means of Equation (15), we have:

$$
\alpha(2-\alpha)\sigma_{sx}^2 + (1-\alpha-\eta)T^2\sigma_{sv}^2 + 2(\alpha-1)(1-\alpha-\eta)T\sigma_{sxv}^2 = \alpha^2 B_x + \eta^2 T^2 B_v.
\tag{A12}
$$

We can obtain the following equations similarly by calculating $E[\Delta v_{s,k}^2]$ and $E[\Delta x_{s,k}\Delta v_{s,k}]$:

$$
\begin{aligned}
& -\beta^2\sigma_{sx}^2 + (2-\beta-\theta)(\beta+\theta)T^2\sigma_{sv}^2 + 2\beta(1-\beta-\theta)T\sigma_{sxv}^2 \\
& = \beta^2 B_x + \theta^2 T^2 B_v,
\end{aligned}
\tag{A13}
$$
$$
\begin{aligned}
& \beta(1-\alpha)\sigma_{sx}^2 + (1-\beta-\theta)(\alpha+\eta-1)T^2\sigma_{sv}^2 \\
& + (\alpha+2\beta+\theta-\alpha\theta-2\alpha\beta-\eta\beta)T\sigma_{sxv}^2 = \alpha\beta B_x + \eta\theta T^2 B_v,
\end{aligned}
\tag{A14}
$$

Substituting the solutions of the linear system involving Equations (A12)–(A14) into (A2) and using Equation (A10), we arrive at Equation (20).

Appendix B. Derivation of Equation (26)

The e_{fin} of the α-β-η-θ filter is derived using the final value theorem in the z-domain. Applying a z-transform to Equations (1)–(4), we obtain:

$$
X_p(z) = X_s(z)/z + TV_s(z)/z,
\tag{A15}
$$
$$
V_p(z) = V_s(z)/z,
\tag{A16}
$$
$$
X_s(z) = X_p(z) + \alpha(X_o(z) - X_p(z)) + T\eta(V_o(z) - V_p(z)),
\tag{A17}
$$
$$
V_s(z) = V_p(z) + (\beta/T)(X_o(z) - X_p(z)) + \theta(V_o(z) - V_p(z)).
\tag{A18}
$$

By simplifying of these equations, the relationship between $X_p(z)$ and $(X_o(z), V_o(z))$ is obtained as:

$$
X_p(z) = \frac{(\alpha+\beta)z + \alpha\theta - \alpha - \eta\beta}{f_{\alpha\beta\eta\theta}(z)}X_o(z) + \frac{(\eta+\theta)z - \eta}{f_{\alpha\beta\eta\theta}(z)}TV_o(z),
\tag{A19}
$$

where:

$$f_{\alpha\beta\eta\theta}(z) = z^2 + (\alpha + \beta + \theta - 2)z + \alpha\theta - \eta\beta - \alpha - \theta + 1. \tag{A20}$$

Because we have assumed that the target has constant acceleration a_c but do not assume the measurement errors,

$$X_o(z) = Z[a_c(kT)^2/2] = \frac{z(z+1)}{2(z-1)^3}a_cT^2, \tag{A21}$$

$$V_o(z) = Z[a_c(kT)] = \frac{z}{(z-1)^2}a_cT, \tag{A22}$$

where $Z[\ ]$ denotes the z-transform. With Equations (A19), (A21) and (A22), the predicted error in the z-domain, $E_p(z) \equiv X_o(z) - X_p(z)$, is:

$$E_p(z) = \frac{z(z+1)(z^2+(\theta-2)z+1-\theta)-2z(z-1)(\eta z+\theta z-\eta)}{2(z-1)^3 f_{\alpha\beta\eta\theta}(z)}a_cT^2. \tag{A23}$$

Thus, applying the final value theorem, Equation (26) is derived.

Appendix C. Derivation of Equations (34)–(36)

The relationship between steady state Kalman gains and Q_{gen} is next derived. The k in Equations (7)–(11) is omitted in the following equations because of the steady state assumption. The i-th row and j-th column of a matrix P are denoted as $P^{i,j}$. With Equations (11) and (33), $\tilde{P}$ is calculated as:

$$\tilde{P} = \begin{pmatrix} \hat{P}^{1,1} + 2T\hat{P}^{1,2} + T^2\hat{P}^{2,2} + a & \hat{P}^{1,2} + T\hat{P}^{2,2} + b \\ \hat{P}^{1,2} & \hat{P}^{2,2} + c \end{pmatrix}. \tag{A24}$$

Equation (9) can also be written as [34]:

$$K = \hat{P}H^T R^{-1}. \tag{A25}$$

Substituting Equations (13)–(15) into Equation (A25), we have:

$$K = \begin{pmatrix} \alpha & T\eta \\ \beta/T & \theta \end{pmatrix} = \begin{pmatrix} \hat{P}^{1,1}/B_x & \hat{P}^{1,2}/B_v \\ \hat{P}^{1,2}/B_x & \hat{P}^{2,2}/B_v \end{pmatrix}, \tag{A26}$$

which means that:

$$\hat{P} = \begin{pmatrix} \alpha B_x & \eta T B_v \\ \beta B_x/T & \theta B_v \end{pmatrix}. \tag{A27}$$

With $\hat{P}^{1,2} = \hat{P}^{2,1}$ and Equation (A27), we obtain:

$$\eta = \beta B_x/T^2 B_v = \beta R_{xv}, \tag{A28}$$

which implies that Equation (17) is satisfied for not only the RA model, but also arbitrary process noise. Substituting Equations (13) and (14) into Equation (11), we obtain:

$$\hat{P} = \begin{pmatrix} (1-\alpha)\tilde{P}^{1,1} - T\eta\tilde{P}^{1,2} & (1-\alpha)\tilde{P}^{1,2} - T\eta\tilde{P}^{2,2} \\ (1-\theta)\tilde{P}^{1,2} - (\beta/T)\tilde{P}^{1,1} & (1-\theta)\tilde{P}^{2,2} - (\beta/T)\tilde{P}^{1,2} \end{pmatrix}. \tag{A29}$$

Substituting Equation (A27) into Equation (A24), substituting Equation (A24) into Equation (A29) and comparing their elements, we have:

$$\alpha B_x = (1-\alpha)(\alpha B_x + 2\eta T^2 B_v + T^2\theta B_v + a) - T\eta(\eta T B_v + \theta T B_v + b), \tag{A30}$$

$$\beta B_x/T = (1-\alpha)(\eta T B_v + \theta T B_v + b) - T\eta(\theta B_v + c), \tag{A31}$$

$$\theta B_v = (1-\theta)(\theta B_v + c) - (\beta/T)(\eta T B_v + \theta T B_v + b). \tag{A32}$$

Solving the linear system composed of Equations (A30)–(A32) with respect to (a, b, c) using Equation (A28), we obtain Equations (34)–(36).

References

1. Wu, K.; Cai, Z.; Zhao, J.; Wang, Y. Target Tracking Based on a Nonsingular Fast Terminal Sliding Mode Guidance Law by Fixed-Wing UAV. *Appl. Sci.* **2017**, *7*, 333.
2. Fan, Y.; Lu, F.; Zhu, W.; Bai, G.; Yan, L. A Hybrid Model Algorithm for Hypersonic Glide Vehicle Maneuver Tracking Based on the Aerodynamic Model. *Appl. Sci.* **2017**, *7*, 159.
3. Li, W.; Sun, S.; Jia, Y.; Du, J. Robust unscented Kalman filter with adaptation of process and measurement noise covariances. *Digit. Signal Process.* **2016**, *48*, 93–103.
4. Saho, K.; Masugi, M. Automatic Parameter Setting Method for an Accurate Kalman Filter Tracker Using an Analytical Steady-State Performance Index. *IEEE Access* **2015**, *3*, 1919–1930.
5. Jin, B.; Jiu, B.; Su, T.; Liu, H.; Liu, G. Switched Kalman filter-interacting multiple model algorithm based on optimal autoregressive model for manoeuvring target tracking. *IET Radar Sonar Navig.* **2015**, *9*, 199–209.
6. Martino, L.; Read, J.; Elvira, V.; Louzada, F. Cooperative Parallel Particle Filters for on-Line Model Selection and Applications to Urban Mobility. *Digit. Signal Process.* **2017**, *60*, 172–185.
7. Drovandi, C.C.; McGree, J.; Pettitt, A.N. A sequential Monte Carlo algorithm to incorporate model uncertainty in Bayesian sequential design. *J. Comput. Graph. Stat.* **2014**, *23*, 324.
8. Crouse, D.F. A general solution to optimal fixed-gain (α-β-γ etc.) filters. *IEEE Signal Process. Lett.* **2015**, *22*, 901–904.
9. Bar-Shalom, Y.; Li, X.R. *Estimation and Tracking: Principles, Techniques, and Software*; Artech House Publishers: Boston, MA, USA, 1998.
10. Kalata, P.R. The Tracking Index: A Generalized Parameter for α-β and α-β-γ Target Trackers. *IEEE Trans. Aerosp. Electron. Syst.* **1984**, *AES-20*, 174–182.
11. Saho, K.; Masugi, M. Performance analysis of alpha-beta-gamma tracking filters using position and velocity measurements. *EURASIP J. Adv. Signal Process.* **2015**, *2015*, 35.
12. Ekstrand, B. Some Aspects on Filter Design for Target Tracking. *J. Control Sci. Eng.* **2012**, *2012*, doi:10.1155/2012/870890.
13. O'Shea, T.P.; Bamber, J.C.; Harris, E.J. Temporal regularization of ultrasound-based liver motion estimation for image-guided radiation therapy. *Med. Phys.* **2016**, *43*, 455–464.
14. Khin, N.H.; Che, Y.F.; Eileen, S.M.L.; Liang, W.X. Alpha Beta Gamma Filter for Cascaded PID Motor Position Control. *Procedia Eng.* **2012**, *41*, 244–250.
15. Lee, Y.S.; Lee, H.J. Multiple object tracking for fall detection in real-time surveillance system. In Proceedings of the International Conference Advanced Communication Technology 2009 (ICACT2009), Phoenix Park, Korea, 15–18 February 2009; pp. 2308–2312.
16. Matsunami, I.; Nakamura, R.; Kajiwara, A. Target State Estimation Using RCS Characteristics for 26 GHz Short-Range Vehicular Radar. In Proceedings of the IEEE 2013 International Conference on Radar, Adelaide, SA, Australia, 9–12 September 2013; pp. 304–308.
17. Jatoth, R.K.; Gopisety, S.; Hussain, M. Performance Analysis of Alpha Beta Filter, Kalman Filter and Meanshift for Object Tracking in Video Sequences. *Int. J. Image Graph. Signal Process.* **2015**, *7*, 24–30.
18. Abdelkrim, M.; Mohammed, D.; Mokhtar, K.; Abdelaziz, O. A simplified alpha-beta based Gaussian sum filter. *AEU Int. J. Electron. Commun.* **2013**, *67*, 313–318.
19. Mohammed, D.; Mokhtar, K.; Abdelaziz, O.; Abdelkrim, M. A new IMM algorithm using fixed coefficients filters (fast IMM). *AEU Int. J. Electron. Commun.* **2009**, *64*, 1123–1127.

20. Ma, K.; Chang, Y.; Li, H.; Gao, J. A new method of target tracking in Ultra-Short-Range Radar. In Proceedings of the International Conference on Computer Science and Network Technology (ICCSNT2013), Dalian, China, 12–13 October 2013; pp. 10–12.

21. Wang, Y. Feature point correspondence between consecutive frames based on genetic algorithm. *Int. J. Robot. Autom.* **2006**, *21*, 35–38.

22. Yoo, J.C.; Kim, Y.S. Alpha-beta-tracking index (α-β-Λ) tracking filter. *Signal Process.* **2003**, *83*, 169–180.

23. Dai, X.; Zhou, Z.; Zhang, J.J.; Davidson, B. Ultra-wideband radar-based accurate motion measuring: Human body landmark detection and tracking with biomechanical constraints. *IET Radar Sonar Navig.* **2015**, *9*, 154–163.

24. Saho, K.; Sakamoto, T.; Sato, T.; Inoue, K.; Fukuda, T. Pedestrian imaging using UWB Doppler radar interferometry. *IEICE Trans. Commun.* **2013**, *E96-B*, 613–623.

25. Wang, Y.; Liu, Q.; Fathy, A.E. CW and pulse-Doppler radar processing based on FPGA for human sensing applications. *IEEE Trans. Geosci. Remote Sens.* **2013**, *51*, 3097–3107.

26. Zhoua, G.; Wub, L.; Xiea, J.; Denga, W.; Quan, T. Constant turn model for statically fused converted measurement Kalman filters. *Signal Process.* **2015**, *108*, 400–411.

27. Jahromi, M.J.; Bizaki, H.K. Target Tracking in MIMO Radar Systems Using Velocity Vector. *J. Inf. Syst. Telecommun.* **2014**, *2*, 150–158.

28. Geetha, B.; Ramachandra, K.V. A Three State Kalman Filter with Range and Range-Rate Measurements. *Int. J. Comput. Appl.* **2013**, *3*, 85–101.

29. Yoon, J.H.; Kim, D.Y.; Bae, S.H.; Shin, V. Joint Initialization and Tracking of Multiple Moving Objects Using Doppler Information. *IEEE Trans. Signal Process.* **2011**, *59*, 3447–3452.

30. Saho, K. Fundamental properties and optimal gains of a steady state velocity measured α-β tracking filter. *Adv. Remote Sens.* **2014**, *3*, 61–76.

31. Sudano, J.J. The alpha-beta-eta-theta tracker with a random acceleration process noise. In Proceedings of the IEEE National Aerospace and Electronics Conference (NEACON2000), Dayton, OH, USA, 10–12 October 2000; pp. 165–171.

32. Jury, E.I. *Theory and Application of the z-Transform Method*; John Wiley and Sons: New York, NY, USA, 1964.

33. Baldi, P. Gradient Descent Learning Algorithm Overview: A General Dynamical Systems Perspective. *IEEE Trans. Neural Netw.* **1995**, *6*, 182–195.

34. Gelb, A. *Applied Optimal Estimation*; The M.I.T. PRESS: Cambridge, MA, USA, 1974.

applied
sciences

MDPI

Article

Dynamic Multiple Junction Selection Based Routing Protocol for VANETs in City Environment

Irshad Ahmed Abbasi [1,2,*] , Adnan Shahid Khan [1] and Shahzad Ali [3]

1   Department of Computer Systems and Communication Technologies, Faculty of Computer Science and
    Information Technology, Universiti Malaysia Sarawak (UNIMAS), Kota Samarahan 94300, Malaysia;
    15010165@siswa.unimas.my (I.A.A.); skadnan@unimas.my (A.S.K.)
2   Department of Computer Science, Faculty of Science and Arts at Balgarn, University of Bisha, P.O. Box 60,
    Sabt Al-Alaya 61985, Saudi Arabia
3   Department of Computer Science, Al Jouf University, Tabarjal 74331, Saudi Arabia; shahzad@ieee.org
*   Correspondence: irshad.upesh@gmail.com or aabasy@ub.edu.sa; Tel.: +966-537-080-524

Received: 23 January 2018; Accepted: 16 February 2018; Published: 28 April 2018

Abstract: VANET (Vehicular Ad-hoc Network) is an emerging offshoot of MANETs (Mobile Ad-hoc Networks) with highly mobile nodes. It is envisioned to play a vital role in providing safety communications and commercial applications to the on-road public. Establishing an optimal route for vehicles to send packets to their respective destinations in VANETs is challenging because of quick speed of vehicles, dynamic nature of the network, and intermittent connectivity among nodes. This paper presents a novel position based routing technique called Dynamic Multiple Junction Selection based Routing (DMJSR) for the city environment. The novelty of DMJSR as compared to existing approaches comes from its novel dynamic multiple junction selection mechanism and an improved greedy forwarding mechanism based on one-hop neighbors between the junctions. To the best of our knowledge, it is the first ever attempt to study the impact of multiple junction selection mechanism on routing in VANETs. We present a detailed depiction of our protocol and the improvements it brings as compared to existing routing strategies. The simulation study exhibits that our proposed protocol outperforms the existing protocols like Geographic Source Routing Protocol (GSR), Enhanced Greedy Traffic Aware Routing Protocol (E-GyTAR) and Traffic Flow Oriented Routing Protocol (TFOR) in terms of packet delivery ratio, end-to-end delay, and routing overhead.

Keywords: intelligent transportation; multiple junctions; position based routing; optimal route

1. Introduction

The emerging Vehicular Ad-hoc Networks (VANETs) is getting a spotlight from entities like academicians, research institutes, and industries because it is envisioned to play a very important role for the future transportation system. It aims at enhancing the driving experience of the users by playing a crucial role in developing Intelligent Transportation System (ITS) and road safety by providing road conditions and vehicular traffic information to the drivers. For the safety of drivers and regulating the flow of vehicles, the drivers have to be alerted to unwanted traffic accidents, traffic congestion, road conditions, and other associated features. The main objective of VANETs is to address these issues by providing the exact and timely information to the drivers [1–12].

In VANETs, nodes are self-organized and capable of communicating in an infrastructure-less environment [1–5,13,14]. The integration of advanced wireless technologies in vehicles helps them to communicate without any infrastructure, which reduces the cost of hefty infrastructure deployment. For the wireless access in a vehicular environment, IEEE 802.11 committee has developed a standard named as IEEE 802.11p. For Vehicle-to-Vehicle (V2V) and Vehicle-to-Infrastructure (V2I) communications, the Federal Communications Commission (FCC) has allocated a bandwidth of 75 MHz in 5.9 GHz band for licensed Dedicated Short Range Communication (DSRC). Each vehicle exchanges information with other vehicles as well as roadside units within their radio ranges by using the allocated bandwidth by FCC [1–3,13,14]. Also, various significant projects like CarTALK2000(Car Talk 2000) [15], C2CCC(Car 2 Car Communication Consortium) [16], California PATH(California Partners for Advanced Transportation Technology) [17], FleetNet [18], DEMO 2000 by Japan Automobile Research Institute (JSK) [19], Chauffeur in EU [20] and Crash Avoidance Metrics Partnership (CAMP) [19] are initiated by different firms for developing vehicular communication.

VANETs are an offshoot of Mobile Ad hoc Networks with certain differentiating characteristics. For instance, the nodes in MANETs (Mobile Ad-hoc Networks) have random movements, but the movement of vehicular nodes is constrained by predefined roads. Traffic control system, speed barriers, and congestion on the road also affect the speed and mobility pattern of vehicles. The vehicles can be equipped with transceivers having longer transmission ranges, broad on-board storage capacities, digital maps, and sensors for revealing vehicle states. Unlike MANETs, limited battery, processing power, and storage do not make an issue in VANET due to the availability of rechargeable sources of energy [11–14,21,22].

There are a lot of technical challenges in the design of effective vehicular communications. One of them is to design an efficient routing strategy capable of computing shortest rich density optimal path. Finding such a path is a challenge because of the high mobility of the nodes, the intermittent connectivity, and the uneven distribution of vehicular nodes, which varies with time and location. The nodes in one area may be sparsely connected while in another area they can be densely connected. While, at the same time, the predictable mobility and mobility restrictions are helpful in establishing routes [2,3,8]. Traditional MANETs routing protocols are unable to handle these issues. For routing in VANETs, researchers have evaluated various well-known MANETs routing protocols but they have concluded that MANETs routing protocols do not fit well in VANETs setting [23–26].

In the existing literature, most of the well-known routing protocols specifically designed for VANETs establish routing path by considering immediate junction/street from source to destination. Considering only immediate junction at one time may result in routing the packets along the streets where there is no carrier or negligible connectivity resulting in an increased delay as well as packet loss. If the packets stay in a buffer for a long time then there is a possibility that they will be discarded due to the expiry of time-to-live. Consequently, this degrades the network performance. To the best of our knowledge, a protocol that is capable of establishing dynamic optimal routing path by considering multiple junctions/streets from source to destination based on shortest path and connectivity is missing in the existing literature. This paper presents a novel position based routing protocol for VANETs called Dynamic Multiple Junction Selection based Routing protocol (DMJSR). It is designed specifically for city scenarios. DMJSR takes into account vehicles speed, their directions and locations, and bi-directional double lanes road. It makes use of GPS and digital maps for getting the vehicle current location and the road information respectively. It considers multiple-neighbor junctions while making routing decisions which maximizes connectivity and enhances networks performance as compared to existing approaches. Moreover, it is also novel in its mechanisms as it combines multi-junction selection mechanism with position and direction based forwarding mechanism for forwarding of packets between the selected junctions. It establishes dynamic optimal connected routing paths, which enhances the performance in terms of packet delivery ratio, end-to-end delay, and routing overhead. It is vital to achieve enhanced performance in terms of the aforementioned parameters for providing safety services (like coordinated communication between two vehicles, real-time traffic management, cooperative message transfer, accidents warnings, road hazard control warnings, post-crash announcements, and traffic watchfulness), convenience services (like parking availability, electronic toll collections, and active prediction) and commercial oriented services (like distant vehicle diagnostics/personalization, downloading digital maps, internet access, advertisements, real-time video relay, etc.).

The rest of the paper is organized as follows. Section 2 describes the existing routing techniques along with their limitations and elaborates the motivation of our work. Section 3 illustrates the proposed routing strategy. Section 4 presents the simulation results and analysis. The paper is concluded in Section 5.

2. Related Work

Routing protocols for VANETs can be categorized into two main types. The first type is topology-based routing protocols and the second type is position-based routing protocols. The topology-based routing protocols can further be categorized into reactive, proactive, and hybrid protocols [27,28]. The reactive protocols like Dynamic Source Routing(DSR) [29] and Ad-hoc on Distance Vector Routing(AODV) [30] are based on demand routing because in these routing protocols only those routes that are currently in use are maintained [14,21,31]. Proactive routing protocols, like Optimized Link State Routing (OLSR) [32], are table driven routing protocols because in such protocols all the available routes in the network topology are maintained [2,3,31,33]. Due to frequently changing network topology in VANETs, the topology based routing protocols are not suitable. Such protocols are not scalable under highly mobile large-scale VANETs. These protocols incur high latency in finding a route because they store all the unused routing paths [3,34,35]. A protocol like AODV consumes additional bandwidth due to transmission of periodic heavy beacon messages in intermittently connected VANETs [14,31]. In addition, the maintenance of routing paths in highly mobile environments results in large routing overhead for topology-based routing protocols [5–8,11]. In order to overcome the restrictions of topology-based routing, position-based routing which is based on location information is introduced [35]. In position-based routing protocols, the vehicular nodes are equipped with GPS for getting accurate information about their geographical locations, speeds, and moving directions. Destination node location can be found by using location services such as GLS (Grid Location Service) [36], RLS (Reactive Location Service) [37] and HLS(Hierarchical Location Service) [34]. Beacon messages are used to obtain one-hop neighbor information. There is no need to maintain the routing paths in position-based routing protocols. Due to all these features, position-based routing protocols overcome the major limitations of topology-based routing protocols [2,3,11,12,31,33,34]. In the literature, many position-based routing protocols are proposed in VANETs. As the focus of this work is city environments, therefore, we mention some of the well-known protocols specifically designed for city scenarios.

Geographic Source Routing Protocol (GSR) [38] was designed for city scenarios. It uses Dijkstra shortest path algorithm [2] to accomplish the shortest path between sender and destination. The shortest path computed by GSR consists of road junctions that are ordered sequentially. The data packets have to transverse these junctions in order to move from source to destination. The source node makes use of Reactive Location Services (RLS) [14] for locating the destination. GSR uses greedy forwarding for relaying packets between the junctions. The simulation outcomes, with the use of realistic vehicular traffic in urban surroundings, illustrated that GSR performs better then topology based routing protocols like DSR and AODV in terms of end-to-end delay and packet delivery ratio [2,31]. However, GSR is not a traffic aware routing protocol. Therefore, it does not make use of traffic information for making the routing decisions [3].

In [33], Greedy Perimeter Coordinator Routing (GPCR) was proposed for city environments. GPCR consists of greedy forwarding and perimeter mode. The unique feature of the protocol is that it makes routing decision at the coordinator node without considering digital maps. The node that is near the junction or at the junction is called coordinator. In GPCR the restricted greedy forwarding bound the packet carrier node to forward the packets to a node at the junction instead of dispatching them across the junction. The local optimum situation is overcome by perimeter mode. An assumption is made in perimeter mode that the graphs are naturally planner in city scenarios. Thus, it avoids graph planarization that may create partitions in the network [3,21,38].

Greedy Perimeter Stateless Routing Junction+ (GPSRJ+) [39] is an improved version of GPCR. Unlike GPCR, it avoids unnecessary packet stop at a junction which raises hop count. It employs two-hop neighbor information to predict which city street its neighboring junction vehicle will take. If prediction identifies that its neighboring junction will forward the packet on to a street with a different direction, it forwards the packet to the junction vehicle; else it avoids the junction and forwards the packet to a vehicle that is nearest to the destination vehicle. The simulation result shows that GPSRJ+ surpasses GPCR in terms of packet delivery ratio and hop count.

Anchor-based Street and Traffic-Aware Routing (A-STAR) [40] identifies anchor path with high connectivity by using information about city bus routes. When a packet faces local optimum situation then route recovery strategy is used to overcome the local optimum problem by calculating a new anchor path. The simulation results verify that A-STAR outperforms Greedy Perimeter Stateless Routing (GPSR) and GSR because it is capable of finding the end-to-end connected path even in case of low vehicular density. However, one problem with this approach is that routes are along the anchor path that may not be optimal resulting in increased end-to-end delay [2,31].

Enhanced Greedy Traffic-Aware Routing Protocol (E-GyTAR) [2] is similar to GyTAR but during junction selection mechanism, it considers only directional density. It establishes the shortest path based on directional density and thereby route the packet toward the destination. It uses greedy packet forwarding strategy to forward the packet between the junctions. It avoids local optimum by using carry and forward approach. The main drawback of this routing protocol is that it selects junctions based on directional density and ignores non-directional density on a multi-lane road. If directional density is absent then this protocol cannot find the way to relay packets towards the destination [3].

Traffic Flow Oriented Routing Protocol (TFOR) [3] is a recently proposed technique which consists of two modules: (a) A junction selection mechanism based on traffic flows and the shortest routing path and (b) A forwarding strategy based on two-hop neighbor information. It accomplishes shortest optimal path based on shortest distance to the recipient node and vehicular traffic density. Simulation outcomes show that TFOR outperforms E-GyTAR and GyTAR in terms of packet delivery ratio and end-to-end delay.

Directional Geographic Source Routing (DGSR) [14]. It is an enhanced version of geographic source routing (GSR) with directional forwarding strategy. In this routing scheme, the source vehicular node uses location services to acquire the position of destination vehicle. It establishes the shortest path from source to destination using Dijkstra Algorithm. The shortest path is consisting of junctions which are ordered sequentially. The packets from source vehicular node follow the sequence of junctions to reach the destination. If packet meets a local optimum, DGSR uses carry and forward approach to overcome the local optimum problem.

Enhanced Greedy Traffic Aware Routing Protocol-Directional (E-GyTARD) [14]. It is an extended version of E-GyTAR [2] with directional forwarding. It consists of two mechanisms: (i) Junction selection; (ii) Directional greedy forwarding strategy. It uses location services to get the position of the destination node. It selects junctions on the basis of directional traffic density and shortest distance to the destination. It forwards packets in between junction using directional greedy forwarding. Simulation outcomes in realistic urban scenarios show that E-GyTARD outperformed GSR and DGSR in terms of packet delivery. Table 1 shows the comparative characteristic of all the aforementioned routing protocols.

Table 1. The comparative analysis of position-based routing protocols.

Charateristics	Location Based Routing Protocols for VANETs								
	GSR [34]	GPCR [33]	GPSRJ+ [39]	A-STAR [40]	GyTAR [31]	E-GyTAR [2]	TFOR [3]	DGSR [14]	E-GyTARD [14]
Dynamic Junction Selection	No	No	No	No	Yes	Yes	Yes	Yes	Yes
Static Junction Selection	Yes	-	No	Yes	No	No	No	No	No
Scenario	City	City	City	City	City	City	City	City	City
Hop Count	Single hop	Single hop	Two hops	Single hop	Single hop	Single hop	Two hops	Single hop	Single hop
Realistic Mobility Flows	Yes	Yes	Yes	Yes	Yes	Yes	Yes	Yes	Yes
Local Optimum Handling Strategy	Fall back on greedy mode	Right hand rule	Perimeter mode	Reconstruct anchor path	Carry and forward	Carry and forward	Carry and forward	Carry and forward	Carry and forward
Map Required	Yes	Yes	Yes	Yes	Yes	Yes	Yes	Yes	Yes
GPS Required	Yes	Yes	Yes	Yes	Yes	Yes	Yes	Yes	Yes
Prediction Based	No	No	Yes	No	Yes	Yes	Yes	No	Yes
Location Service Required	Yes	Yes	Yes	Yes	Yes	Yes	Yes	Yes	Yes
Traffic Aware	No	No	No	No	Yes	Yes	Yes	No	Yes

[1] VANETs, Vehicular Ad-hoc Networks; GSR, Geographic Source Routing Protocol; GPCR, Greedy Perimeter Coordinator Routing Protocol; GPSRJ+, Greedy Perimeter Stateless Routing Junction+ Protocol; A-STAR, Anchor-based Street and Traffic Aware Routing Protocol; E-GyTAR, Enhanced Greedy Traffic Aware Routing Protocol; TFOR, Traffic Flow Oriented Routing Protocol; DGSR, Directional Greedy Source Routing Protocol; E-GyTARD, Enhanced Greedy Traffic Aware Routing Protocol-Directional.

3. Dynamic Multiple Junction Based Source Routing Protocol

3.1. Limitatiotions of Existing Protocols

The protocols that are considered for performance comparison in this work are GyTAR, E-GyTAR and TFOR. The similarity among all these protocols is that all of these protocols are based on dynamic junction selection mechanism. In these routing protocols, one junction is selected at a time based on traffic density and shortest path to the destination. GSR is based on static junction selection mechanism while GyTAR, E-GyTAR and TFOR are based on dynamic junction selection mechanism. Existing literature shows that the protocols that are based on dynamic junction selection mechanism perform better as compared to protocols using static junction selection [2,3,11]. The dynamical junction selection protocols (GyTAR, E-GyTAR and TFOR) select one junction at a time for sending packets towards the destination. However, there are a few limitations of this approach. For instance, let us consider one of the scenarios shown in Figure 1. Figure 1 is added to this paper just to explain the possible shortcomings of the existing protocols and how the proposed protocol can solve these shortcomings. It is worth noting that the proposed routing protocol performs well for all the different simulation settings that were used for performance evaluation during our work.

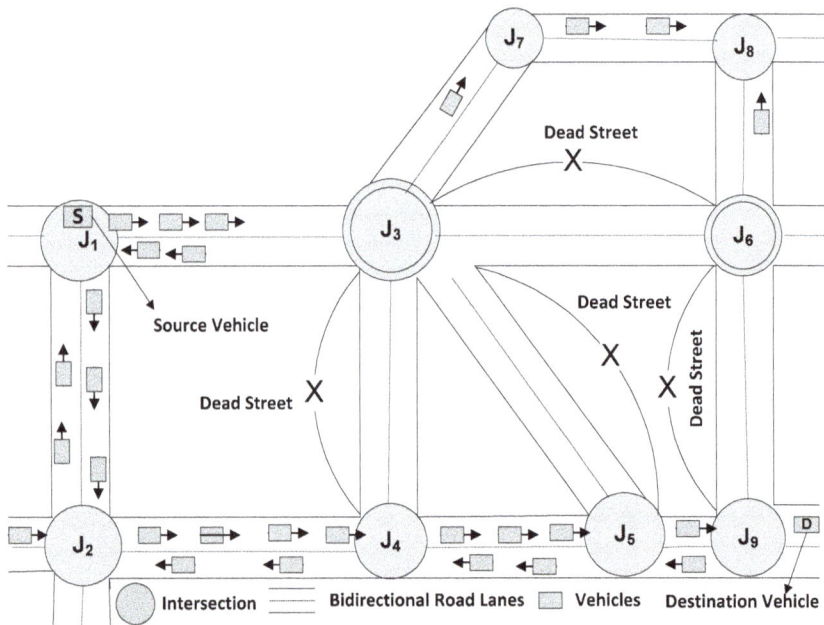

Figure 1. Limitations of Greedy Traffic Aware Routing Protocol (GyTAR), Enhanced Greedy Traffic Aware Routing Protocol (E-GyTAR), and Traffic Flow Oriented Routing Protocol (TFOR).

Suppose source vehicle S is at junction J_1 and wants to forward a packet to the destination vehicle D. TFOR, GyTAR and E-GyTAR choose next junction on the basis of vehicular traffic density and shortest path to the destination node, therefore J_3 will be selected as the next junction by all the protocols. At J_3 GyTAR, E-GyTAR and TFOR will be unable to find next appropriate junction as a next junction through which packet can be routed toward the destination. Because the next shortest path providing streets that lead to destination contain no vehicular density to forward the packet towards the destination. When the packet is routed towards J_3, all the protocols are unable to decide next path because from J_3 to J_6 there is no traffic density. Similarly, J_3 to J_5 and J_3 to J_4 there is no traffic density. All these streets are out of traffic as there is no traffic density for routing packet further. In this case, routing performance is degraded here because of inappropriate selection of junctions due to consideration of a single junction only. The packet delivery ratio decrease and end-to-end delay increases despite the optimal paths (J_1–J_2, J_2–J_4, J_4–J_5, J_5–J_9) are available with rich traffic density. Consequently, all these protocols are inefficient in junction selection mechanism to decide optimal routing paths on the basis of vehicular traffic density and shortest distance to the destination. Furthermore, their current junction selection mechanism is very limited to move the packet progressively closer to the destination due to the consideration of just the immediate junction from the current junction. The probability of incurring a dead street (street without packet carriers/vehicles) along a selected path to the destination increases while considering just the immediate junction.

We need a routing protocol that selects junctions efficiently by considering connectivity in such a way that it maximizes packet delivery ratio, minimizes routing overhead and end-to-end delay. We propose a novel routing strategy called dynamic multiple junction selection based routing mechanism, which selects multiple junctions based on shortest path and traffic density for tackling aforementioned like scenarios.

3.2. DMJSR Protocol Overview

It is a novel position based routing protocol capable of finding robust routes in city environment by considering traffic density and shortest distance to the destination. The proposed protocol in this paper uses multi-hop communication for routing of data in VANETs. The main objective of this routing protocol is to ensure the selection of paths having more connectivity within city environments to enhance the performance of the network.

3.2.1. Protocol Assumptions

DMJSR is a junction-based geographic routing protocol. Certain assumptions are made for the working of this protocol. Similar assumptions are made in [2,3,31]. GPS is utilized by a vehicle in order to locate its position. The destination vehicular node can be located using location services such as GLS [36]. Every vehicle contains an onboard navigation system, which provides the location of neighboring junctions and useful street level information by using preloaded digital maps. Moreover, it is also assumed that every vehicular node is familiar with its direction and velocity. We also assume that each vehicular node has information about vehicular traffic density between two junctions. This information can be made available by traffic sensors deployed beside junctions or simple distributed mechanism for road traffic estimation realized by all vehicles [41]. In the presence of aforementioned assumptions, the detailed description of our routing protocol is given below. It consists of two modules: (i) Dynamic multiple junction selection mechanism and (ii) improved greedy forwarding mechanism between the junctions on the basis of one-hop neighbor information.

3.2.2. Dynamic Multiple Junction Selection

Similar to the existing routing approaches GyTAR, E-GyTAR and TFOR, DMJSR routing protocol uses anchor based routing approach with city streets information. It utilizes street map topology for routing of data packets between vehicles. The major difference between the considered protocols and our protocol is the junction selection mechanism. DMJSR selects the next appropriate junction by considering the next two immediate junctions dynamically from current junction based on vehicular traffic density and shortest curve metric distance to the destination. Now a big question about the proposed protocol can be that why two junctions are considered and not three, four and so on? The answer to this question is given in Section 4.3.4. By considering two junctions, it decreases the probability of facing connectivity problem that occurs during existing junction selection mechanisms. Also, it decreases the probability of incurring dead-street (street without packet carriers/vehicles) along a selected path to the destination. It also moves packet progressively closer to the destination by using those streets that are rich in traffic/connectivity as compared to existing approaches. When choosing next two-hop neighbor junction, the sender vehicular node or intermediate vehicular node looks for the positions of two-hop neighboring junctions by using digital city streets map and a score is calculated and assigned to all candidate two-hop neighbor junctions based on traffic density and curve metric distance of candidate junctions to the destination. The two-hop neighbor junction with the highest score is selected as the next destination junction. Candidate junctions are allocated score using Algorithm 1.

In Algorithms 1 and 2, alpha (α) and beta (β) are the weighting factors for distance and traffic density respectively between the junctions. By adjusting the value of α and β, we can make a tradeoff between distance and traffic density when selecting next junction. Traffic density is vital for providing connectivity for relaying packet toward the destination. H_1 and H_2 are the weighting factors for candidate one-hop neighbor junction and two-hop neighbor junction respectively. An adjustment in the value of H_1 and H_2 can make a tradeoff between the importance of candidate one-hop neighbor junction and two-hop neighbor junctions. In our simulations, equal weights are assigned to all aforementioned parameters. Lines 1 to 10 set the values to the parameters used in our algorithm. Line 11 checks if candidate one-hop neighbor has next neighbor junction that leads to the destination then Algorithm 1 invokes Algorithm 2 using line 12 which is given below.

Algorithm 1. The Dynamic Multiple Junction Selection Mechanism

---

Input: Area, α, H_2
Output: The next destination junction NDj
1. begin
2.   set score $\leftarrow 0$
3.   set $\beta \leftarrow 1 - \alpha$
4.   set $H_1 \leftarrow 1 - H_2$
5.   for each candidate junction j do
6.   set $D_n \leftarrow$ the curve metric distance between NC_j and destination
            /* NC_j is the next candidate junction (one-hop) */
7.   set $Dc \leftarrow$ the curve metric distance between C_j and destination
            /* C_j is the current junction */
8.   set $Dp_1 \leftarrow D_n/D_c$ /* Closeness of candidate junction to destination */
9.   set $TD_1 \leftarrow$ no. of vehicles between NC_j and C_j in both directions
10.   if score $< (\alpha \times (1 - Dp_1) + \beta \times TD_1)$
11.   if NC_j contains next candidate neighbour junction NC_k //two-hop
12.   invoke Algorithm 2 //GetnextneighbourjuctionKofJ (ND_k, Score)
13.   set ScoreK = Score
14.   set $NC_j \leftarrow ND_k$
15.   set $ND_j \leftarrow NC_j$
16.   set score $\leftarrow H_1.(\alpha \times (1 - Dp_1) + \beta \times TD_1) + H_2. (ScoreK)$
17.   else
18.   set $ND_j \leftarrow NC_j$
19.   set score $\leftarrow \alpha \times (1 - Dp_1) + \beta \times TD_1$
20.   end
21.   end
22.   end
23.   return ND_j
24. end

Algorithm 2 is used to assign a score to each of second-hop neighbor junctions. It returns the junction with the highest score and control switches back to Algorithm 1. Algorithm 1 uses Line no 16 to compute the scores of the two-hop neighbor junctions. The junction having the highest score will be selected as the next two-hop candidate neighbor junction through which packet is relayed toward the destination. Any candidate two-hop neighbor junction that is closest to the destination and provides higher traffic density will be selected as the next destination junction through which packet moves toward the destination. The working of the proposed protocol is illustrated in Figure 2.

Algorithm 2. Second-Hop Neighbor Junction Score Computation

Input: Area, α

Output: The next destination junction ND_k with Score
   GetnextneighbourjuctionKofJ (NC_k, Score)

1.   begin
2.   for each candidate junction K of J do
3.   set $Dn_k \leftarrow$ the curve metric distance between NC_k and destination
             /* NC_k is the next candidate junction */
4.   set $Dc_k \leftarrow$ the curve metric distance between C_k and destination
             /* C_k is the current junction */
5.   set $Dp_2 \leftarrow Dn_k / Dc_k$ //closeness of second hop w.r.t destination
6.   set $TD_2 \leftarrow$ no. of vehicles between NC_k and C_k in both directions
7.   if score < ($\alpha \times (1 - Dp_2) + \beta \times TD_2$) then
8.   set $ND_k \leftarrow NC_k$
9.   set score $\leftarrow \alpha \times (1 - Dp_2) + \beta \times TD_2$
10.   end
11.   end
12.   return (ND_k, $Score_k$)
13. end

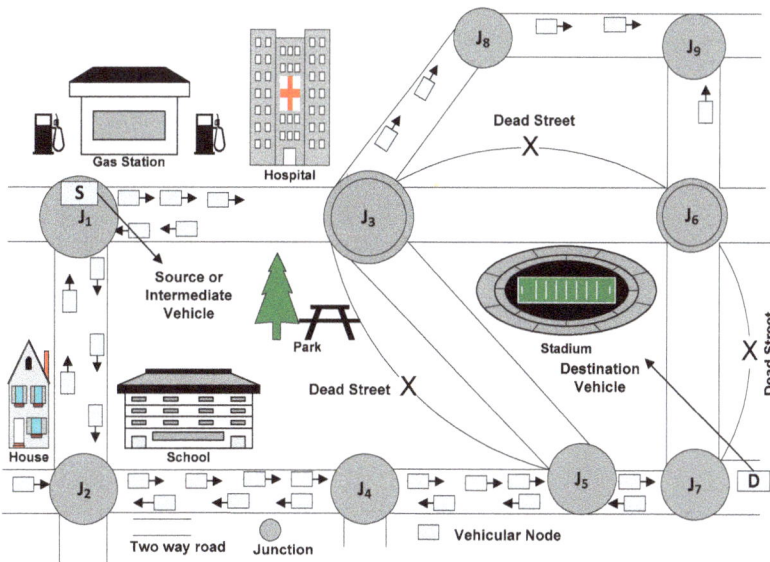

Figure 2. Working of Dynamic Multiple Junction Selection based Routing Protocol (DMJSR).

3.2.3. Illustrative Example for DMJSR Working

Consider the scenario of Figure 2 where the source vehicle S is at current junction J_1 and is dispatching packet towards the destination D using DMJSR. In this case, there are 4 candidate two-hop neighbor junctions of current junction J_1 available through which packet can be routed toward the destination. These include J_4, J_5, J_6 and J_8.

Our proposed algorithm will assign weights to each of two-hop neighbor junctions of J_1 according to the traffic density and shortest curve metric distance to the destination. City streets that are connecting J_1 to J_4 have higher traffic density than J_1–J_5, J_1–J_8 and J_1–J_6. Consequently, J_1 and J_4 have higher connectivity than J_1–J_6, J_1–J_8 and J_1–J_5.Therefore, DMJSR will assign more weight to J_4 as compared to J_5, J_6 and J_8 and it will be chosen as next destination junction. In this way, our routing approach will remove the limitations of GyTAR, E-GyTAR, and TFOR. In the case of such routing protocols, J_3 would have been selected instead of J_4, which will not be considered because all these protocols based on the selection of one-hop junction, which results in a sub-optimal choice in this case. The packet would have been stuck in local maximum because after selecting J_3, junctions J_5 and J_6 have no traffic density as these are along dead streets (those streets that contain no vehicle for carrying packet towards the destination). Our routing protocol route the packet from current junction J_1 to J_4 through J_2 and in this way it will relay the packet towards the destination node. Therefore, in DMJSR, a packet will travel successively closer towards the destination node along the urban streets where there are adequate vehicular nodes to provide connectivity. It is worth mentioning that Figure 2 is just an illustration of one of the possible cases where the considered protocols would suffer in terms of performance and the proposed protocol will work well. The proposed protocol works well for all the considered simulation scenarios as discussed in Section 4.

3.2.4. Forwarding between Junctions

After the determination of destination junction, DMJSR uses a greedy approach with one-hop neighbor information for moving forward data packets between junctions. It is achieved by marking all data packets with the location of the next destination junctions. Every vehicular node maintains a neighbor table that contains the direction, speed and position of each neighbor vehicular node. Neighbor table is updated through periodic beacon messages exchange. The packet carrier vehicular node consults its neighbor table for the most recent predicted locations of neighboring vehicles prior to forwarding the data packet. It forwards the packet to a one-hop neighbor that is closest to the destination.

Protocols like GyTAR [31] and E-GyTAR [2] make use of one-hop neighbor information based predictive strategy. This strategy prefers to forward the packet to the next node based on the speed of the vehicular node instead of preferring a node that lies closest to the destination. The one-hop neighbor having the highest speed is selected as next forwarding node. However, this approach has a few limitations. Firstly, in the presence of appropriate one-hop neighbors, it may end up selecting an inappropriate one-hop neighbor for forwarding packets which may incur more delay and hop count. For instance, Figure 3 presents a scenario mentioned in TFOR [3].

In this Scenario, suppose the neighbor A of F (forwarding vehicle) has greater speed among all of its one-hop neighbors such as B and E. The forwarding vehicular node F chooses vehicle A instead of vehicular node B because of its greater speed as compared to other neighbors and forward packet to it at time t_1. Vehicular node A is unable to forward packet to vehicular node C at time t_2, because vehicular node C is outside the transmission range of vehicular node A. But if the forwarding vehicular node F makes use of two-hop neighbor information which is accomplished by beacons exchange, then at time t_2, F will be able to dispatch the packet to vehicular node C through neighbor vehicular node B instead of vehicular node A. This is because B is nearer to C and C is the closest vehicular node to destination vehicle D. By doing so, it minimizes the delay while reducing the number of intermediate hops while relaying the packet from source to destination.

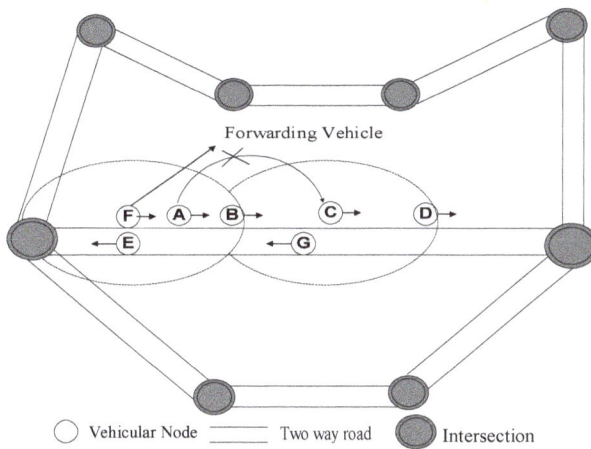

Figure 3. Limitation of speed based improved greedy forwarding strategy used in GyTAR, and E-GyTAR.

Secondly, speed-based forwarding strategy can cause an increase in a number of hops due to the variable high speed of different nodes [3]. By considering these facts, TFOR proposed two-hop neighbor based greedy forwarding that prefers position and direction of two-hop neighbor nodes instead of speed.

Two-hop neighbor based forwarding strategy used in TFOR minimizes the hop count and diminishes the end-to-end delay as compared to one-hop speed based improved greedy forwarding of GyTAR [31]. Our proposed protocol DMJSR employs one-hop neighbor information based greedy forwarding technique in which forwarding node considers position and direction of one-hop neighbor as a substitute of two-hop neighbor based greedy forwarding of TFOR and one-hop speed based improved greedy forwarding of GyTAR and E-GyTAR which helps it to achieve improved performance as compared to existing strategies.

The reason for this selection is because first of all it is simpler as compared to other variants and secondly it generates less routing overhead in the highly dense environment as compared to maintaining two-hop neighbor information.

Despite the forwarding strategy based on one-hop neighbor information, the risk remains that a packet carrier vehicle is unable to locate a next candidate forwarding vehicle to forward the packet, in that case, it will make use of carry and forward approach [31]. The packet carrier vehicle will hold the packet until an appropriate forwarding node in its vicinity is found. However, our approach minimizes the use of carry and forward as compared to existing approaches because of its emphasis on connectivity during junction selection which reduces end-to-end delay and packet loss. As frequent use of carry and forward causes more delays due to lack of connectivity and also packet loss because of expiry of the time-to-live duration of the packet. In general, if our protocol during junction selection phase finds that there is no vehicular density around the next candidate junctions, it will simply employ carry and forward technique to move packet toward next destination junction.

4. Simulation Setup and Result Analysis

For performance evaluation, GloMoSim (Global Mobile System Simulator) (2.03, UCLA (University of California Los Angeles), Los Angeles, CA, USA, 2008) [3,35,42] is used to carry out simulations.

4.1. Mobility Model

It is a very vital step to choose an appropriate mobility model for VANETs simulation. The mobility model must exhibit the real vehicular characteristics. The mobility model also affects the performance of protocols [2,3]. The mobility models for vehicular environment usually have two categories. These are macroscopic and microscopic mobility models. The macroscopic mobility is actually a reflection of mobility constraints such as roads, number of road lanes, the city streets, speed limits, traffic lights, traffic density, and traffic flows. The microscopic mobility exhibits the characteristics such as vehicles conduct with each other and with the infrastructure [35,43]. Both micro and macro mobility categories are made available by the VanetMobiSim. It is an extended version of CANU (Communication in Ad-hoc Networks for Ubiquitous Computing) mobility simulation model [35,42,43].

4.2. Simulation Scenario

In our simulation, VanetMobiSim (1.1, Institut Eurecom, Sophia Antipolis, Valbonne, France, 2007) [43] is used to generate the vehicular mobility traces in an area of 3000 × 2700 m². This area contains 18 intersections and 30 bidirectional multilane roads in which vehicular nodes mobility is simulated. At the start of the simulation, the vehicular nodes are randomly distributed over the multilane roads and move in both directions. The movements of the vehicles on the roads are based on car following model or intelligent driving model [43]. The vehicles speed depends on the kind of the vehicles (like car, bus, truck, etc.) and nature of the roads. The number of vehicles is varied from 100 to 350. The simulation results are averaged over 10 runs. The rest of the parameters are summarized in Table 2.

Table 2. Simulation setup.

Simulation/Scenario		MAC/Routing	
Simulation time	250 min	MAC (Medium Access Control) protocol	802.11 DCF (Distributed Coordination Function)
Map size	3000 × 2700 m²	Channel capacity	54 Mbps
Mobility model	VanetMobiSim	Transmission range	266 m
Number of intersections	23	Traffic model	15 CBR connections
Number of double lane roads	36	Packet sending rate	(1–10 packet(s)/second)
Number of vehicles	100–350	Vehicle Speed	35–60 Km/h
Number of simulation runs	10	Packet size	128 bytes
Weighting factors	$\alpha = 0.5$, $\beta = 0.5$, $H_1 = 0.5$, $H_2 = 0.5$	Beacon Interval	1 s

4.3. Simulation Results and Discussion

For performance comparison, we have compared three other routing protocols with our proposed protocol. These protocols are GSR [38], E-GyTAR [2] and TFOR [3]. The performance metrics used for comparison include packet delivery ratio, end-to-end delay, and routing overhead. Packet delivery ratio is defined as the percentage of packets that are successfully delivered to their destination vehicles. End-to-end delay is defined as the average delay incurred by the packet from its source to its destination. Routing overhead is the ratio of total control packets produced to the total data packets received at the destinations during the complete simulation [2,3,31].

4.3.1. Packet Delivery Ratio

Figure 4 depicts that packet delivery ratio for all the considered protocols increases with increase in vehicular density. It can be observed from the figure that packet delivery ratio of DMJSR is higher as compared to GSR, E-GyTAR and TFOR. This is primarily because, in DMJSR, the routing path is established progressively based on multiple junctions ensuring that the junctions with higher vehicular traffic density are selected. Consequently, a packet will travel successively closer towards the destination along the city streets where there are adequate vehicular nodes to ensure high network connectivity. Whereas in case of GSR, the source vehicle computes series of junction statically before transmission of the packet without taking into account the vehicular traffic density. As a result, some data packets are unable to find their destination node because of lack of connectivity on some segments of city streets. However, the problem with E-GyTAR and TFOR is that they establish path based on one junction at a time. As a result, sometimes those junctions whose next streets contain no vehicular traffic are selected. Selection of such streets results in local optimum and as a result, packet delivery ratio decreases. The new dynamic multiple junction selection mechanism and one-hop position and direction based forwarding mechanism minimize the occurrence of such cases and this helps DMJSR to bring considerable improvement in terms of packet delivery ratio as compared to GSR, E-GyTAR and TFOR.

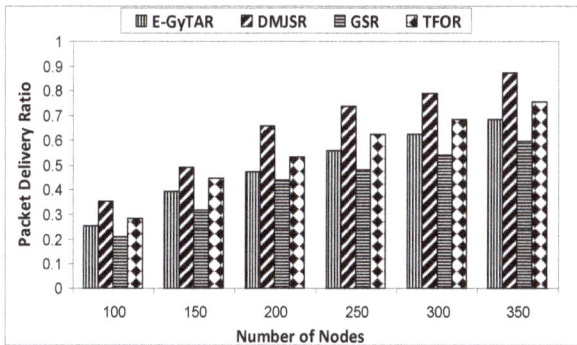

Figure 4. Packet delivery ratio vs. the number of nodes (@5 packets/second). GSR, Geographic Source Routing Protocol; DMJSR, Dynamic Multiple Junction Selection based Routing Protocol; TFOR, Traffic Flow Oriented Routing Protocol; E-GyTAR, Enhanced Greedy Traffic Aware Routing Protocol.

Figure 5 shows a comparison of packet delivery ratio vs. packet sending rates. Increasing packet-sending rate to certain values for all protocols decreases packet delivery ratio. It brings congestion which compels the network to drop some packets. In GSR, certain nodes on the preselected paths instigate more packet transmission which results in more decrease in packet delivery ratio as compared to others routing protocols. The packet-sending rate does not spoil too much the performance of E-GyTAR, TFOR and DMJSR in terms of delivery ratio.

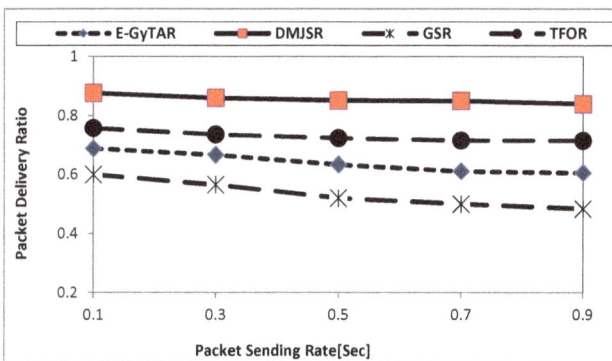

Figure 5. Delivery ratio as a function of the packet-sending rate (350 nodes).

4.3.2. End-To-End Delay

Figure 6 illustrates that DMJSR outperforms GSR, E-GyTAR and TFOR in terms of end-to-end delay as well. This is because DMJSR avoids pre-determining an end-to-end routing path before dispatching data packets without considering connectivity as GSR does: the route in DMJSR is discovered progressively based on multiple junction mechanism by considering network connectivity when relaying data packets from source to destination. In E-GyTAR only directional density is used to find the path, but in urban scenarios with a two-lane road, there are a lot of streets having non-directional density and shortest distance to the destination that can play a key role for relaying packets towards the destination during routing. The speed based greedy forwarding approach of E-GyTAR results in incurring more hops while relaying the packet from source to destination resulting in an increase in end-to-end delay. In DMJSR, there is a reduction in the number of hops required to deliver data packets due to use of the forwarding mechanism that maintains one-hop neighbor information based on position and direction instead of the speed of vehicular node to forward packets between two junctions. Although, TFOR uses directional and non-directional traffic density to accomplish routing path but still suffer from the local optimum problem at junction level which causes more delay. DMJSR selects those multiple junctions that provide enough connectivity as compared to E-GyTAR and TFOR and is less likely to be affected by local optimum. For GSR, E-GyTAR and TFOR, the likelihood of packets staying more time in suspension buffer of a vehicular node is more as compared to DMJSR, which results in an increase in end-to-end delay for GSR, E-GyTAR and TFOR. The novel combination of multiple junction selection mechanism with position and direction based one-hop forwarding mechanism in DMJSR leads to a considerable reduction of end-to-end delay in comparison to the other protocols.

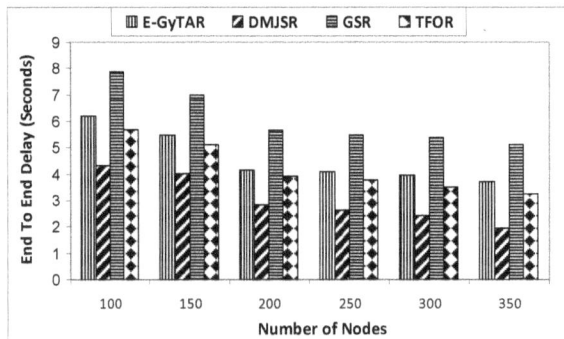

Figure 6. End-to-end delay vs. the number of nodes (@5 packets/second).

4.3.3. Routing Overhead

Figure 7 shows the routing-overhead of all the three protocols against the number of the nodes. With an increase in vehicular density, there is an increase in routing overhead for all the protocols. This is because the amount of control messages depends on the number of vehicles. DMJSR has least routing overhead as compared to GSR, E-GyTAR and TFOR. GSR incurs higher routing overhead because of generation of more beacon messages. It obtains the locations of its neighbor by generating a higher amount of beacons messages as compared to other routing protocols. While in E-GyTAR the routing overhead increases because of its improper junction selection mechanism that causes local optimum in the absence of directional density of vehicles. TFOR also produces more overhead because of its junction selection mechanism. Additionally, maintaining two-hop neighbor information in greedy forwarding brings more routing overhead for TFOR in highly dense environments like traffic jams.

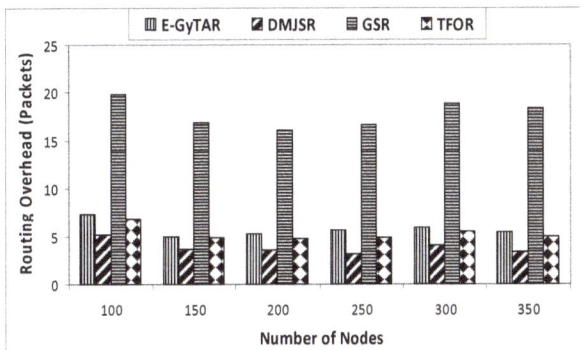

Figure 7. Routing overhead vs. the number of nodes (@5 packets/second).

4.3.4. Impact of Increasing Number of Considered Junctions Dynamically

One important question that needs to be answered is that what is the optimal number of junctions that should be considered for achieving the best performance in terms of various parameters? For answering this question, we have compared the performance in terms of packet delivery ratio with respect to increasing the number of considered junctions. The simulation result shows that considering two junctions results in a better performance as compared to considering more than two junctions.

Figure 8 shows that increasing the number of considered junctions from 1 to 2 results in an increase in packet delivery ratio. However, as we start increasing the number of junctions beyond 2, the performance of DMJSR starts to degrade in terms of packet delivery ratio. The major reason behind this observation lies in one of the very basic characteristics of VANETs, i.e., the network in VANETs is very dynamic and the network topology changes very rapidly. Because of this reason, considering more than 2 junctions while making a decision for packet forwarding degrades the performance instead of achieving an enhanced performance. Moreover, keeping all the information that lies between multiple junction such as density, direction of vehicles etc., in a very dynamic topology also leads to an overhead in terms of processing and storage. The simulation results revealed that the proposed protocol is capable of achieving better performance as compared to existing routing protocols. As a next step, performance evaluation using a more complex mobility model based on real traffic traces that fully capture the complexity involved in the movement of the nodes can be considered.

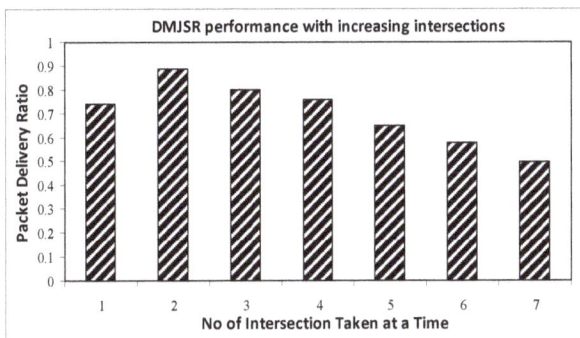

Figure 8. Packet delivery ratio vs. increasing no of junctions dynamically at a time @ 350 vehicular nodes.

5. Conclusions

In this paper, a novel multiple junction selection based routing protocol (DMJSR) is proposed. We first presented the major problems faced by some of the already well-known protocols in VANETs. Then we explained our new proposed protocol and explained how it overcomes the limitations faced by existing protocols. DMJSR selects two junctions at a time based on traffic density and shortest curve metric distance to the destination. DMJSR employs a novel multiple junction approach, an improved greedy forwarding strategy based on one-hop neighbor information to relay packets in between junctions, and a recovery strategy, which can incorporate urban surroundings challenges in a better way. The simulation results indicated that our proposed protocol convincingly outperforms TFOR, E-GyTAR, and GSR in terms of packet delivery ratio end-to-end delay because of its ability to incorporate in better ways the main challenges faced in the city environments as compared to the existing routing protocols.

Acknowledgments: This work is fully funded by the Research and Innovation Management Center University of Malaysia Sarawak (RIMC-UNIMAS) under the grant number F08/SpSG/1403/16/4.

Author Contributions: Irshad Abbasi is the main author of this manuscript. All the authors have contributed to this manuscript. All authors have read and approved the final manuscript. Irshad Abbasi conceived the novel idea and designed the experiments; Irshad Abbasi performed the experiments; Irshad Abbasi analyzed the data; Irshad Abbasi contributed materials/analysis tools; Irshad Abbasi wrote the entire paper. Irshad Abbasi, Adnan Shahid Khan and Shahzad Ali checked, reviewed and revised the paper. Irshad Abbasi accomplished final proofreading. Adnan Shahid Khan supervised the research.

Conflicts of Interest: The authors declare no conflict of interest.

References

1. Abdalla, G.M.T.; Rgheff, M.A.A.; Senouci, S.M. Current trends in vehicular ad hoc networks. *Ubiquitous Comput. Commun. J.* **2008**, *11*, 1–9.
2. Bilal, S.M.; Madani, S.A.; Khan, I.A. Enhanced junction selection mechanism for routing protocol in VANETs. *Int. Arab J. Inf. Technol.* **2011**, *8*, 422–429.
3. Abbasi, I.A.; Nazir, B.; Bilal, S.M.; Madani, S.A. A traffic flow oriented routing protocol for VANET. *EURASIP J. Wirel. Commun. Netw.* **2014**, *2014*, 121. [CrossRef]
4. Sharma, R.; Choudhry, A. An extensive survey on different routing protocols and issue in VANETs. *Int. J. Comput. Appl.* **2014**, *106*, 1–18.
5. Shah, S.A.A.; Shiraz, M.; Nasir, M.K.; Noor, R.B.M. Unicast routing protocols for urban vehicular networks: Review, taxonomy, and research issues. *J. Zhejiang Univ. Sci. (Comput. Electron.)* **2014**, *15*, 489–513. [CrossRef]
6. Silva, A.C. Applicability of position-based routing for VANET in highways and urban Environment. *Port. J. Netw. Comput. Appl.* **2013**, *36*, 961–973.
7. Dhankhar, S.; Agrawal, S. A survey on routing protocols and issues. *Int. J. Innov. Res. Sci. Eng. Technol.* **2014**, *3*, 13427–13434.
8. Husain, A.; Shringar, R.; Kumar, B.; Doegar, A. Performance comparison topology based and position based Routing protocols in vehicular network environments. *Int. J. Wirel. Mob. Netw. (IJWMN)* **2011**, *3*, 289–303. [CrossRef]
9. Jimenez, F.; Clavijo, M.; Jose, E.N.; Gomez, O. Improving the Lane Reference Detection for Autonomous Road Vehicle Control. *Sensors* **2016**, *13*, 9497524. [CrossRef]
10. Naranjo, J.E.; Jimenez, F.; Anaya, J.J.; Talavera, E.; Gomez, O. Application of vehicle to another entity (V2X) communications for motorcycle crash avoidance. *J. Intell. Transp. Syst.* **2017**, *21*, 285–295. [CrossRef]
11. Abbasi, I.A.; Khan, A.S. A Review of Vehicle to Vehicle Communication Protocols for VANETs in the Urban Environment. *J. Future Internet* **2018**, *10*, 14. [CrossRef]
12. Lin, Y.W.; Chen, Y.S.; Lee, S.L. Routing protocols in vehicular ad hoc networks: A survey and future perspectives. *J. Inf. Sci. Eng.* **2010**, *26*, 913–932.
13. Gupta, N.; Prakash, A.; Tripathi, R. Adaptive Beaconing in mobility aware clustering based MAC protocol for safety message dissemination in VANET. *Wirel. Commun. Mob. Comput.* **2017**, *2017*, 1246172. [CrossRef]
14. Bilal, S.M.; Khan, A.R.; Ali, S. Review and performance analysis of position based routing protocols. *Wirel. Pers. Area Commun.* **2017**, *94*, 559–578. [CrossRef]

15. Car-Talk 2000. Available online: http://www.cartalk2000.net/ (accessed on 5 April 2017).

16. CAR 2 CAR Communication Consortiums. Available online: http://www.car-to-car.org (accessed on 27 April 2017).

17. California Partners for Advanced Transit and Highways. Available online: http://www.path.berkeley.edu (accessed on 3 January 2018).

18. FleetNet. Available online: https://fleetnetamerica.com (accessed on 11 June 2017).

19. Crash Avoidance Metrics Partnership. Available online: https://www.mentor.com (accessed on 29 August 2017).

20. Chauffeur in EU. Available online: http://www.autoeurope.com/go/chauffeur-services (accessed on 3 January 2018).

21. Karp, B.; Kung, H.T. Greedy perimeter stateless routing for wireless networks. In Proceedings of the 6th Annual International Conference on Mobile Computing and Networking (ICMCN '6), Boston, MA, USA, 6–11 August 2000; ACM: New York, NY, USA, 2000; pp. 243–254.

22. Rahim, A.; Ahmad, I.; Khan, Z.S.; Sher, M.; Javed, A.; Shoaib, M.; Mahmood, R. A Comparative study of mobile and vehicular adhoc networks. *Int. J. Recent Trends Eng.* **2009**, *2*, 195–197.

23. Hartenstein, H.; Laberteaux, K.P. *Vehicular Applications and Internetworking Technologies*; Wiley Online Library: Hoboken, NJ, USA, 2010; pp. 1–14.

24. Jaap, S.; Bechler, M. Evaluation of routing protocols for vehicular ad hoc networks in city traffic scenarios. In Proceedings of the 5th International Conference on Intelligent Transportation Systems Telecommunications (ITST '5), Brest, France, 27–29 June 2005; pp. 1–9.

25. Cheng, J.; Cheng, J.; Zhou, M.C.; Liu, F.Q.; Gao, S.C.; Liu, C. Routing in Internet of vehicles: A Review. *IEEE Trans. Intell. Transp. Syst.* **2015**, *16*, 2339–2352. [CrossRef]

26. Sharef, B.T.; Alsaqour, R.A.; Ismail, M. Review: Vehicular communication ad hoc routing protocols: A survey. *J. Netw. Comput. Appl.* **2014**, *40*, 363–396. [CrossRef]

27. Lee, K.C.; Lee, U.; Gerla, M. *Advances in Vehicular Ad-Hoc Networks: Developments and Challenges*; Chapter Survey of Routing Protocols in Vehicular Ad Hoc Networks; IGI Global: Harrisburg, PA, USA, 2009.

28. Li, F.; Wang, Y. Routing in vehicular ad hoc networks: A survey. *IEEE Veh. Technol. Mag.* **2007**, *15*, 12–22. [CrossRef]

29. Johnson, D.B.; Maltz, D.A. *Dynamic Source Routing in Ad Hoc Wireless Networks*; Kluwer Academic Publishers: Norwell, MA, USA, 1996; pp. 153–181.

30. Perkins, C.E.; Royer, E.M. Ad hoc on demand distance vector routing. In Proceedings of the Second IEEE Workshop on Wireless Mobile Computing Systems and Applications (WMCSA '2), New Orleans, LA, USA, 25–26 February 1999; pp. 90–100.

31. Jerbi, M.; Senouci, S.M.; Meraihi, R.; Doudane, Y.G. An improved vehicular ad hoc routing protocol for city Environments. In Proceedings of the IEEE International Conference on Communications (ICC '07), Glasgow, UK, 24–28 June 2007; pp. 1–8.

32. Clausen:, T.; Jacquet, P. *Optimized Link State Routing Protocol*; RFC 3626; Internet Engineering Task Force: Fremont, CA, USA, 2003.

33. Lochert, C.; Mauve, M.; Fubler, H.; Hartenstein, H. Geographic routing in city scenarios. *SIGMOBILE Mob. Comput. Commun. Rev.* **2005**, *5*, 69–72. [CrossRef]

34. Kieb, W.; Fubler, H.; Widmer, J.; Mauve, M. Hierarchical location service for mobile ad hoc networks. *SIGMOBILE Mobile Comput. Commun. Rev.* **2004**, *8*, 47–58.

35. Politecnico, M.F.; Abruzzi, C.D.D.; Harri, J.; Filali, F. *Vehicular Mobility Simulation for VANETs*; Christian Bonnet Institute Eurecom, Department of Mobile Communications 06904: Sophia, France, 2011; pp. 1–14.

36. Li, J.; Jannotti, J.; Decouto, D.S.J.; Karger, D.R.; Morris, R. A scalable location service for geographic ad hoc routing. In Proceedings of the 6th Annual International Conference on Mobile Computing and Networking (MobiCom '6), Boston, MA, USA, 6–11 August 2000; ACM: New York, NY, USA, 2000; pp. 120–130.

37. Kasemann, M.; Fubler, H.; Hartenstein, H.; Mauve, M. *A Reactive Location Service for Mobile Ad Hoc Networks*; Technical Report TR-02-014; Department of Computer Science, University of Mannheim: Mannheim, Germany, 2002.

38. Lochert, C.; Hartenstein, H.; Tian, J.; Fubler, H.; Hermann, D.; Mauve, M. A routing strategy for vehicular ad hoc network in city environment. In Proceedings of the IEEE Intelligent Vehicles Symposium (IVS), Columbus, OH, USA, 9–11 June 2003; pp. 156–161.

39. Lee, K.C.; Haerri, J.; Lee, U. Enhanced perimeter routing for geographic forwarding protocols in urban vehicular scenarios. In Proceedings of the IEEE Globecom Workshops, Washington, DC, USA, 26–30 November 2007; pp. 1–9.

40. Seet, B.C.; Liu, G.; Lee, B.S.; Foh, C.H.; Wong, K.J.; Lee, K.K. A-STAR: A mobile ad hoc routing strategy for metropolitan area in vehicular communications. In *NETWORKIN2004, Networking Technologies, Services, and Protocols; Performance of Computer and Communication Networks, Mobile and Wireless Communications*; Lecture Notes in Computer Science; Springer: Berlin/Heidelberg, Germany, 2004; pp. 989–999.

41. Jerbi, M.; Senouci, S.M.; Rasheed, T.; Ghamri, Y. An infrastructure free intervehicular communication based traffic information. In Proceedings of the First IEEE International Symposium on Wireless Vehicular Communication (WiVec '07), Baltimore, MD, USA, 30 September–1 October 2007.

42. GLOMOSIM Simulator. Available online: https://networksimulationtools.com/glomosim-simulator-projects (accessed on 15 March 2017).

43. Harri, J.; Filali, F.; Bonnet, C.; Fiore, M. VanetMobiSim: Generating realistic mobility patterns for VANETs. In Proceedings of the 3rd ACM International Workshop on Vehicular Ad Hoc Networks (ACM Wkshps '06), Los Angeles, CA, USA, 24–29 September 2006; pp. 86–97.

*applied
sciences*

MDPI

Article

A Comparative Study of Clustering Analysis Method for Driver's Steering Intention Classification and Identification under Different Typical Conditions

Yiding Hua [1] ⓘ, Haobin Jiang [1,2,*], Huan Tian [1], Xing Xu [2] and Long Chen [1,2]

[1] School of Automobile and Traffic Engineering, Jiangsu University, Zhenjiang 212013, China;
    dingyihua0209@163.com (Y.H.); 2211604073@stmail.ujs.edu.cn (H.T.); chenlong@ujs.edu.cn (L.C.)
[2] Automotive Engineering Research Institute, Jiangsu University, Zhenjiang 212013, China;
    xuxingujs@gmail.com
* Correspondence: jianghb@ujs.edu.cn; Tel.: +86-0511-8878-2845

Received: 22 August 2017; Accepted: 28 September 2017; Published: 30 September 2017

Abstract: Driver's intention classification and identification is identified as the key technology for intelligent vehicles and is widely used in a variety of advanced driver assistant systems (ADAS). To study driver's steering intention under different typical operating conditions, five driving school coaches of different ages and genders are selected as the test drivers for a real vehicle test. Four kinds of typical car steering condition test data with four different vehicles are collected. Test data are filtered by the Butterworth filter and are used for extracting the driver steering characteristic parameters. Based on Principal Component Analysis (PCA), the three kinds of clustering analysis methods, including the Fuzzy C-Means algorithm (FCM), the Gustafson–Kessel algorithm (GK) and the Gath–Geva algorithm (GG), considered are proposed to classify and identify driver's intention under different typical operating conditions. Results show that the three approaches can successfully classify and identify drivers' intention respectively despite some accuracy error by FCM. Meanwhile, compared with FCM and GK, GG was the best performing in classification and identification of the driver's intention. In order to verify the validity of the identification method designed by this article, five different drivers were selected. Five tests were carried out on the driving simulator. The results show that the results of each identification are exactly the same as the actual driver's intention.

Keywords: driver's steering intention; real vehicle test; Principal Component Analysis (PCA); cluster analysis

---

1. Introduction

In recent years, with the rapid integration of high-tech and advanced automotive technologies such as computers, the Internet, communications and navigation, automatic control, artificial intelligence, machine vision, precision sensors, high-precision maps and smart cars (or unmanned vehicles), smart driving has become one of the world's automotive engineering research hotspots and a new impetus of the automotive industry's growth. According to authoritative media at home and abroad, in the future of the automotive industry, more than 90% of scientific and technological innovation will focus on the field of automotive intelligence. Therefore, smart vehicles are safe, efficient, energy-efficient next-generation vehicles [1,2], and the study of smart cars has a very important significance, which has become the focus of the global automotive industry.

The driver intention is reflected by his/her own inner state in the driving process. It cannot be obtained directly during driving and is only predicted by the driver's movements, vehicle status and traffic environment information. Driver's intention classification and identification are identified as comprising the key technology for intelligent vehicles and are widely used in a variety of Advanced Driver Assistant Systems (ADAS) [3,4], such as the Adaptive Cruise Control System (ACC) [5,6], the Active Front Steering System (AFS) [7–10], the Parking Assistance Systems (PAS) [11,12], the steer-by-wire systems [13] and Man-machine Co-driving Electric Power Steering (MCEPS) system [14,15]. The classification and identification of driver's intention are based on the real-time acquisition of the driver's operating signal and car state, or at the same time, by monitoring the driver's head movement range and facial expressions, the driver's behavior is distinguished and identified to obtain the driver's driving intention [16].

Many scholars are committed to study the classification and identification of driver's intention. Liang Li et al. [17] proposed a novel method based on an artificial error back-propagation neural network to identify the driver's starting intention. Takano et al. [18] proposed an intelligent cognitive method for driver's intention identification based on the Hidden Markov Model (HMM). The method mainly includes data segmentation, time series data labeling and the identification and generation of the driving mode. Raksin et al. [19] proposed an algorithm based on the driver's intention to identify the direct yaw moment control, in which the driver steering intention identification is through the Hidden Markov Model (HMM). The use of the dynamic Bayesian network was combined with the past driving state and the current driving state to predict the driver's intention of parking at the crossroads. Tesheng Hsiao [20] used the maximum posterior probability assessment method to obtain the driver's steering model parameters. He established a steering model that can effectively improve the recognition accuracy and that has the function of predicting the driver's driving strategy. The previous research works mainly concentrated on a single traffic environment, such as the straight road or the crossroads intentions, and not on variety of typical steering conditions under the driving intention identification study.

The driving intention under each condition requires multiple drivers' characteristic steering parameters, such as driving parameters (steering angle, angular velocity and torque) and vehicle status parameters (roll angle, lateral acceleration and yaw rate). However, if all the characteristic parameters are used for classification and identification, the computational complexity is increased due to the large number of characteristic parameters. Additionally, analysis of the situation becomes much more difficult. Although each feature parameter provides some information, some of the characteristic parameters are correlated, and the characteristic parameters are not independent of each other. Therefore, the information provided by these characteristic parameters overlaps to some extent. Therefore, we need to use a kind of theoretical algorithm to reduce the dimension of the data and to decorrelate the input variables. Principal Component Analysis (PCA) is used to reduce the data dimension, which is used in various applications such as error recognition [21], pedestrian identification [22] and image tracking [23]. However, PCA is not used in the driving intention identification study under typical steering conditions.

In essence, the driver's intention is a pattern recognition process. The cluster analysis is a typical method of pattern recognition. Compared with the traditional classification and identification of driver's intention using the neural network and fuzzy mathematics, the clustering algorithm only needs a small amount of data, which eliminates the need to construct the nonlinear recognizer and ensures that the accuracy is stable. In order to classify and identify driver's intention, three clustering analysis methods are studied: the Fuzzy C-Means algorithm (FCM), the Gustafson–Kessel algorithm (GK) and the Gath–Geva algorithm (GG). Due to its flexibility and robustness for ambiguity, the FCM algorithm is currently an active topic [24] and has been widely applied in the areas of pattern recognition [25], function approximation [26], image processing [27], machine learning [28], and so on. The GK algorithm can generate a fuzzy partition that provides the degree of membership of each data point to a given cluster [29]. The GG algorithm can make the parameters of the univariate membership functions be directly derived from the parameters of the clusters [30].

Therefore, this paper selected five driving school coaches of different driving experiences and genders as real vehicle test driver, and four typical car steering conditions' test data with four different vehicles were collected. Additionally, this paper used principal component analysis and clustering analysis to classify and identify the driver's intention. The paper analyzed the advantages and disadvantages of the three different clustering methods (FCM, GK and GG) in the direction of driver's steering intention.

The paper is organized as follows. In the second section, five driving school coaches of different ages and genders are selected as the test drivers for the real vehicle test. In the third section, driver's characteristic steering parameters under different conditions are proposed. The fourth section uses principal component analysis and clustering analysis to classify and identify the driver's intention. In the fifth section, the clustering results and analysis are presented. Finally, the conclusions are drawn in the last section.

2. Experiment

2.1. Experimental Devices

The real vehicle experiment of driver's steering in different conditions consists of the following components: S-Motion biaxial optical speed sensor (Kistler, Winterthur, Switzerland) (signal delay only 6 ms), biaxial optical speed sensor mounting bracket, KiMSW Force steering wheel sensor 250 Nm (Kistler, Winterthur, Switzerland) (steering angle accuracy: 0.015°), universal mounting bracket (Kistler, Winterthur, Switzerland), power distribution box (Kistler, Winterthur, Switzerland) and SDI-600GI Model GPS/INS (SDI, Beijing, China) (accuracy: 10 cm error). The experimental devices are shown in Figure 1.

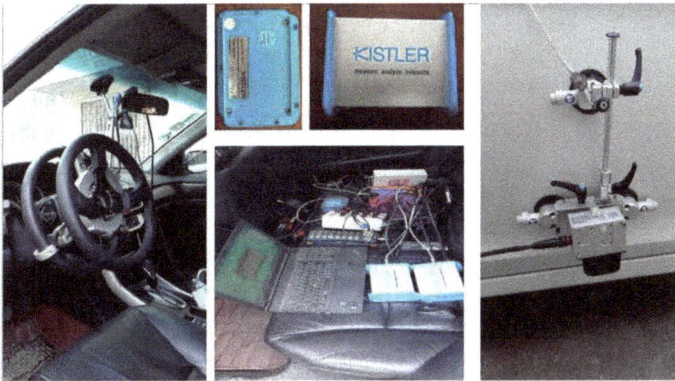

Figure 1. Experimental devices and the actual installation.

2.2. Experimental Design and Experimental Vehicles

This experiment selected 5 driving school coaches of different driving ages and genders as the test drivers, which are shown in Table 1. The turn right/left steering condition, U-turn condition, lane keeping condition and lane changing condition are proposed in the real vehicle test. Additionally, the speed of the vehicle is certain during the test. The test vehicles were four passenger cars: GM GL8, Skoda Octavia, Honda Accord and SAIC MG. The experimental vehicles are shown in Figure 2.

Table 1. Driver information.

No.	Ages (Years)	Driving Experience (Years)	Gender
Driver 1	55	33	female
Driver 2	28	10	male
Driver 3	53	31	male
Driver 4	46	22	male
Driver 5	53	21	male

Figure 2. Experimental vehicles (**1**) GM GL8, (**2**) Skoda Octavia, (**3**) Honda Accord, (**4**) SAIC MG.

3. The Method of the Driver's Intention Identification

Driver's Characteristic Steering Parameter under Different Conditions

In order to accurately describe the steering characteristics of the driver under each steering condition, it is ensured that there will be no loss or distortion of the driver's intention information. According to GB/T 6323-1994 "Vehicle Handling Stability Test Method", this paper selects the typical steering conditions of the driver's characteristic parameters, as shown in Table 2. In order to identify the driver's steering intention under different steering conditions, this paper chooses the driver steering parameters and vehicle dynamics parameters for the first two seconds under different operating conditions. The purpose of this is to use the initial operation of the driver to identify the driver's steering intention in the next period. It is helpful to lay the foundation for dynamic control for further advanced driver assistant systems. Furthermore, this can improve the accuracy of the intelligent vehicle active safety control system. However, various uncertainties exist due to the large amount of interference signal in the test data, as shown in Figure 3. Especially, torque, angular velocity, yaw rate and lateral acceleration need to be filtered.

Table 2. Driver's characteristic steering parameters.

Symbol	Meaning	Units
δ_m	Average steering angle	deg
δ_{max}	Maximum steering angle	deg
$\dot{\delta}_m$	Average angular velocity	deg/s
$\dot{\delta}_{max}$	Maximum angular velocity	deg/s
T_m	Average torque	Nm
T_{max}	Maximum torque	Nm
γ_{max}	Maximum yaw rate	deg/s
ϕ_{max}	Maximum roll angle	deg
$a_{y_{max}}$	Maximum lateral acceleration	m/s^2

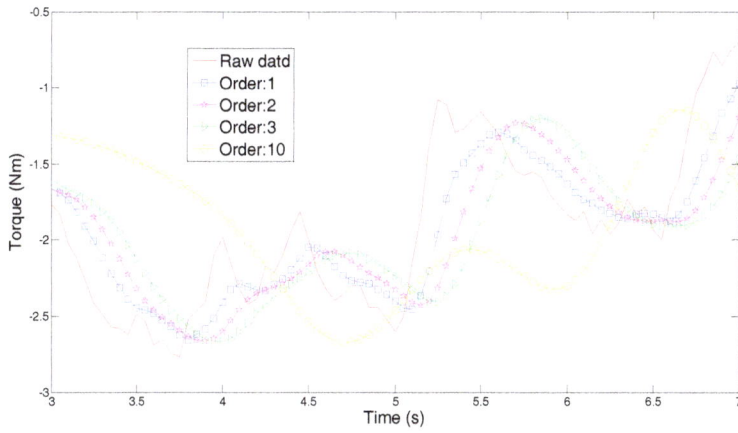

(a)

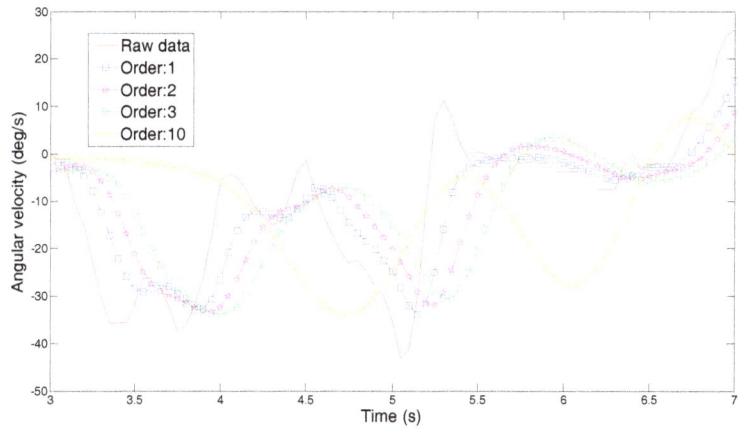

(b)

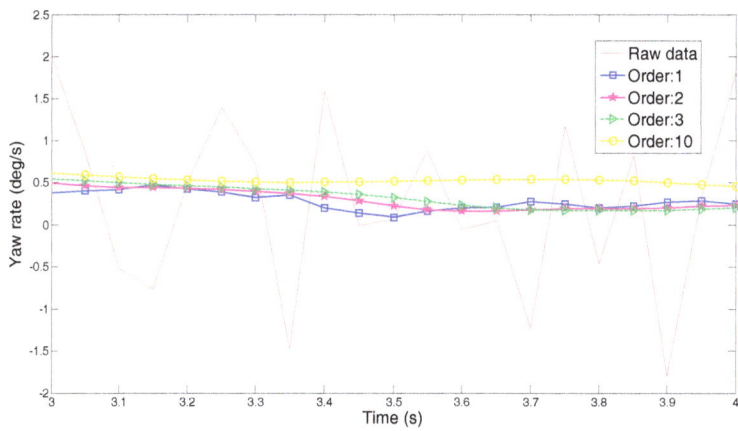

(c)

Figure 3. *Cont.*

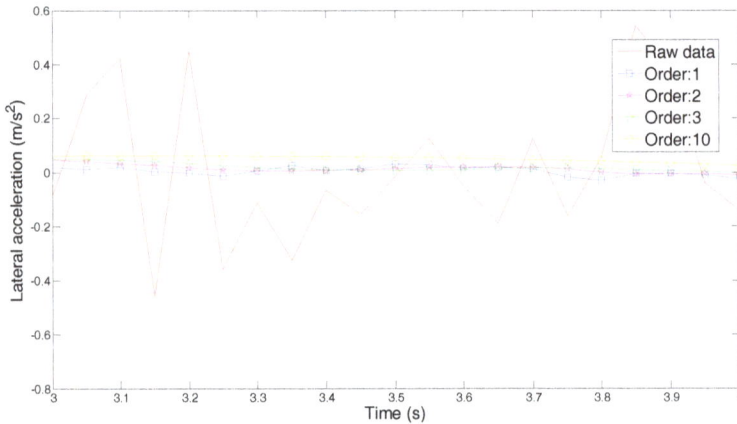

(d)

Figure 3. Filter results with the Butterworth filter: (**a**) torque; (**b**) angular velocity; (**c**) yaw rate; and (**d**) lateral acceleration.

The Butterworth filter is a kind of electronic filter whose frequency response curve is the smoothest. The filter was first proposed by a British engineer, Stephen Butterworth, in a paper published in the British journal Radio Engineering in 1930. The attenuation rate of the first-order Butterworth filter is 6 dB per octave. The second-order Butterworth filter has a decay rate of 12 dB per octave, and the third-order Butterworth filter has an attenuation rate of 18 dB per octave, and so on. The amplitude of the Butterworth filter is monotonically decreasing, and it is also the only filter that maintains the same shape regardless of the order of the amplitude of the diagonal frequency curve. The higher the order of the filter is, the faster the amplitude attenuation in the resistive band is. The difference with the Chebyshev, Bessel and elliptical filters is that the attenuation of the Butterworth filter is slower than the other filters, but is very flat and does not vary.

The Butterworth low-pass filter can be expressed by the square of the amplitude:

$$|H(\omega)|^2 = \frac{1}{1 + \left(\frac{\omega}{\omega_c}\right)^{2n}} \tag{1}$$

where n is the order of the filter and ω_c is the cut-off frequency.

Therefore, we choose the Butterworth filter with different orders to process the test data. Filter results can be seen in Figure 3. In Figure 3a, the first-order Butterworth filter obtains the closest data to the raw data; simultaneously, the value of torque is also very smooth. Therefore, the first-order Butterworth filter is the most suitable to handle the torque signal. In Figure 3b, we can reach the same conclusion about the angular velocity signal. In Figure 3c,d, although the first-order Butterworth filter obtains the closest data to the raw data, the first-order Butterworth filter is less smooth than the second-order Butterworth filter. Therefore, the second-order Butterworth filter is the most suitable to handle the yaw rate and the lateral acceleration signal.

Since the number of experimental data is very large, which is shown in Table 3, only about 20% (133) of the experimental data are selected (after removing the wrong data). This is because, compared with the traditional classification and identification of the driver's intention using the neural network and fuzzy mathematics, the clustering algorithm only needs a small amount of data, which eliminates the need to construct the nonlinear recognizer and ensures that the accuracy is stable. Additionally, the characteristic parameters of these sets of test data are extracted and analyzed, which lays the foundation for the driver's steering intention identification.

Table 3. Actual experimental data and speed conditions.

The Number of Experiment and the Speed Limit	Typical Steering Conditions				
	Turning Right/Left Condition	U-Turn Condition	Lane Keeping Condition	Lane Changing Condition	The Sum of All the Conditions
Number of Experimental Data	320	82	160	162	724
Speed Conditions (km/h)	20–50	20–30	30–60	30–40	20–60

The representative samples of the experimental data of the real vehicle tests are shown in Figures 4–9, which contain four conditions of the data, respectively, turning right condition, U-turn condition, lane keeping condition and lane changing condition.

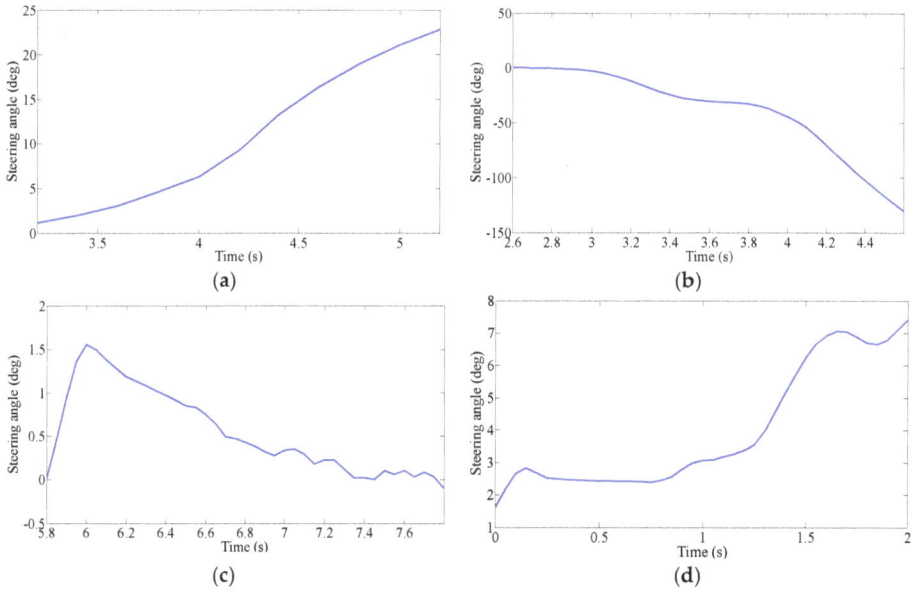

Figure 4. Steering angle under different steering conditions: (**a**) turning right; (**b**) U-turn; (**c**) lane keeping; (**d**) lane changing.

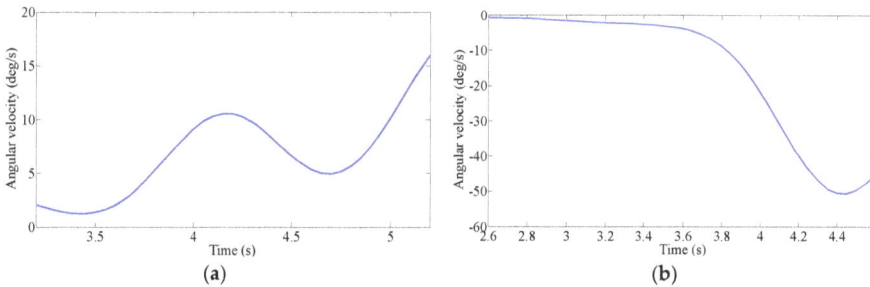

Figure 5. *Cont.*

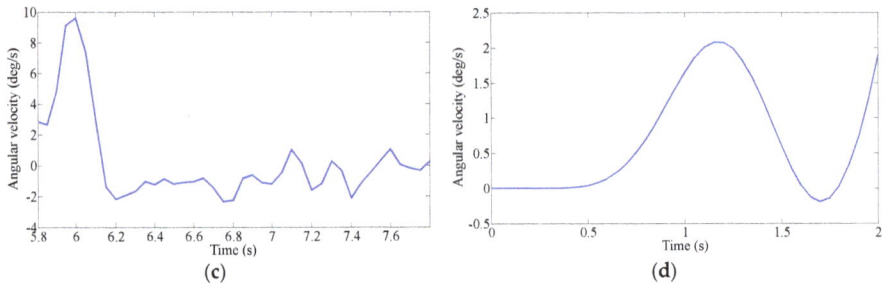

Figure 5. Angular velocity under different steering conditions: (**a**) turning right; (**b**) U-turn; (**c**) lane keeping; (**d**) lane changing.

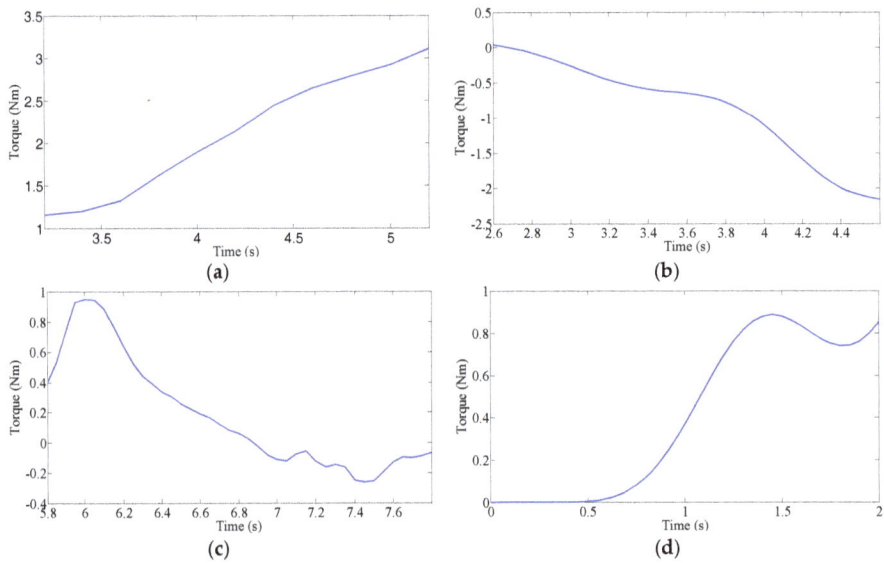

Figure 6. Torque under different steering conditions: (**a**) turning right; (**b**) U-turn; (**c**) lane keeping; (**d**) lane changing.

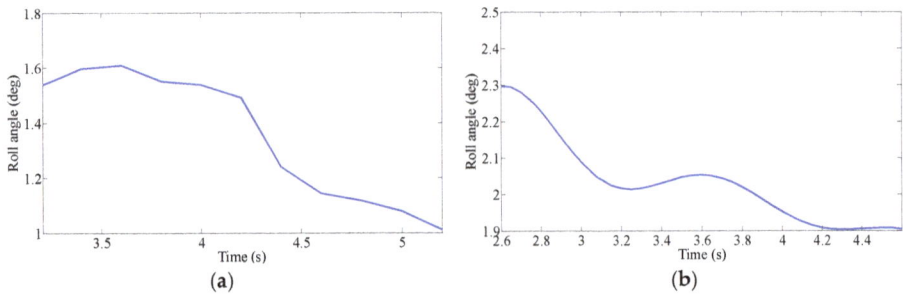

Figure 7. *Cont.*

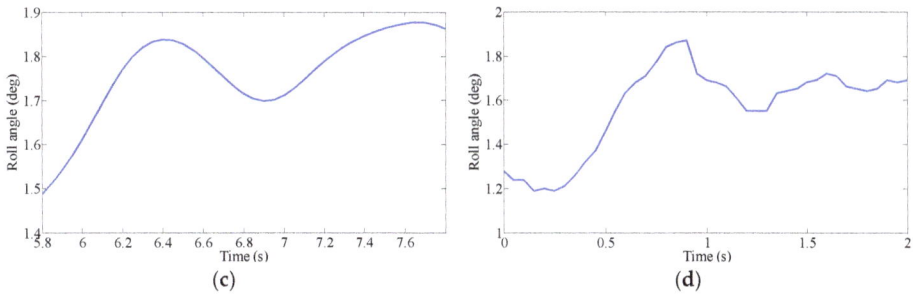

(c)

(d)

Figure 7. Roll angle under different steering conditions: (**a**) turning right; (**b**) U-turn; (**c**) lane keeping; (**d**) lane changing.

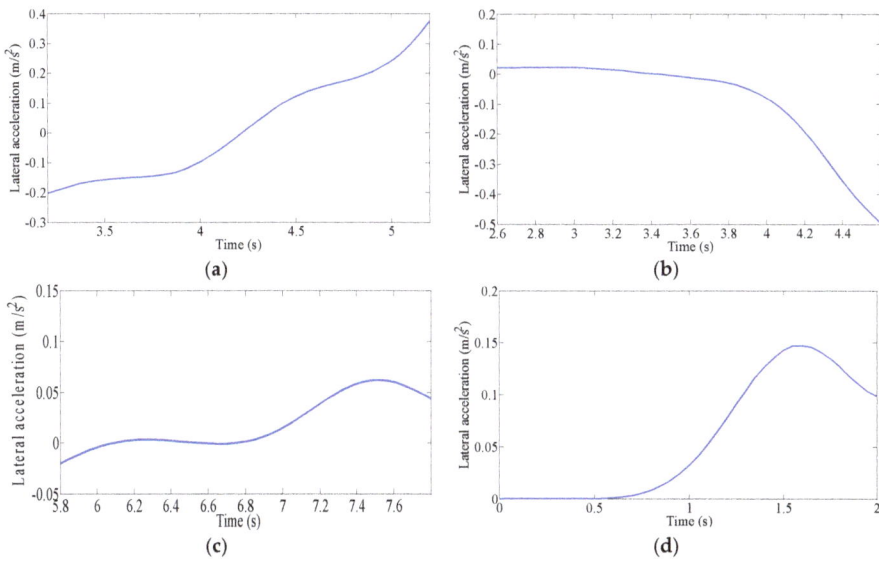

(a)

(b)

(c)

(d)

Figure 8. Lateral acceleration under different steering conditions: (**a**) turning right; (**b**) U-turn; (**c**) lane keeping; (**d**) lane changing.

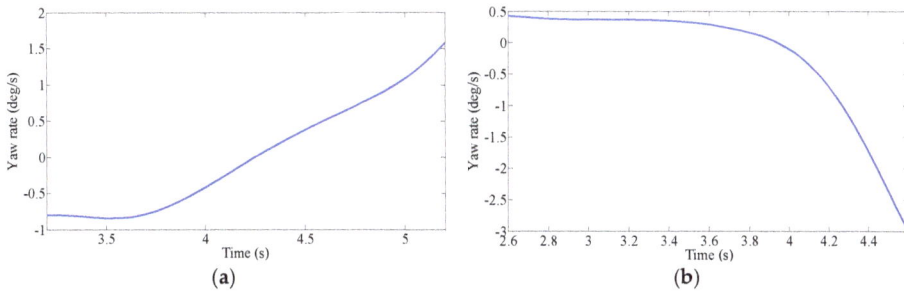

(a)

(b)

Figure 9. *Cont.*

71

Figure 9. Yaw rate under different steering conditions: (**a**) turning right; (**b**) U-turn; (**c**) lane keeping; (**d**) lane changing.

4. Principal Component Analysis of Steering Parameters

In solving practical problems and research, it is often possible to collect more information about the research object in order to have a comprehensive understanding of the problem. However, due to the theoretical development and application of technical constraints, having too many variables to be processed and too much information have become analysis obstacles. To solve this problem, Principal Component Analysis (PCA) should be used to analyze data. PCA is a statistical analysis method that simplifies multiple indicators into a small number of comprehensive indicators, with as few as possible to reflect the original variable information [31], to ensure that the original loss of information and the number of variables is as small as possible. Let $X = (X_1, X_2, \cdots, X_p)'$ be a p-dimensional random vector, and its linear variation is as follows:

$$
\begin{aligned}
PC_1 &= a_1'X = a_{11}X_1 + a_{21}X_2 + \ldots + a_{p1}X_p \\
PC_2 &= a_2'X = a_{12}X_1 + a_{22}X_2 + \ldots + a_{p2}X_p \\
&\cdots\cdots\cdots\cdots\cdots\cdots\cdots\cdots\cdots\cdots\cdots\cdots\cdots \\
PC_p &= a_p'X = a_{1p}X_1 + a_{2p}X_2 + \ldots + a_{pp}X_p
\end{aligned}
\tag{2}
$$

Using the new variable PC_1 to replace the original p variables X_1, X_2, ..., X_p, PC_1 should reflect as much as possible the original variable information. If the first principal component is not enough to represent the vast majority of the original variables' information, two main components PC_2, and so on, will be used. The main purpose of principal component analysis is to simplify the data, so in practical applications, we will not take p principal components and usually use m ($m < p$) principal components. The number of principal components m should be based on the cumulative contribution of the variance of each principal component to the final decision.

$$
p_r = \lambda_k / \sum \lambda_i
\tag{3}
$$

where λ is the eigenvalue corresponding to each principal component, k is the number of selected main components and I is the total number of components.

The principal component analysis of 133 sets of experimental data is carried out by MATLAB software, and nine principal components (Y_1, Y_2, ..., Y_9) were obtained. The eigenvalue, contribution rate and cumulative contribution rate of each principal component are shown in Table 4. According to the principal component analysis principle, the first four principal components are selected, and the correlation between the characteristic parameters and the principal components is analyzed. The representative average angular velocity, average steering angle, maximum yaw rate and maximum lateral acceleration, the four parameters of acceleration, are used for cluster analysis. The method of principal component analysis is shown in Figure 10.

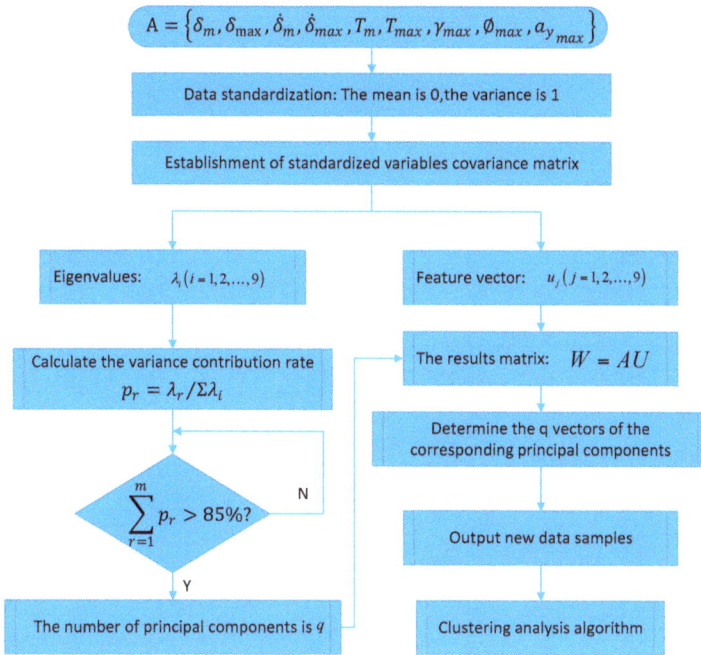

Figure 10. Method of principal component analysis.

Table 4. The principal component eigenvalue, contribution rate and cumulative contribution rate.

Main Ingredient	Eigenvalues	Contribution Rate (%)	Cumulative Contribution Rate (%)
Y1	4.279	41.33	41.33
Y2	1.439	17.42	58.75
Y3	1.235	14.53	73.28
Y4	1.005	11.95	85.23
Y5	0.699	5.12	90.35
Y6	0.325	4.91	95.26
Y7	0.032	3.61	98.87
Y8	0.006	1.09	99.96
Y9	0.005	0.04	100

5. Comparison of Clustering Analysis Methods

5.1. Fuzzy C-Means Algorithm

The fuzzy C-means clustering is defined as:

$$\bar{J}(X; U, V, \lambda) = \sum_{i=1}^{c} \sum_{k=1}^{N} (\mu_{ik})^m D_{ikA}^2 + \sum_{k=1}^{N} \lambda_k \left(\sum_{i=1}^{c} \mu_{ik} - 1 \right) \tag{4}$$

and by setting the gradients of $(\bar{J})$ with respect to U, V and λ to zero. If $D_{ikA}^2 > 0$, $\forall i, k$ and $m > 1$, then $(U, V) \in M_{fc} \times R^{n \times c}$ may minimize only if:

$$\mu_{ik} = \frac{1}{\sum_{j=1}^{c} \left(D_{ikA} / D_{jkA} \right)^{2/(m-1)}}, 1 < i \le c, 1 \le k \le N \tag{5}$$

Additionally:

$$v_i = \frac{\sum\limits_{k=1}^{N} \mu_{ik}^m x_k}{\sum\limits_{k=1}^{N} \mu_{ik}^m}, 1 \le i \le c \tag{6}$$

5.2. Gustafson–Kessel Algorithm

The Gustafson–Kessel algorithm obtains the objective function by introducing the covariance matrix, which is suitable for the clustering analysis of the correlation between the variables and is suitable for the distribution of irregular data [32]. The clustering algorithm uses the adaptive distance of the clustering covariance matrix to measure. By obtaining the objective function to achieve the membership matrix of the fuzzy clustering, , and the clustering center $V = (v_1, v_2, v_3, \cdots, v_c)^T$, where: c is the number of samples; u_{ij} is the clustering center membership relative to the data point; and they meet $u_{ij} \in [0,1]$, $\sum\limits_{i=1}^{c} u_{ij} = 1, 1 \le j \le N$. The data sequence $X = (x_1, x_2, \cdots, x_N)$ is given, and its minimized objective function is:

$$J(X, V, U) = \sum\limits_{i=1}^{c} \sum\limits_{j=1}^{N} (u_{ij})^m D_{ij}^2 \tag{7}$$

$$D_{ij}^2 = \|x_j - v_i\|_{A_i}^2 = (x_j - v_i)^T A_i (x_j - v_i) \tag{8}$$

$$A_i = \det(F_i)^{\frac{1}{n}} F_i^{-1} \tag{9}$$

where: m is the fuzzy index, representing the degree of fuzzy clustering, and the greater the value of m, the greater the degree of overlap between the major clusters; usually, m takes one or two; D_{ij}^2 is the distance from any data point x_j to the cluster center v_i, and it is a square inner product norm; A_i is a positive definite symmetric matrix, and it is determined by the clustering matrix covariance matrix F_i.

However, Formula (9) cannot be minimized directly considering its linear features. In order to obtain a viable solution, A_i must be constrained in some way. The usual way is to constrain the determinant of A_i. The allowable matrix A_i varies with its determinant, corresponding to the shape of the optimized cluster, while its volume remains unchanged:

$$\|A_i\| = \rho_i, \rho > 0 \tag{10}$$

where ρ_i is certain for each cluster. Using the Lagrange multiplier method, A_i is obtained:

$$A_i = [\rho_i \det(F_i)]^{1/n} F_i^{-1} \tag{11}$$

where F_i is defined by:

$$F_i = \frac{\sum\limits_{k=1}^{N} (\mu_{ik})^m (x_k - v_i)(x_k - v_i)^T}{\sum\limits_{k=1}^{N} (\mu_{ik})^m} \tag{12}$$

5.3. Gath–Geva Algorithm

The Gath–Geva algorithm was proposed by Bezdek and Dunn [33]:

$$D_{ik}(x_k, v_i) = \frac{\sqrt{\det(F_{\omega i})}}{\alpha_i} \exp\left(\frac{1}{2}\left(x_k - v_i^{(l)}\right)^T F_{\omega i}^{-1} \left(x_k - v_i^{(l)}\right)\right) \tag{13}$$

The difference between the GG algorithm and the GK algorithm is that the distance norm involves an exponential term. F_{wi} is the fuzzy covariance matrix of the i-th cluster, given by:

$$F_{wi} = \frac{\sum\limits_{k=1}^{N} (\mu_{ik})^{\omega} (x_k - v_i)(x_k - v_i)^T}{\sum\limits_{k=1}^{N} (\mu_{ik})^{\omega}}, 1 \le i \le c \tag{14}$$

The prior probabilities formula for each classification are given by:

$$\alpha_i = \frac{1}{N} \sum_{k=1}^{N} \mu_{ik} \tag{15}$$

6. Clustering Results and Analysis

In order to further analyze the driver's steering characteristics under the steering conditions after clustering, the results are shown by the average angular velocity and the average steering angle. As shown in Figures 11–13, the fuzzy C-means algorithm, the Gustafson–Kessel algorithm and the Gath–Geva algorithm were used to divide the driver's steering test results into four classes, and in each algorithm, the cluster center for each class is marked. (0.18, 0.59), (0.33, 0.18), (0.55, 0.785) and (0.785, 0.625) are the four clustering centers given by the fuzzy C-means algorithm; the Gustafson–Kessel algorithm gives four clustering centers: (0.155, 0.62), (0.27, 0.305), (0.46, 0.53) and (0.76, 0.67). The four cluster centers (0.115, 0.63), (0.255, 0.365), (0.44, 0.48) and (0.742, 0.649) are given by the Gath–Geva algorithm. According to the fourth cluster center, we can see that the average angular velocity and the average steering angle of the drivers' steering are larger. This reflects the driver's steering intention under the U-turn condition. Analysis of the first and second cluster centers revealed that these two types of conditions exist; the lower average angular velocity and larger average steering angle. It can be judged at this time that the first cluster reflects the drivers' steering intention under the lane change condition, and the second cluster centers reflects drivers' steering intention under the lane keeping condition. The remaining third cluster center between the fourth and first two cluster centers usually belongs to the ordinary driver's intentions for the turn right/left steering conditions.

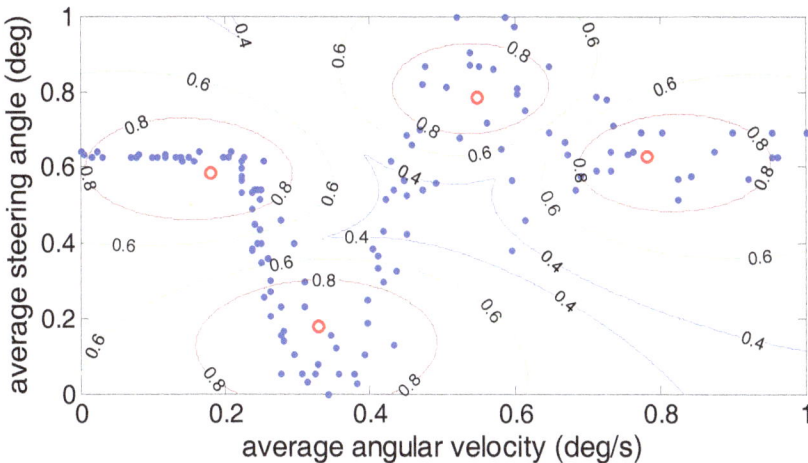

Figure 11. Visualization of the classification results of the driver's steering intention based on the fuzzy C-means algorithm.

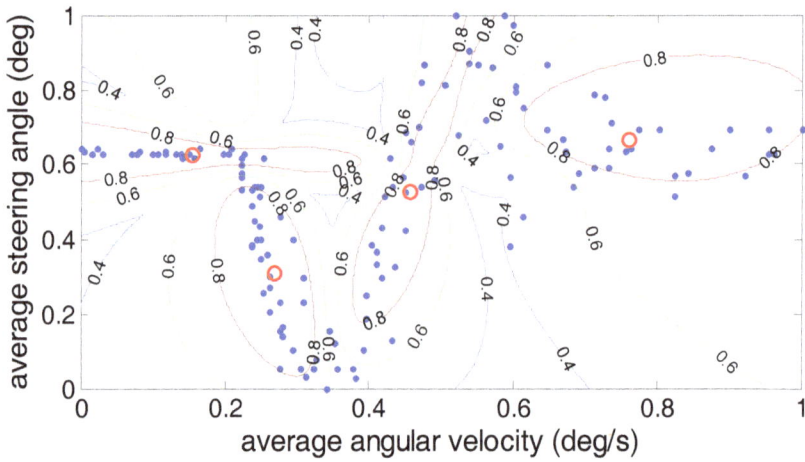

Figure 12. Visualization of the classification results of the driver's steering intention based on the Gustafson–Kessel algorithm.

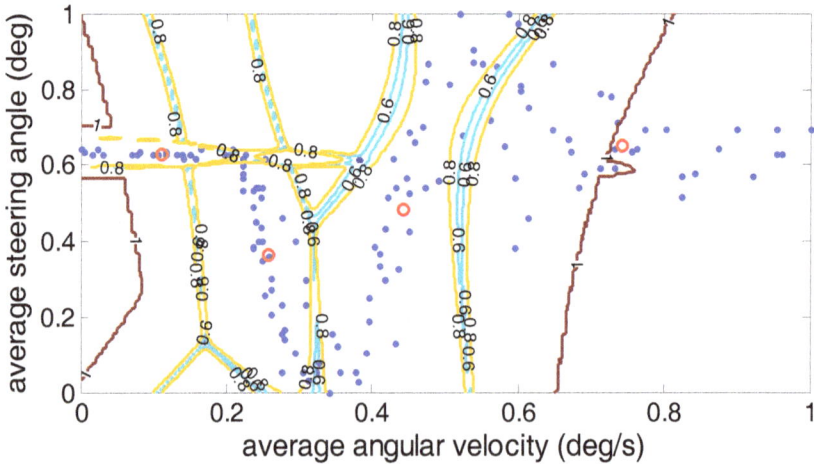

Figure 13. Visualization of the classification results of the driver's steering intention based on the Gath–Geva algorithm.

Three kinds of clustering analysis methods can be used to separate the experimental data under different steering conditions into four different types. However, by comparing the three clustering methods, we find that for the third cluster center, the average steering angle is greater than the average steering angle of the center of the fourth cluster using the fuzzy C-means algorithm, which is contrary to the fact that the average steering angle under the normal right/left turn is less than the average steering angle of the U-turn. Therefore, the accuracy of the Gustafson–Kessel algorithm and the Gath–Geva algorithm is superior to the fuzzy C-means algorithm.

In order to analyze the clustering effect more scientifically, in the clustering method, the most representative criteria for evaluating the clustering effect are the Partition Coefficient (*PC*) and the Classification Entropy (*CE*), which are defined as follows:

The Partition Coefficient (*PC*) measures the amount of "overlapping" between clusters. It is defined by Bezdek as follows:

$$PC(c) = \frac{1}{N} \sum_{i=1}^{c} \sum_{j=1}^{N} (\mu_{ij})^2 \tag{16}$$

The Classification Entropy (*CE*) measures the fuzziness of the cluster partition only, which is similar to the partition coefficient.

$$CE(c) = -\frac{1}{N} \sum_{i=1}^{c} \sum_{j=1}^{N} \mu_{ij} \log(\mu_{ij}) \tag{17}$$

In this paper, *C* is the number of clusters and *N* is the total number of experiments. When the two validity evaluation functions reach the optimal value, that is *PC* (*c*) reaches the maximum value and *CE* (*c*) reaches the minimum value, the clustering analysis effect is better. The comparisons of *PC* (*c*) and *CE* (*c*) and the required time among different clustering algorithms under different working conditions are shown in Table 5.

Table 5. Recorded data of the evaluation in three ways.

Algorithm	PC (c)	CE (c)	Time Consumed (s)
Fuzzy C-means	0.6935	0.6096	5.1685
Gustafson–Kessel	0.7356	0.4892	5.7304
Gath–Geva	0.9493	0.0377	5.5268

By analyzing Table 6, we can see that the Gath–Geva algorithm achieves the maximum value for the Partition Coefficient (*PC*). At the same time, the Gath–Geva algorithm achieves the minimum value for the Classification Entropy (*CE*). Due to the complexity of the Gath–Geva algorithm, the time consumed area is somewhat more than the fuzzy C-means, but it is also within acceptable limits. To sum up, the Gath–Geva algorithm is better than the other two methods for the classification and identification of driver's intention.

Table 6. The results of the identification.

Distances	Condition 1	Condition 2	Condition 3	Condition 4	Condition 5
The distance to Clustering Center 1	3.4	2.3	0.28	2.8	3.64
The distance to Clustering Center 2	6.8	6.1	4.03	0.65	7.39
The distance to Clustering Center 3	0.25	0.74	2.79	2.7	5.71
The distance to Clustering Center 4	2.2	3.0	4.99	2.47	0.16
The result of identification	Clustering 3	Clustering 3	Clustering 1	Clustering 2	Clustering 4

7. The Results of the Identification of Driver's Steering Intention

The identification process of the driver's steering intention is shown in Figure 14 by the method of principal component analysis and the Gath–Geva algorithm, including the offline part, online part and identification. The offline part is to analyze the driver's steering intention data under different steering conditions through the principal component analysis and Gath–Geva algorithm analysis and get the clustering center. The online and identification parts are the processes of the real-time identification of the driver's steering intention. Firstly, the first 2 s of the test data are obtained under a certain condition in the driving simulator. Secondly, the characteristics of the data parameters are extracted, and then, the distances between the characteristic parameters and the center of each clustering center are calculated. Finally, according to the principle of the smallest distance, the driver's steering intention is determined. The distance calculation formula is:

$$d_i = \|x - c_i\|, i = 1, 2, 3, 4 \tag{18}$$

where x represents the characteristic parameters of one condition, $x = (x_1, x_2, \cdots, x_p)$; c_i represents the clustering center parameters for clustering, $c_i = (c_{i1}, c_{i2}, \cdots, c_{ip})$.

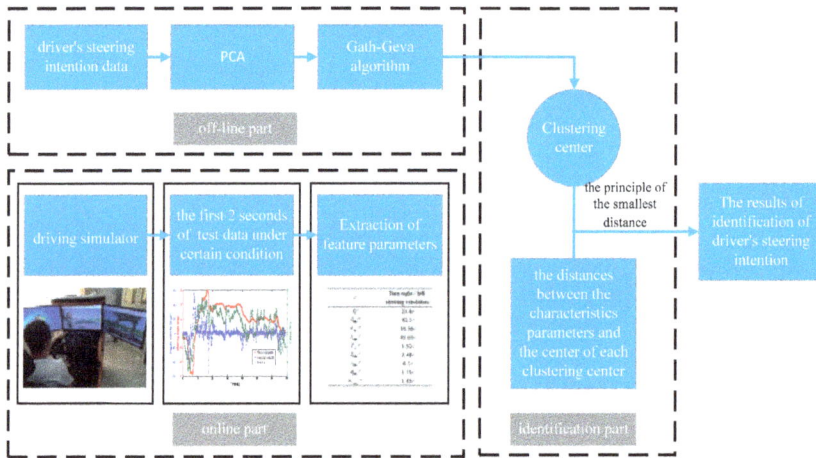

Figure 14. The flowchart for the identification of the driver's steering intention. PCA: Principal Component Analysis.

In order to verify the validity of the identification method designed by this article, five different drivers were selected. Five tests were carried out on the driving simulator, namely turning right, turning left, lane changing, lane keeping and U-turn condition. (0.45, 0.65), (0.44, 0.48) and (0.742, 0.649) are given by the Gath–Geva algorithm according to the above, respectively representing lane changing, lane keeping, turning right and U-turn condition. As shown in Table 6, in Conditions 1–5, the distances between the characteristic parameters and the center of each clustering center were calculated. It is shown that the results of each identification are exactly the same as the actual driver's intention. Therefore, the effectiveness of the identification method designed by this article is verified.

8. Conclusions

In this paper, driver's characteristic steering parameters under different conditions were proposed. Real vehicle tests under four kinds of typical operating conditions were implemented by five excellent driving school coaches with different ages and genders. The test vehicles covered four different countries' passenger cars. Then, the principal component analysis and clustering analysis were combined to classify and identify the driver's steering intention. By comparison and analysis, the Gath–Geva algorithm was significantly better than the other two clustering algorithms under different typical operating conditions to classify and identify the driver's steering intention. In order to verify the validity of the identification method designed by this article, five different drivers were selected. Five tests were carried out on the driving simulator. It was show that the results of each identification were exactly the same as the actual driver's intention. Therefore, the effectiveness of the identification method designed by this article was verified.

Acknowledgments: This work is financially supported by The National Natural Science Fund (No. U1564201 and No. 51675235).

Author Contributions: Yiding Hua and Xing Xu conceived of and designed the method. Yiding Hua and Huan Tian performed the experiments and analyzed the experimental data. Finally, Yiding Hua wrote the paper with the help of Haobin Jiang and Long Chen.

Conflicts of Interest: The authors declare no conflict of interest.

References

1. Sun, Y.; Xiong, G.G.; Chen, H.Y. Evaluation of Intelligent Behavior of Unmanned Vehicles Based on Fuzzy-EAHP. *Automot. Eng.* **2014**, *36*, 22–27.
2. Sai, S. Current and future ITS. *IEICE Trans. Inf. Syst.* **2013**, *96*, 176–183. [CrossRef]
3. Mueller, C.; Siedersberger, K.H.; Faerber, B. Active roll motion as feedback for decoupled steering interventions in Advanced Driver Assistance Systems (ADAS). *Forsch. Ingenieurwesn* **2017**, *81*, 41–55.
4. Anaya, J.J.; Ponz, A.; Garcia, F. Motorcycle detection for ADAS through camera and V2V Communication, a comparative analysis of two modern technologies. *Expert Syst. Appl.* **2017**, *77*, 148–159. [CrossRef]
5. Zhao, R.C.; Wong, P.K.; Xie, Z.C. Real-time weighted multi-objective model predictive controller for adaptive cruise control systems. *Int. J. Automot. Technol.* **2017**, *18*, 279–292. [CrossRef]
6. Li, S.B.; Guo, Q.Q.; Xin, L. Fuel-Saving Servo-Loop Control for an Adaptive Cruise Control System of Road Vehicles with Step-Gear Transmission. *IEEE Trans. Veh. Technol.* **2017**, *66*, 2033–2043. [CrossRef]
7. Zheng, B.; Anwar, S. Yaw stability control of a steer-by-wire equipped vehicle via active front wheel steering. *Mechatronics* **2009**, *19*, 799–804. [CrossRef]
8. Sun, X.D.; Chen, L.; Jiang, H.B.; Yang, Z.B.; Chen, J.C.; Zhang, W.Y. High-performance control for a bearingless permanent magnet synchronous motor using neural network inverse scheme plus internal model controllers. *IEEE Trans. Ind. Electron.* **2016**, *63*, 3479–3488. [CrossRef]
9. Sun, X.D.; Shi, Z.; Chen, L.; Yang, Z.B. Internal model control for a bearingless permanent magnet synchronous motor based on inverse system method. *IEEE T. Energy Convers.* **2016**, *31*, 1539–1548. [CrossRef]
10. Sun, X.D.; Su, B.K.; Chen, L.; Yang, Z.B.; Xu, X.; Shi, Z. Precise control of a four degree-of-freedom permanent magnet biased active magnetic bearing system in a magnetically suspended direct-driven spindle using neural network inverse scheme. *Mech. Syst. Signal Process.* **2017**, *88*, 36–48. [CrossRef]
11. Cheng, L.; Qiao, T.Z. Localization in the Parking Lot by Parked-Vehicle Assistance. *IEEE Trans. Intell. Transp.* **2016**, *17*, 3629–3634. [CrossRef]
12. Sun, X.D.; Chen, L.; Yang, Z.B. Overview of bearingless permanent magnet synchronous motors. *IEEE Trans. Ind. Electron.* **2013**, *60*, 5528–5538. [CrossRef]
13. Cheon, D.S.; Nam, K.H. Steering torque control using variable impedance models for a steer-by-wire system. *Int. J. Automot. Technol.* **2017**, *18*, 263–270. [CrossRef]
14. Kim, W.H.; Son, Y.S.; Chung, C.C. Torque-Overlay-Based Robust Steering Wheel Angle Control of Electrical Power Steering for a Lane-Keeping System of Automated Vehicles. *IEEE Trans. Veh. Technol.* **2016**, *65*, 4379–4392. [CrossRef]
15. Sun, X.D.; Chen, L.; Yang, Z.B.; Zhu, H.Q. Speed-sensorless vector control of a bearingless induction motor with artificial neural network inverse speed observer. *IEEE ASME Trans. Mech.* **2013**, *18*, 1357–1366. [CrossRef]
16. Windridge, D.; Shaukat, A.; Hollnagel, E. Characterizing Driver Intention via Hierarchical Perception-Action Modeling. *IEEE Trans. Hum. Mach. Syst.* **2013**, *43*, 17–31. [CrossRef]
17. Li, L.; Zhu, Z.B.; Wang, X.Y. Identification of a driver's starting intention based on an artificial neural network for vehicles equipped with an automated manual transmission. *Proc. Inst. Mech. Eng. D J. Automob.* **2016**, *230*, 1417–1429. [CrossRef]
18. Takano, W.; Matsushita, A.; Iwao, K. Recognition of human driving behaviors based on stochastic symbolization of time series signal. In Proceedings of the IEEE/RSJ International Conference on Intelligent Robots and Systems, Nice, France, 22–26 September 2008; pp. 167–172.
19. Raksin, C.P.; Mizushima, T.; Nagai, M. Direct yaw moment control system based on driver behavior recognition. *Veh. Syst. Dyn.* **2008**, *46*, 911–921.
20. Tesheng, H. Time-varying system identificationvia maximum a posteriori estimation and its application to driver steering models. In Proceedings of the American Control Conference, Westin Seattle Hotel, Seattle, WA, USA, 11–13 June 2008; pp. 11–13.
21. Harmouche, J. Incipient fault detection and diagnosis based on Kullback Leibler divergence using PCA: Part I. *Signal Process.* **2014**, *94*, 278–287. [CrossRef]
22. Nguyen, T.H.; Kim, H. Novel and efficient pedestrian detection using bidirectional PCA. *Pattern Recogn.* **2013**, *46*, 2220–2227. [CrossRef]
23. Ding, M. Adaptive KPCA. *Signal Process.* **2010**, *90*, 1542–1553. [CrossRef]

24. Zhou, K.; Yang, S. Exploring the uniform effect of FCM clustering: A data distribution perspective. *Knowl. Based Syst.* **2016**, *96*, 76–83. [CrossRef]
25. Andrea, B.; Palma, B. A survey of fuzzy clustering algorithms for pattern recognition I. *IEEE Trans. Syst. Man. Cybern. Soc.* **1999**, *29*, 778–785.
26. Zarita, Z.; Pauline, O. Design of wavelet neural networks based on symmetry fuzzy c-means for function approximation. *Neural Comput. Appl.* **2013**, *23*, 247–259.
27. Zhou, D.G.; Zhou, H. A modified strategy of fuzzy clustering algorithm for image segmentation. *Soft Comput.* **2015**, *19*, 3261–3272. [CrossRef]
28. Saber, S.; Selamat, A.; Fujita, H. Systematic mapping study on granular computting. *Knowl. Based Syst.* **2015**, *80*, 78–97.
29. Abonyi, J.; Babuska, R.; Szeifert, F. Modified Gath-Geva fuzzy clustering for identification of Takagi-Sugeno fuzzy models. *IEEE Trans. Syst. Man Cybern. B* **2002**, *32*, 612–621. [CrossRef] [PubMed]
30. Sarbu, C.; Zehl, K.; Einax, J.W. Fuzzy divisive hierarchical clustering of soil data using Gustafson-Kessel algorithm. *Chemom. Intell. Lab.* **2007**, *86*, 121–129. [CrossRef]
31. Price, A.L.; Patterson, N.J.; Plenge, R.M. Principal components analysis corrects for stratification in genome-wide association studies. *Nat. Genet.* **2006**, *38*, 904–909. [CrossRef] [PubMed]
32. Gustafson, D.E.; Kessel, W.C. Fuzzy clustering with fuzzy covariance matrix. In Proceedings of the 1978 IEEE Conference on Decision and Control including the 17th Symposium on Adaptive Processes, San Diego, CA, USA, 10–12 January 1979; pp. 761–766.
33. Bezdek, J.C. *Pattern Recognition with Fuzzy Objective Function Algorithms*; Plenum Press: New York, NY, USA, 1981.

applied
sciences

MDPI

Article

Localized Space-Time Autoregressive Parameters Estimation for Traffic Flow Prediction in Urban Road Networks

Jianbin Chen [1,2] , Demin Li [1,2,*], Guanglin Zhang [1,2] and Xiaolu Zhang [1,2]

[1] College of Information Science & Technology, Donghua University, Shanghai 201620, China;
    chen_jianbin@mail.dhu.edu.cn (J.C.); glzhang@dhu.edu.cn (G.Z.); xiaoludhu@mail.dhu.edu.cn (X.Z.)
[2] Engineering Research Center of Digitized Textile & Fashion Technology, Ministry of Education,
    Donghua University, Shanghai 201620, China
*   Correspondence: deminli@dhu.edu.cn; Tel.: +86-21-6779-2325

Received: 26 October 2017; Accepted: 9 February 2018; Published: 12 February 2018

Abstract: With the rapid increase of private vehicles, traffic congestion has become a worldwide problem. Various models have been proposed to undertake traffic prediction. Among them, autoregressive integrated moving average (ARIMA) models are quite popular for their good performance (simple, low complexity, etc.) in traffic prediction. Localized Space-Time ARIMA (LSTARIMA) improves ARIMA's prediction accuracy by extending the widely used STARIMA with a dynamic weight matrix. In this paper, a localized space-time autoregressive (LSTAR) model was proposed and a new parameters estimation method was formulated based on the LSTARIMA model to reduce computational complexity for real-time prediction purposes. Moreover, two theorems are given and verified for parameter estimation of our proposed LSTAR model. The simulation results showed that LSTAR provided better prediction accuracy when compared to other time series models such as Shift, autoregressive (AR), seasonal moving average (Seasonal MA), and Space-Time AR (STAR). We found that the prediction accuracy of LSTAR was a bit lower than the LSTARIMA model in the simulation results. However, the computational complexity of the LSTAR model was also lower than the LSTARIMA model. Therefore, there exists a tradeoff between the prediction accuracy and the computational complexity for the two models.

Keywords: LSTAR; STARIMA; parameters estimation; traffic flow prediction; urban road network

1. Introduction

In recent decades, the number of vehicles in urban areas has increased rapidly and the urban road network is becoming larger and more complex. Due to this, traffic congestion has become a major problem in big cities, which has led to more fuel consumption and environment pollution. Statistics show that the average annual traffic congestion cost in the US in 2014 was 1433 dollars per auto commuter, or over 5 billion dollars per city for very large urban areas [1]. In order to improve the efficiency of the urban road network and reduce traffic congestion, intelligent transportation systems (ITS) [2] have been developed by integrating information technology, automatic control technology, and geographic information systems (GIS). With the introduction of ITS, real-time traffic information is available to the vehicles in the road network for trip planning through vehicular navigation systems or dynamic route guidance systems. Unlike the computer network, which can transmit data package from source to destination in a very short time, vehicles in urban road networks need much longer times to travel to their destinations. Thus, trip planning should consider not only current traffic information, but also future traffic conditions. Therefore, short-term traffic flow prediction with real-time traffic information as prior knowledge is quite essential and has attracted increasing attention [3].

The earliest traffic prediction method was based on the macroscopic traffic simulation model proposed by Lighthill and Whitham [4] and Richards [5], known as the Lighthill-Whitham-Richards (LWR) model. In this model, vehicles in the highway are treated as "traffic flow" and their dynamics can be analyzed with the continuous fluid conservation equation in fluid mechanics. The microscopic traffic modeling method of cellular automaton (CA) [6] simulates the traffic flow dynamics by analysis of the interaction of individual vehicles. Traffic simulation models focus on traffic flow dynamics using only current traffic information; no historic information is needed. The weakness of the traffic simulation model is that it needs the origination-destination (OD) matrix of all vehicles to simulate the traffic dynamics, which is normally hard to collect.

The autoregressive moving average (ARMA) model or autoregressive integrated moving average (ARIMA) model [7], also called the Box-Jenkins model, is an important prediction model in economics and other areas. Furthermore, it is considered as the standard of time series prediction. ARIMA and its variations as seasonal ARIMA (SARIMA) [8], vector ARMA (VARMA) models [9], and so on have been widely used for traffic prediction. The space-time ARIMA (STARIMA) [10] model has a long historical background which is based on the ARMA with exogenous inputs (ARMAX) model. Since the 1980s, STARIMA has been applied to different areas such as river flow, spread of disease, spatial econometrics, and so on. In 2005 [11], the STARIMA methodology was first proposed for the spatiotemporal behavior of traffic flow. In the STARIMA model, traffic flow data is in the form of a spatial time series which is collected at specific locations at constant intervals of time to be used for the short-term forecasting of space-time stationary traffic-flow processes. Furthermore, the model can be used for assessing the impact of traffic-flow changes on other parts of the network through the use of weight matrices estimated on the basis of the distances among the various locations. Unlike the VARMA model, which is a generic model without any known information, the number of parameters to be estimated for STARIMA is much less, as the road network topology is considered. However, in the proposed STARIMA model [11], the weight matrices are assigned equally without considering the traffic condition differences between the directly connected first order spatial neighbors and the not directly connected higher order neighbors in the whole road network.

In the past 10 years, there have been many research studies on STARIMA. Min et al. [12] presented a dynamic form of the STARIMA that accounted for temporal dynamics. They replaced the traditional distance-weighted spatial weight matrix with a temporally dynamic matrix that reflected the current traffic turn ratios observed at each road intersection. The weight matrix can be updated in real time based on current conditions, but the method was limited to intersection-based flow data and was fixed spatially.

Tao Cheng et al. [13] extended the standard STARIMA model to a Localized STARIMA (LSTARIMA) model, which described the modeling of dynamic and heterogeneous autocorrelation in network data with improved traditional models. The constructed model provided an improvement over the traditional space-time series models. Their paper showed that the performance of prediction was improved when compared to standard STARIMA models. The LSTARIMA model captured the autocorrelation of traffic data locally and dynamically in the road network with a dynamic spatial weight matrix. The LSTARIMA model has also shown good performance in traffic prediction without the need for data pre-processing (e.g., a logarithmic transformation and differencing). Compared with other ARIMA variations, the LSTARIMA model has a simpler structure (because the LSTARIMA has smaller AR, and MA order values p and q). As the future traffic state of the current road depends not only on its prior states, but also on its neighbor roads, the weight matrix of roads is key to traffic flow prediction.

In this paper, our contributions are as follows:

1. An LSTAR model with lower computational complexity based on the LSTARIMA was proposed. In the LSTARIMA model of Cheng et al. [13], the same weight matrix W was used for AR and MA components of the whole road network. We used different matrices, W and U, for AR and MA components. And individual observation $z_i(t)$ was used instead of the N-dimension column

vector $Z(t)$ to allow each road to have its own weight matrix W, U. Since the ARMA model can be properly approximated by a high-order AR model, we further developed the reconstructed LSTARIMA model into our proposed LSTAR model.

2. A more reasonable weight matrix and new traffic information collection with the Vehicular Ad hoc Networks (VANET) approach was proposed. As the number of vehicles output from upstream roads has more impact on the future traffic condition compared to speed difference, it was used to determine the dynamic spatial weights instead of the speed difference. To obtain the traffic information needed for weight matrix determination, the vehicles stopped at red lights were used to collect traffic information via VANET.

3. Two theorems were given and verified for parameter estimation of our proposed LSTAR model. When the distribution of traffic flow is stable, the weight matrix can be treated as time invariant. When the traffic flow distribution is not stable, the weight matrix is time variant. For these two different cases, we provided two theorems to determine the parameters.

4. Related simulations were performed. Through the simulation results, we observed that the prediction accuracy of LSTAR was a bit lower than the LSTARIMA model. However, the computational complexity of the LSTAR model was also lower than the LSTARIMA model. Therefore, there existed a tradeoff between the prediction accuracy and the computational complexity for the two models.

The rest of this paper is organized as follows. In Section 2, state-of-the-art traffic information collection, traffic prediction, and traffic applications are reviewed. In Section 3, we introduce the LSTAR model, the construction of the weight matrix of the LSTAR model, and a new traffic information collection method. In Section 4, parameters estimation methods of the LSTAR model are given and proven. In Section 5, the experimental evaluation is presented. Finally, Section 6 provides the conclusions and identifies future research directions.

2. State-of-the-Art and Related Topics

Traffic information collection provides the data input for traffic prediction, and there are many applications that use traffic prediction results to improve traffic conditions. In this section, the state-of-the-art traffic information collection, traffic prediction, and urban traffic applications are reviewed.

2.1. Traffic Information Collection

As traffic information is the base data of ITS, how to collect traffic information efficiently is an important area of research. Loop detectors are pressure, magnetic, and other sensors buried underground to detect if there are vehicles passing over them. They are widely deployed in urban areas to count the number of vehicles passing through fixed points of roads. Compared to loop detector technology, machine-vision-based traffic monitoring is a state-of-the-art approach with the advantages of easy maintenance, real-time visualization, and high flexibility [14]. With properly installed cameras, traffic information such as speed, volume, and even traffic accidents can be detected. However, it is expensive to establish these systems as well as to maintain a huge number of fixed devices on the road side, and they can only gather the traffic information of fixed points.

With the equipment of global position system (GPS) receivers on vehicles or mobile phones, vehicles can detect their own real-time location and speed. An alternative traffic information collection approach has been proposed to estimate the traffic state by checking the location and speed of some vehicles running on the road [15,16]. These probe vehicles are known as floating vehicles, and are normally buses and taxis. Due to the low cost of GPS receivers, the overall cost of the floating vehicles system is low. The shortcoming of floating vehicles is that their distribution in the urban traffic network is not even in space and time, which means that they may not be able to provide complete traffic states of the whole road network.

Vehicular Ad hoc Networks (VANET) [17–19] are an emerging technology developed for traffic security and data transformation [20]. In recent years, many research studies have used VANET to collect traffic information. The first type of system uses VANET to only estimate the traffic density of the road by detecting the number of vehicles in the VANET communication range [21]. The other type of system assumes that each vehicle is also equipped with a GPS receiver so more detailed traffic information can be collected [22,23]. As GPS receivers are more commonly equipped when compared with VANET and GPS can provide more information, more research has focused on the approach with GPS. With the communication capability of VANET, the collected traffic information can be easily shared and used by other ITS applications such as traffic prediction, route planning, and so on.

After traffic information is collected, it can be used as an input to other ITS applications as base data. Since VANET can obtain the traffic information of the whole network without infrastructure and can be easily integrated with other ITS systems, it was used to collect traffic information in our paper. Section 3.3 discussed our traffic information collection method via VANET in detail.

2.2. Traffic Prediction

Short-term traffic prediction is one of the most important topics in ITS research and practice. Aside from the ARIMA series prediction method, there have also been many other methodologies engaged for this purpose.

The Kalman filtering method [24,25], which is based on historical data and present data to predict a future state, has been widely used in the forecasting of traffic flow. However, its computational complexity is too high for complex urban traffic flow prediction. Neural Networks (NNs) models [26,27] have also been utilized to predict traffic flow for their high prediction accuracy. The weak point of NNs is the long model training time. Other research studies have referred to Support Vector Machines (SVM) [28]. However, the error of SVM is high under the circumstance of peak periods and blocking traffic accidents compared with the Bayesian network [29].

The K-NN method (K-nearest neighborhood) [30] performs well in short-term forecasting even when accidents have occurred. However, this algorithm has high complexity and needs a large amount of calculation when searching for class neighbors. Markov-based models [31] have also shown good performance on traffic flow prediction since the traffic condition at the next interval is closely related to the recent states. However, in these models, there are many states to consider. New technologies such as big data [32] and particle filtering [33,34] have also been used in traffic predication and other urban mobility applications.

Although these traffic prediction approaches have provided good prediction results in some scenarios, most of them are too complex or require a long training time. Thus, there have been respectable efforts put towards improving various ARIMA prediction models.

2.3. Urban Traffic Applications

Current traffic information and predicted traffic information are the base data of ITS, but they are meaningless without practical applications. Route planning and vehicle navigation systems are some of the most popular applications that use real-time and predicted traffic information [35]. The first-generation route planning system only considers the static features of the road network to obtain the shortest path. With the development of real-time traffic information collection technology, dynamic route planning has been proposed to re-calculate the new shortest path with the updated real-time data at each intersection. Such a system provides better travel planning when compared with the static route. However, using only current data may lead to frequent route changes in complex traffic conditions. The newest route planning systems use the predicted traffic information to arrive at the best and most stable route to the destination.

The other main applications using current and predicted traffic information are traffic management applications. The most common usage is to display current traffic states, the predicted traveling time to land mark locations, and traveling proposals on the traffic information board. With this information, drivers can re-plan their travel accordingly. Adaptive traffic signal control is the key technology of traffic management. The reactive traffic signaling control system adjusts the signal phase and cycle lengths according to current traffic data. The predictive traffic signal systems retime the traffic sign according to the predicted traffic information. With the predictive approach, the total waiting time of vehicles is reduced and the efficiency of the traffic network is improved [36].

Predicted traffic information is not only used by navigation systems and traffic management systems, but also by other urban applications such as parking management and so on. For the wide usage of traffic prediction, there are continuous research interests in this field.

3. Model and Preliminaries

3.1. LSTAR Model Construction

According to the STARIMA model defined in Reference [11], both space and time are considered.

$$Z_t = \sum_{k=1}^{p} \sum_{l=0}^{\lambda_k} \phi_{kl} W_l z_{t-k} - \sum_{k=1}^{q} \sum_{l=0}^{m_k} \theta_{kl} W_l \varepsilon_{t-k} + \varepsilon_t. \tag{1}$$

Z_t is an N-dimensional column vector of road i, while $i = 1, 2, \ldots, N$, and W_l is an $N \times N$ matrix with element w_{ij}^l. ε_t is the residual vector. ϕ_{kl}, θ_{kl} are the AR and MA parameters, respectively. Then, the observation of road i, $z_i(t)$ can be described as:

$$z_i(t) = \sum_{k=1}^{p} \sum_{l=0}^{\lambda_k} \sum_{j=1}^{N} \phi_{kl} w_{ij}^{(l)} z_j(t-k) - \sum_{k=1}^{q} \sum_{l=0}^{m_k} \sum_{j=1}^{N} \theta_{kl} w_{ij}^{(l)} \varepsilon_j(t-k) + \varepsilon_i(t). \tag{2}$$

In 2014, a new space-time model, the localized STARIMA (LSTARIMA) model, was proposed by Cheng et al. [13] to consider spatial heterogeneity and temporal non-stationarity. The model is described by the following form:

$$Z_i(t) = \sum_{k=1}^{p_i} \sum_{h=0}^{\lambda_k(t-k,i)} \phi_{i,kh} W^{(h,t-k,i)} Z_i(t-k) - \sum_{l=1}^{q_i} \sum_{h=0}^{n_l(t-l,i)} \theta_{i,lh} W^{(h,t-l,i)} \varepsilon_i(t-l) + \varepsilon_i(t). \tag{3}$$

$Z_i(t)$ is an N-dimensional column vector of the observation value on link $1, \ldots, N$ with tag i at time t, which can be any prediction variable of roads such as speed, traffic flow, density, and so on. The term $\varepsilon_i(t)$ is a residual on link $1, \ldots, N$ at time t. The first term in Equation (3) is the AR component, whereas the second term is the MA. The parameters p_i and q_i are the AR and MA orders, respectively. h is the spatial order that represents the order of spatial separation between two locations. The parameters $\lambda_k(t-k,i)$ and $n_l(t-l,i)$ are the dynamic spatial orders associated with the kth and lth temporally lagged terms in the AR and MA components, respectively. They specify the size of the spatial neighborhood that could influence the link of interest i within temporal lags k and l. The parameters $W^{(h,t-k,i)}$ and $W^{(h,t-l,i)}$ are the dynamic spatial weight matrices $W^{(h,t,i)}$ pertaining to link i at temporal lags k and l. $\phi_{i,kh}$ and $\theta_{i,lh}$ are the AR and MA parameters for each link i ($i = 1, 2, \ldots, N$).

Although spatial tag i was added to $Z_i(t)$ in the LSTARIMA model construction, the spatial heterogeneity was not fully considered. As the $Z_i(t)$ here is an N-dimension column vector which covers all of the roads (road $1, 2, \ldots, N$) in the network, all of the roads will share the same $\phi_{i,kh}$ and $\theta_{i,lh}$. In this LSTARIMA model, the same matrix W is used for both AR and MA components. As the weight matrix of AR and MA components is not always the same, using only one weight matrix W is not proper.

In this paper, different weight matrices W, U were used for AR and MA components and individual road traffic flow observation $z_i(t)$, according to Equation (2), was defined to allow every road to have its own weight matrix W_i, U_i according to spatial location, but not sharing the same weight matrix among all roads, as in Equation (3).

Then, the LSTARIMA can be rewritten as follows:

$$z_i(t) = \sum_{k=1}^{p_i} \sum_{h=0}^{\lambda_{i,k}} \sum_{j=1}^{N_i} \phi_{i,kh} w_{ij}^{(h)}(t-k) z_j(t-k) - \sum_{k=1}^{q_i} \sum_{h=0}^{m_{i,k}} \sum_{j=1}^{N_i} \theta_{i,kh} u_{ij}^{(h)}(t-k) \varepsilon_j(t-k) + \varepsilon_i(t). \tag{4}$$

The parameters $w_{ij}^{(h)}(t-k)$ and $u_{ij}^{(h)}(t-l)$ are the elements of dynamic spatial weight matrices $W_i^{(h)}(t-k)$ and $U_i^{(h)}(t-l)$ pertaining to link i at temporal lags k and l. Like the traditional STARIMA model, LSTARIMA makes use of spatial weight matrices W and U to model the influence of the spatiotemporal neighborhoods. However, it relaxes the globally fixed temporal dependence for all locations by using different AR and MA parameters according to location. Furthermore, it accounts for the temporal non-stationarity by allowing the matrix elements value and size of the spatial neighborhoods to vary with time.

According to Bo [37], the ARMA model can be properly approximated by the high-order AR model. As AR only has one type of parameter to be estimated, ARMA and ARIMA have two or three types of parameters to be estimated, and the parameters estimation of AR is easy even when the order is a little bit higher. There exist plenty of studies that have used a high-order AR model to approximate many processes of interest [37]. Furthermore, traffic flow has complex dynamics and may not exactly match an ARIMA model. In addition, many studies have removed MA and used only different AR models to conduct traffic prediction and obtain good results with limited AR order [38]. For reducing computational complexity and real-time prediction purpose, in this paper, we proposed the LSTAR model for traffic flow prediction.

With the MA component removed, the LSTARIMA prediction model (Equation (4)) is changed into the following LSTAR model:

$$z_i(t) = \sum_{k=1}^{p_i} \sum_{h=0}^{\lambda_{i,k}} \sum_{j=1}^{N_i} \phi_{i,kh} w_{ij}^{(h)}(t-k) z_j(t-k) + \varepsilon_i(t) \tag{5}$$

where $\varepsilon_i(t)$ is white noise, $\phi_{i,hk}$ is the parameter for each link i ($i = 1, 2, \ldots, N$), and $w_{ij}^{(h)}(t-k)$ are the elements of the dynamic spatial weight matrix $W_i^{(h)}(t-k)$ pertaining to link i at temporal lag k.

3.2. Weight Matrix Construction

Weight matrix construction is an essential topic in STARIMA models. In the STARIMA model, the weight matrix is time invariant and equal for the same neighbor order. In the LSTARIMA model, a time variant weight matrix was introduced to improve traffic prediction accuracy with lower AR and MA orders. Furthermore, the speed difference was used to construct the weight matrix in the LSTARIMA model [13]. The speed of a road is an important character of traffic flow, but is not the essential one in terms of impact to surrounding roads. It is obvious that a road that outputs only one vehicle will not impact neighbor roads at the same level as saturated roads with the same speed. The traffic output amount to the neighbor roads in a time slot has more impact on the future traffic state of neighbor roads. Thus, the output vehicle number during a time slot was used in the weight matrix construction instead of speed in this paper.

For all pairwise road sections (i, j) with spatial lag h, the corresponding $w_{ij}^{(h)}(t)$ is defined as follows:

$$w_{ij}^{(h)}(t) = \begin{cases} 1 & h = 0, i = j \\ \frac{Q_{ij}(t)}{\sum Q(t)} & h = 1, \sum Q(t) \neq 0 \\ \frac{Q_{ij}^{(h)}(t)}{\sum Q^{(h)}(t)} & h \neq 1, \sum Q^{(h)}(t) \neq 0 \\ 0 & \sum Q^{(h)}(t) = 0 \end{cases} \tag{6}$$

where $Q_{ij}(t)$ is the number of vehicles running from road j towards road i at time slot t and $\sum Q(t)$ is the sum of vehicle numbers from directly connected roads to road i at time slot t. $Q^{(h)}{}_{ij}(t)$ is the number of vehicles on the h order neighbor road j towards $h - 1$ order neighbor of road i at time slot t, and $Q^{(h)}{}_{ij}(t) = 0$ if j is not the h order neighbor of i. $\sum Q^{(h)}(t)$ is the sum of $Q^{(h)}{}_{ij}(t)$.

3.3. Traffic Information Collection

In this paper, we proposed a traffic information collection system via VANET for urban areas. In this system, each vehicle was assumed to have a GPS receiver and VANET equipped to report its location. This assumption is reasonable as currently more and more vehicles are equipped with such devices. Considering that there are always traffic lights at the intersection of urban roads, we used vehicles stopped at red lights to collect traffic information by checking the location of all vehicles inside their VANET communication range periodically.

As shown in Figure 1, when the traffic light turned red for the east-west direction, the first vehicle stopped at the west side was selected as the traffic information collector (TIC). If there were no vehicle stops at the west side, the first vehicle stopped at the east side would be the TIC. In the example of Figure 1, vehicle V11 is the TIC and collects the traffic information during the red light period.

Step 1. When the traffic light turns to red at T0, V11 broadcasts traffic information collection request.

Step 2. All of the vehicles in the communication range R of V11 will report their locations to V11 after receiving the request from V11.

Step 3. V11 catalogs the vehicles to four vehicle sets according to the location. They are marked as $Vset(E, T0)$, $Vset(S, T0)$, $Vset(W, T0)$, and $Vset(N, T0)$.

Step 4. After time $\tau < R/V_{max}$, V11 collects the traffic information again according to Steps 1–3 and obtains $Vset(E, T0 + \tau)$, $Vset(S, T0 + \tau)$, $Vset(W, T0 + \tau)$, $Vset(N, T0 + \tau)$. V_{max} is the maximum allowed velocity. Time $\tau < R/V_{max}$ will let all vehicles running towards the intersection be detectable at time $T0 + \tau$.

For example:

If $R = 150$ m and $V_{max} = 20$ m/s, $\tau = 5$ s can be used as $5 < \frac{150}{20} = 7.5$. The maximal length a vehicle can run during τ is $\frac{20 \text{ m}}{\text{s}} \times 5 \text{ s} = 100$ m < 150 m. Then, no vehicle entering the intersection at T0 can run outside the communication range of V11 and be detectable at $T0 + \tau$.

Step 5. The vehicles' set run from road A to road B is calculated by formula: $Vset(A \rightarrow B, T0 + \tau) = Vset(A, T0) \cap Vset(B, T0 + \tau)$.

For example:

As shown in Figure 1, the traffic output from S to E from T0 to $T0 + \tau$ is:

$$Vset(S \rightarrow E, T0 + \tau) = Vset(S, T0) \cap Vset(E, T0 + \tau)$$
$$= \{V1, V2, V3, V4\} \cap \{V1, V8, V9, V10\} = \{V1\}$$

Step 6. The TIC calculates the traffic output of each road with time interval τ until the traffic light for the east-west direction turns green. As the traffic light for the south-north direction turns red, the first vehicle stopped at the north or south side will be selected as the TIC and collect traffic information continuously.

As a part of the advance travel information system (ATIS), the TIC will send out the collected traffic information via VANET for applications such as traffic prediction. In normal urban traffic conditions, there should always be vehicles stopped at the red light to act as the TIC. If there is no vehicle stopped at the red light to be the TIC, the traffic information of the last time slot will be used. This is acceptable as it normally happens in very low traffic density cases and real-time traffic information is not important.

With the collected traffic information, the traffic output of each road section j can be calculated by:

$$Q^{(h)}{}_{ij}(t) = \sum_n \sum_m Vset(j \to j_m, T + n\tau),$$

where $T + n\tau \in (t - 1, t]$, j is the h order neighbor of road i, j_m is the $h - 1$ order neighbor of road i (j_m is i when $h = 1$), m is the total number of j_m which are downstream j.

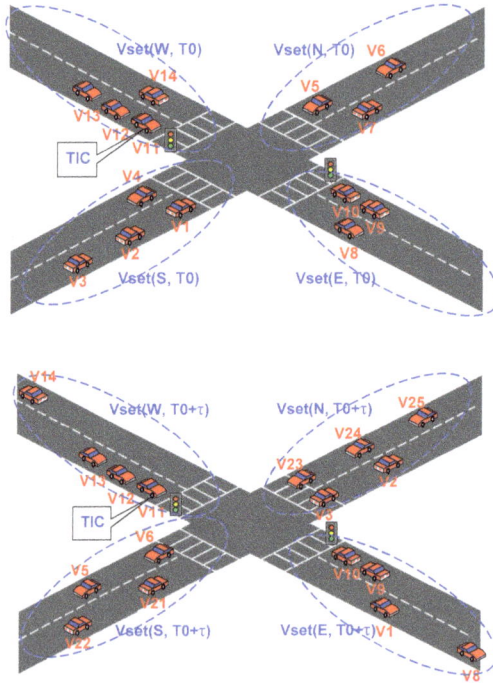

Figure 1. Traffic information collection.

4. Main Results

For the LSTAR model, in this section we discuss how to determine the parameter $\phi_{i,kh}$ with a given weight matrix $W_i^{(h)}(t)$. With the determined weight matrix $W_i^{(h)}(t)$ and $\phi_{i,kh}$, the future traffic observation $\hat{z}(t)$ of road i can be predicted. Normally, $w_{ij}^{(h)}(t)$ is time variant. When the traffic flow distribution is stable in the road network, $w_{ij}^{(h)}(t)$ will be time invariant since $\dfrac{Q^{(h)}{}_{ij}(t)}{\sum Q^{(h)}(t)}$ is constant. Considering the weight matrix differences of being time invariant or not, two theorems are discussed, and the LSTAR parameters estimation can be conducted accordingly.

Definitions 1. $r_{ij}(m)$ *is the correlation coefficient of road i and j, $r_i(m) = r_{ii}(m)$ is the autocorrelation coefficient of road i. We define* $\Phi_i = \left[\phi_{i,10}, \phi_{i,11}, \cdots \phi_{i,1\lambda_{i,0}}, \phi_{i,20}, \phi_{i,21}, \ldots, \phi_{i,p_i\lambda_{i,k}} \right]^T$,

let $n = \sum_{k=1}^{P_i} \lambda_{i,k}$ to be the dimension of Φ_i. $\quad R_i \quad = \quad [r_i(0), r_i(1), \dots, r_i(n-1)]^T,$
$\Sigma = [\sigma_i^2, 0, \dots, 0]^T$, $S(h,m) = \sum_{j=1}^{N_i} w_{ij}^{(h)} r_{ij}(m),$

$$A = \begin{bmatrix} S(0,1) & S(1,1) & \cdots & S(\lambda_{i,k}, p_i) \\ S(0,0) & S(1,0) & \cdots & S(\lambda_{i,k}, p_i - 1) \\ S(0,1) & S(1,1) & \cdots & S(\lambda_{i,k}, p_i - 2) \\ \vdots & \vdots & \vdots & \vdots \\ S(0, n-1) & S(1, n-1) & \cdots & S(\lambda_{i,k}, n-1-p_i) \end{bmatrix}, \overline{A} = \begin{bmatrix} A & R_i - \Sigma \end{bmatrix}.$$

Theorem 1 follows:

Theorem 1. *If* $W_i^{(h)}(t)$ *is time invariant and* $Rank(A) = Rank(\overline{A}) = n$, *the LSTAR model can be uniquely determined.*

Proof. As $W_i^{(h)}(t)$ is time invariant, we can observe that $W_i^{(h)}(t) = W_i^{(h)}(t-k)$ and it can be rewritten to $W_i^{(h)}$.

The model of Equation (5) will be a time invariant system. Since $z_i(t)$ is a stationary random signal, Equation (5) will be:

$$z_i(t) = \sum_{k=1}^{p_i} \sum_{h=0}^{\lambda_{i,k}} \sum_{j=1}^{N_i} \phi_{i,kh} w_{ij}^{(h)} z_j(t-k) + \varepsilon_i(t). \tag{7}$$

Pre-multiplying both sides of the model (Equation (7)) by $z_i(t-m)$:

$$z_i(t)z_i(t-m) = \sum_{k=1}^{p_i} \sum_{h=0}^{\lambda_{i,k}} \sum_{j=1}^{N_i} \phi_{i,kh} w_{ij}^{(h)} z_j(t-k)z_i(t-m) + \varepsilon_i(t)z_i(t-m). \tag{8}$$

Taking the expected values in both sides, we obtain an equation similar to the Yule-Walker equation:

$$r_i(m) = \sum_{k=1}^{p_i} \sum_{h=0}^{\lambda_{i,k}} \sum_{j=1}^{N_i} \phi_{i,kh} w_{ij}^{(h)} r_{ij}(m-k) + \sigma_i^2 \delta(m) \tag{9}$$

where the expected value is:

$$E(\varepsilon_i(t)z_i(t-m)) = \sigma_i^2 \delta(m). \tag{10}$$

As $\phi_{i,hk}$ does not have tag j, we obtain:

$$r_i(m) = \sum_{k=1}^{p_i} \sum_{h=0}^{\lambda_{i,k}} \phi_{i,kh} \sum_{j=1}^{N_i} w_{ij}^{(h)} r_{ij}(m-k) + \sigma_i^2 \delta(m). \tag{11}$$

For Equation (11), rewrite $\phi_{i,kh}$ to a column vector as $\Phi_i = \left[\phi_{i,10}, \phi_{i,11}, \dots, \phi_{i,1\lambda_{i,0}}, \phi_{i,20}, \phi_{i,21}, \dots \phi_{i,k\lambda_{i,k}} \right]^T$. In addition, $n = \sum_{k=1}^{p_i} \lambda_{i,k}$ is the number of parameters $\phi_{i,kh}$ to be estimated.

As $S(h, m) = \sum_{j=1}^{N_i} w_{ij}^{(h)} r_{i,j}(m)$, we obtain:

$$
\begin{bmatrix} r_i(0) \\ r_i(1) \\ r_i(2) \\ \vdots \\ r_i(n-1) \end{bmatrix} = \begin{bmatrix} S(0,1) & S(1,1) & \cdots & S(\lambda_{i,k}, p_i) \\ S(0,0) & S(1,0) & \cdots & S(\lambda_{i,k}, p_i - 1) \\ S(0,1) & S(1,1) & \cdots & S(\lambda_{i,k}, p_i - 2) \\ \vdots & \vdots & \vdots & \vdots \\ S(0, n-1) & S(1, n-1) & \cdots & S(\lambda_{i,k}, n-1-p_i) \end{bmatrix} \begin{bmatrix} \phi_{i,10} \\ \phi_{i,11} \\ \phi_{i,12} \\ \vdots \\ \phi_{i,k\lambda_{i,k}} \end{bmatrix} + \begin{bmatrix} \sigma_i^2 \\ 0 \\ 0 \\ \vdots \\ 0 \end{bmatrix}. \tag{12}
$$

Let $A = \begin{bmatrix} S(0,1) & S(1,1) & \cdots & S(\lambda_{i,k}, p_i) \\ S(0,0) & S(1,0) & \cdots & S(\lambda_{i,k}, p_i - 1) \\ S(0,1) & S(1,1) & \cdots & S(\lambda_{i,k}, p_i - 2) \\ \vdots & \vdots & \vdots & \vdots \\ S(0, n-1) & S(1, n-1) & \cdots & S(\lambda_{i,k}, n-1-p_i) \end{bmatrix}$, $R_i = \begin{bmatrix} r_i(0) \\ r_i(1) \\ r_i(2) \\ \vdots \\ r_i(n-1) \end{bmatrix}$, $\Sigma = \begin{bmatrix} \sigma_i^2 \\ 0 \\ 0 \\ \vdots \\ 0 \end{bmatrix}$.

Then, the augmented matrix of Equation (12) is:

$$
\overline{A} = \begin{bmatrix} S(0,1) & S(1,1) & \cdots & S(\lambda_{i,k}, p_i) & r_i(0) - \sigma_i^2 \\ S(0,0) & S(1,0) & \cdots & S(\lambda_{i,k}, p_i - 1) & r_i(1) \\ S(0,1) & S(1,1) & \cdots & S(\lambda_{i,k}, p_i - 2) & r_i(2) \\ \vdots & \vdots & \vdots & \vdots & \vdots \\ S(0, n-1) & S(1, n-1) & \cdots & S(\lambda_{i,k}, n-1-p_i) & r_i(n-1) \end{bmatrix} = \begin{bmatrix} A & R_i & -\Sigma \end{bmatrix}.
$$

If $Rank(A) = n$, then $Rank(\overline{A}) = n$ and we will have a unique solution of parameters Φ_i:

$$
\Phi_i = A^{-1}(R_i - \Sigma) \tag{13}
$$

Then, we can uniquely define the LSTAR model and predict traffic flow with it.

If $R(A) \neq R(\overline{A})$, there is no solution for Equation (12).

If $Rank(A) = Rank(\overline{A}) < n$, there are infinite solutions for Φ_i. $\square$

Remark 1. *In case the spatial weight matrix $W_i^{(h)}(t)$ is time invariant, we can determine the LSTAR prediction model by the correlation of roads. We can uniquely define the LSTAR model when $Rank(A) = Rank(\overline{A}) = n$. When $Rank(A) = Rank(\overline{A}) < n$, there will be many solutions for parameter Φ. This means that we have defined more parameters than are actually needed. We can reduce $\lambda_{i,k}$ and/or p_i to obtain a unique LSTAR model.*

Definitions 2. *Let $r_i'(m)$ be the autocorrelation coefficient of ith element of $U(t) = W(t)Z(t)$, where the matrix $W(t) = \sum_{h=0}^{\lambda_{i,k}} W_i^{(h)}(t)$, which combines all spatial effect defined in $W_i^{(h)}(t)$ to one matrix, and $Z(t) = [z_1(t), z_2(t), \ldots, z_{N_i}(t)]^T$ is the vector form of $z_i(t)$. We define $\Phi_i = [\phi_{i,1}, \phi_{i,2}, \ldots, \phi_{i,p_i}]^T$, $\overline{R}_i' = [\overline{r}_i'(0), \overline{r}_i'(1), \ldots, \overline{r}_i'(p_i - 1)]^T$, $\Sigma = [\sigma_i^2, 0, \ldots, 0]^T$,*

$$
A' = \begin{bmatrix} r_i'(1) & r_i'(2) & \cdots & r_i'(p_i) \\ r_i'(0) & r_i'(1) & \cdots & r_i'(p_i - 1) \\ \vdots & \vdots & \vdots & \vdots \\ r_i'(p_i - 2) & r_i'(p_i - 3) & \cdots & r_i'(1) \end{bmatrix}, \overline{A'} = \begin{bmatrix} A' & \overline{R}_i' - \Sigma \end{bmatrix}.
$$

We then present Theorem 2.

Theorem 2. *If $W_i^{(h)}(t)$ is time variant, the combined weight matrix $W(t)$ is full ranked, and $Rank(A') = Rank(\overline{A'}) = p_i$, the LSTAR model can be uniquely determined.*

Proof. According to the weight matrix $W_i^{(h)}(t)$ construction in the LSTAR model, the element $w_{ij}^{(h)}(t)$ will always be zero when i, j is not at spatial order h. We combine all of the $\lambda_{i,k}$ weight matrix into one weight matrix, $W(t) = \sum_{h=0}^{\lambda_{i,k}} W_i^{(h)}(t)$. This simplification is reasonable as (1) time variant $w_{ij}(t)$ can somehow give an effect similar to spatial order h; and (2) $w_{ij}(t)$ is always equal to the only non-zero $w_{ij}^{(h)}(t)$:

$$w_{ij}(t) = 0 + 0 + \ldots + w_{ij}^{(h)}(t) + \ldots + 0 = w_{ij}^{(h)}(t) \text{ number of zero is } \lambda_{i,k} - 1$$

Considering the combined spatial weight matrix $W(t) = \sum_{h=0}^{\lambda_{i,k}} W_i^{(h)}(t)$, $Z(t) = [z_1(t), z_2(t), \ldots, z_{N_i}(t)]^T$ is an N_i dimension column vector that includes all neighbor roads within the spatial order to be considered by road i, and $\Sigma = [\sigma_i^2, 0, \ldots, 0]^T$, we can obtain a matrix from the LSTAR model according to Equation (5).

$$Z(t) = \sum_{k=1}^{p_i} \phi_{i,k} W(t-k) Z(t-k) + \Sigma(t) \tag{14}$$

when the rank of $W(t)$ is N_i, Equation (14) can be rewritten as:

$$[W(t)]^{-1}[W(t)Z(t)] = \sum_{k=1}^{p_i} \phi_{i,k} W(t-k) Z(t-k) + \Sigma(t). \tag{15}$$

Let $U(t) = [W(t)Z(t)]$, we obtain:

$$[W(t)]^{-1}U(t) = \sum_{k=1}^{p_i} \phi_{i,k} U(t-k) + \Sigma(t). \tag{16}$$

$\left[W_i^{(h)}(t)\right]^{-1}$ can be treated as an instantaneous window to $U(t)$, so $U(t)$ is stationary in the short term. We have:

$$\overline{U}(t) = \sum_{k=1}^{p_i} \phi_{i,k} U(t-k) + \Sigma(t). \tag{17}$$

Let $u_i(t)$ be the element of $U(t)$, then:

$$\overline{u}_i(t) = \sum_{k=1}^{p_i} \phi_{i,k} u_i(t-k) + \varepsilon_i(t). \tag{18}$$

Pre-multiplying both sides of Equation (18) by $u_i(t-m)$:

$$\overline{u}_i(t) u_i(t-m) = \sum_{k=1}^{p_i} \phi_{i,k} u_i(t-k) u_i(t-m) + \varepsilon_i(t) u_i(t-m). \tag{19}$$

Taking expected values in both sides, we obtain:

$$\overline{r}_i'(m) = \sum_{k=1}^{p_i} \phi_{i,k} r_i'(m-k) + \sigma_i^2 \delta(m) \tag{20}$$

where the expected value $r_i'(m) = E(u_i(t-m)u_i(t-k))$, $\overline{r}_i'(m) = E([W_i^{(h)}(t)]^{-1} u_i(t)u_i(t-m))$, $E(\varepsilon_i(t)u_i(t-m)) = \sigma_i^2 \delta(m)$.

We can then obtain:

$$
\begin{bmatrix} \bar{r}'_i(0) \\ \bar{r}'_i(1) \\ \vdots \\ \bar{r}'_i(p_i-1) \end{bmatrix} = \begin{bmatrix} r'_i(1) & r'_i(2) & \cdots & r'_i(p_i) \\ r'_i(0) & r'_i(1) & \cdots & r'_i(p_i-1) \\ \vdots & \vdots & \vdots & \vdots \\ r'_i(p_i-2) & r'_i(p_i-3) & \cdots & r'_i(1) \end{bmatrix} \begin{bmatrix} \phi_{i,1} \\ \phi_{i,2} \\ \vdots \\ \phi_{i,p_i} \end{bmatrix} + \begin{bmatrix} \sigma_i^2 \\ 0 \\ \vdots \\ 0 \end{bmatrix} \tag{21}
$$

Let $\Phi_i = \begin{bmatrix} \phi_{i,1} \\ \phi_{i,2} \\ \vdots \\ \phi_{i,p_i} \end{bmatrix}$, $A' = \begin{bmatrix} r'_i(1) & r'_i(2) & \cdots & r'_i(p_i) \\ r'_i(0) & r'_i(1) & \cdots & r'_i(p_i-1) \\ \vdots & \vdots & \vdots & \vdots \\ r'_i(p_i-2) & r'_i(p_i-3) & \cdots & r'_i(1) \end{bmatrix}$, $\bar{R}'_i = \begin{bmatrix} \bar{r}'_i(0) \\ \bar{r}'_i(1) \\ \vdots \\ \bar{r}'_i(p_i-1) \end{bmatrix}$, $\Sigma = \begin{bmatrix} \sigma_i^2 \\ 0 \\ \vdots \\ 0 \end{bmatrix}$.

Then the augmented matrix of Equation (21) is:

$$
\overline{A'} = \begin{bmatrix} r'_i(1) & r'_i(2) & \cdots & r'_i(p_i) & \bar{r}'_i(0)-\sigma_i^2 \\ r'_i(0) & r'_i(1) & \cdots & r'_i(p_i-1) & \bar{r}'_i(1) \\ \vdots & \vdots & \vdots & \vdots & \vdots \\ r'_i(p_i-2) & r'_i(p_i-3) & \cdots & r'_i(1) & \bar{r}'_i(p_i-1) \end{bmatrix} = \begin{bmatrix} A' & \bar{R}'_i - \Sigma \end{bmatrix}.
$$

If $Rank(A') = p_i$, then $Rank\left(\overline{A'}\right) = p_i$ and we will have a unique solution of parameters Φ:

$$
\Phi_i = \left[A'\right]^{-1}(\bar{R}'_i - \Sigma). \tag{22}
$$

If $R(A') \neq R\left(\overline{A'}\right)$, there is no solution for Equation (21).

If $Rank(A') = Rank\left(\overline{A'}\right) < p_i$, there are infinite solutions for Φ. $\square$

Remark 2. *Unlike $W_i^{(h)}(t)$ with most of its elements being zero and normally not being full ranked, most elements of the combined weight matrix $W(t)$ are not zero. So $W(t)$ is normally a full rank matrix. For some special cases when $W(t)$ is not a full rank matrix, we can reduce the size of $W(t)$ to make it fully ranked. When $W(t)$ is a full rank matrix, we can uniquely define the LSTAR model when $Rank(A) = Rank(\overline{A}) = n$. Similar to Remark 1, we can reduce p_i to obtain a unique LSTAR model if $Rank(A) = Rank(\overline{A}) < n$.*

In this section, two theorems were given and proven according to the weight matrix determined. When the traffic flow distribution is stable, $W_i^{(h)}(t)$ can be treated as time invariant and Theorem 1 can be used. When the traffic flow distribution is not stable, $W_i^{(h)}(t)$ is time variant and Theorem 2 should be used. With the measured weight matrix $W_i^{(h)}(t)$ and estimated Φ_i, future traffic state $\hat{z}_i(t+1)$ can be predicted according to the LSTAR model (Equation (5)) by one-time slot shifting.

5. Practical Example and Experimental Evaluation

5.1. Practical Example

In this paper, we provide a practical example on how to use our LSTAR model to predict future traffic flow of the Shanghai Century Park area. To evaluate the prediction approach of LSTAR, we adopted the widely used traffic simulation tools Simulation of Urban Mobility (SUMO) [39] and OpenStreetMap (OSM) [40], which are recognized as promising candidates for traffic simulations, and the simulation results are commonly accepted as a replacement of real data. Additionally, plenty of works exist that have adopted SUMO and OSM as tools to generate traffic data for research [41,42].

In this example, we demonstrate the model-building procedure for our proposed LSTAR model in the context of traffic flow prediction on a road network. First, we downloaded the OSM format road network map of the area near Shanghai Century Park, as shown in Figure 2. The OSM format map not only included the geography topology of the road network, but also the road type, lane number, speed limitation, traffic light duration, and so on, according to real-world information. Then, the SUMO NetConvert tool was used to convert the OSM format map to a SUMO format map. SUMO was then used to simulate the traffic flow of this area according to the road network information converted from the OSM map.

Figure 2. OpenStreetMap (OSM) map of Shanghai Century Park area.

In the simulation, trip demands were generated randomly every two seconds according to the edge length. The "Fringe factor" was set to 4, which means that roads with no successor or no predecessor had four times the possibility of being selected as the start or end of a trip when compared to other roads. The speed limitations, traffic light durations of each road, and so on were obtained from the real-world data of the OSM map. The simulation duration was one week. The detailed simulation parameters are listed in Table 1.

After we obtained traffic flow data generated from SUMO, they were used to conduct traffic flow prediction with different prediction models. The prediction intervals were five minutes, 15 min, and 30 min, as normally a prediction interval over 30 min has less significance to real-time route planning or vehicle navigation.

The SUMO format map converted from the OSM map is shown in Figure 3. The roads were renamed as Rn-m for easy usage in the following discussion. In the following section, road R7-3 in a north-to-south direction was selected as the example road to demonstrate the LSTAR prediction procedure. Furthermore, we conducted traffic flow prediction for roads R7-2, R3-3, and R3-4 with the same procedure used for road R7-3.

Table 1. Simulation parameters.

Parameters	Value
Trip Generation Method	Random
Trip Possibility Weight	Edge Length
New Trip Start Interval	2 s
Fringe Factor	4
Max Vehicle Number	300
Traffic Light Duration	OSM Map data
Speed Limitation	OSM Map data
Simulation Duration	604,800 s (1 Week)

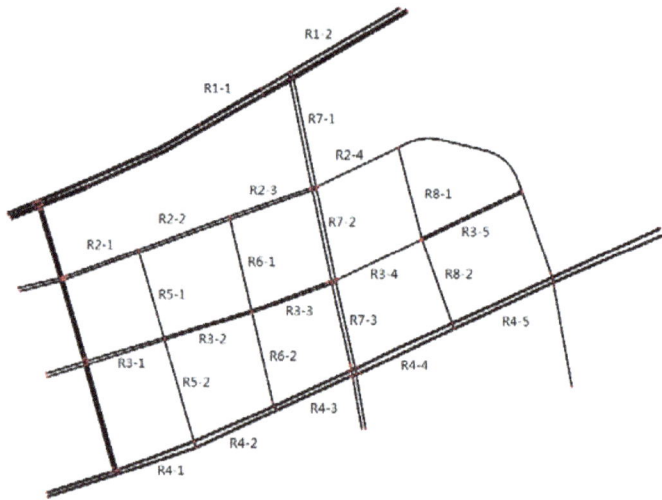

Figure 3. Simulation of Urban Mobility (SUMO) road network.

Construction of a Dynamic Spatial Weight Matrix

Step 1. Build a spatial adjacency matrix.

The first step was to build a spatial adjacency matrix based on the topological structure of the network, which appears in Figure 1. In this paper, spatial adjacency matrices of spatial orders up to three were constructed as per Reference [13]. The spatial neighborhood information can be found in Table 2 with the first, second, and third order neighbors separated.

Step 2. Determine the dynamic spatial order and weights.

The second step was to determine the dynamical spatial order and weights for every road link with the method proposed in this paper (Section 3.2. Weight Matrix Construction). In this simulation, only road R7-3 in a north-to-south direction was selected to show how the weight matrix was determined. According to weight matrix definition, only upstream road sections of R7-3 in the north-to-south direction were considered. The dynamic spatial weights calculation results of road R7-3 in a north-to-south direction with a five-minute time step are shown in Table 2.

Table 2. The dynamic spatial weights.

| Spatial | First | | | | | Second | | | | | | | Third | | |
Temporal Order k	R7-2	R3-3	R3-4	R7-1	R2-3	R2-4	R3-2	R6-1	R3-5	R8-1	R3-1	R5-1	R2-2	R1-1	R1-2
5	0.74	0.11	0.16	0.48	0.22	0.13	0.13	0.00	0.04	0.00	0.17	0.09	0.22	0.26	0.26
10	0.69	0.31	0.00	0.29	0.21	0.11	0.25	0.07	0.04	0.04	0.43	0.07	0.13	0.33	0.03
15	0.61	0.35	0.04	0.59	0.07	0.07	0.21	0.03	0.00	0.03	0.29	0.13	0.10	0.42	0.06
20	0.65	0.23	0.13	0.49	0.19	0.08	0.16	0.03	0.03	0.03	0.26	0.12	0.06	0.44	0.12
25	0.72	0.07	0.21	0.63	0.13	0.08	0.13	0.00	0.04	0.00	0.43	0.00	0.09	0.35	0.13
30	0.36	0.27	0.36	0.32	0.09	0.09	0.32	0.09	0.05	0.05	0.17	0.08	0.00	0.58	0.17
35	0.38	0.38	0.25	0.59	0.29	0.06	0.00	0.00	0.00	0.06	0.23	0.03	0.16	0.48	0.10
40	0.78	0.11	0.11	0.48	0.14	0.07	0.21	0.03	0.07	0.00	0.36	0.04	0.04	0.48	0.08
45	0.74	0.19	0.07	0.50	0.23	0.08	0.19	0.00	0.00	0.00	0.26	0.07	0.04	0.52	0.11
50	0.71	0.14	0.14	0.49	0.16	0.08	0.19	0.05	0.00	0.03	0.19	0.11	0.08	0.56	0.06
55	0.53	0.33	0.13	0.46	0.17	0.13	0.13	0.04	0.08	0.00	0.11	0.05	0.21	0.58	0.05
60	0.47	0.27	0.27	0.29	0.24	0.05	0.24	0.10	0.10	0.00	0.19	0.24	0.29	0.29	0.00
65	0.55	0.27	0.18	0.39	0.18	0.09	0.12	0.12	0.00	0.09	0.25	0.16	0.19	0.38	0.03
70	0.48	0.33	0.19	0.41	0.19	0.15	0.15	0.00	0.07	0.04	0.26	0.06	0.13	0.52	0.03
75	0.56	0.25	0.19	0.50	0.17	0.08	0.13	0.08	0.04	0.00	0.33	0.06	0.11	0.50	0.00
80	0.71	0.29	0.00	0.43	0.19	0.00	0.33	0.05	0.00	0.00	0.24	0.16	0.12	0.36	0.12
85	0.76	0.10	0.14	0.59	0.09	0.05	0.18	0.05	0.05	0.00	0.29	0.04	0.04	0.54	0.08
90	0.67	0.13	0.20	0.46	0.14	0.07	0.25	0.04	0.04	0.00	0.33	0.00	0.21	0.42	0.04
95	0.23	0.69	0.08	0.40	0.20	0.05	0.15	0.00	0.10	0.10	0.22	0.13	0.04	0.39	0.22
100	0.79	0.10	0.10	0.24	0.19	0.10	0.38	0.05	0.05	0.00	0.13	0.13	0.08	0.65	0.03

With the dynamic spatial weights estimated in Table 2, we can see that the weights are time variant in this case as the traffic flow was time variant. Then, we used Theorem 2 to conduct a parameters estimation and traffic flow prediction.

After the future traffic states are predicted, the information can be used to conduct route planning or predictive traffic signal control applications, and so on.

5.2. Experimental Evaluation

The traffic flow prediction accuracy results of the different prediction methods by means of Root Mean Square Error (RMSE) are shown in Figure 4. Figure 5 shows the average RMSE and Root Mean Square Percentage Error (RMSPE). The average of Figures 4 and 5 is the average RMSE, RMSPE values of roads R7-3, R7-2, R3-3, and R3-4 per the prediction models. The definition of RMSE and RMSPE are shown below.

$$\text{RMSE} = \sqrt{\frac{\sum_{i=1}^{n}(x_i - \hat{x}_i)^2}{n}} \tag{23}$$

$$\text{RMSPE} = \frac{\sqrt{\frac{\sum_{i=1}^{n}(x_i - \hat{x}_i)^2}{n}}}{\frac{\sum_{i=1}^{n} x_i}{n}} \times 100\% \tag{24}$$

where n is the prediction interval number, x_i is the actual value, and $\hat{x}_i$ is the prediction value.

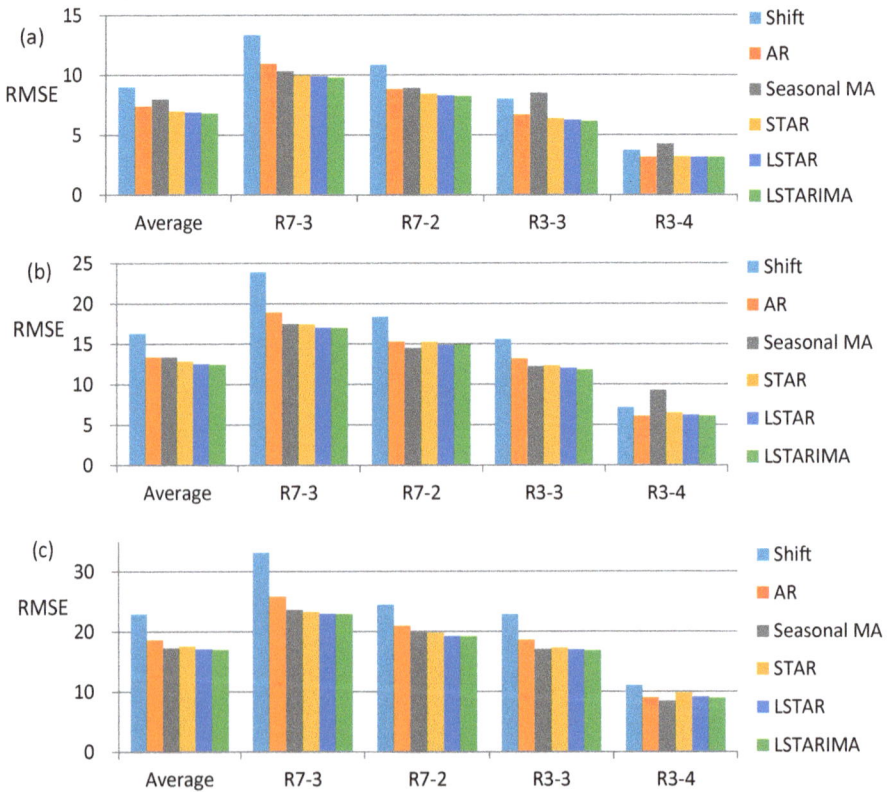

Figure 4. Prediction accuracy comparison of different models with: (**a**) 5-min prediction interval; (**b**) 15-min prediction interval; and (**c**) 30-min prediction interval.

The average RMSE and RMSPE values of all road sections are shown in Figure 5.

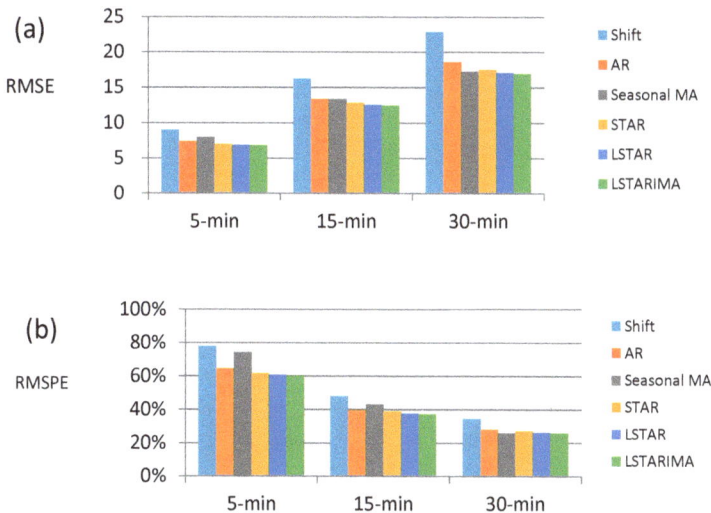

Figure 5. Average prediction accuracy comparison of different models: (**a**) Root Mean Square Error (RMSE); and (**b**) Root Mean Square Percentage Error (RMSPE).

From Figure 4, the results showed that on most roads, the prediction accuracies of the different prediction models were similar for all intervals. The predication accuracy from low to high was Shift, AR, Seasonal MA, STAR, LSTAR, and LSTARIMA, with some exceptions on R3-4 and R3-3. From Figure 5, we can see that the RMSE increased as the prediction intervals increased for all prediction methods, while the RMSPE decreased when the prediction intervals increased. This indicates that although the absolute error increased as the prediction intervals increased, the actual prediction accuracy increased with larger prediction intervals as the percentage form errors decreased. Figure 5 also shows that, regardless of the prediction interval, the average prediction accuracy of LSTAR was always better than Shift, AR, Seasonal MA, and STAR. Moreover, LSTARIMA always had a little higher accuracy when compared to LSTAR in all prediction intervals.

According to Diebold [43], only comparing values such as RMSE is not sufficient to declare that one prediction model is better than another without a statistics significance check. There are many hypothesis tests designed for prediction accuracy comparison and the Diebold-Mariano (DM) test [43] is the most popular one. To further evaluate the prediction performance of LSTAR, the DM test was used to check if LSTAR was better than other statistically significant prediction models. The forecast package of R [44] was used to conduct the DM test for the prediction results of road R7-3. As the DM test can only compare the prediction accuracy of two models, we did the DM test for LSTAR and the other models one by one. The DM test hypothesis was that LSTAR had better performance than all of the other methods subjected to the test. The p-values of each DM test are shown in Table 3.

Table 3. Diebold-Mariano (DM) test results; AR: autoregressive; Seasonal MA: seasonal moving average; STAR: Space-Time AR.

Prediction Model	5 min	15 min	30 min
Shift	0.0000	0.0000	0.0000
AR	0.0351	0.0225	0.0000
Seasonal MA	0.0611	0.0822	0.07814
STAR	0.1023	0.0884	0.0929
LSTARIMA	0.8985	0.7828	0.6797

The DM test results showed that the LSTAR prediction accuracy was significantly better than Shift at p-value < 1%, better than AR at p-value < 5%, and almost better than Seasonal MA and STAR at p-value < 10% (one p-value of STAR > 10%). LSTARIMA was not significantly better than LSTAR at p-value < 10%, as no p-value was greater than 90% with the hypothesis that LSTAR is better than LSTARIMA.

6. Conclusions

This paper discussed the application of a local space-time autoregressive (LSTAR) model for traffic flow prediction. In this paper, we showed the prediction process of the LSTAR model in detail. The LSTAR model appears to be the best model among the Shift, AR, Seasonal MA, and STAR models given its greater parameter flexibility (dynamic spatial neighborhood and dynamic spatial weight). According to the DM test results, the LSTAR prediction accuracy was significantly better than Shift and AR, and was better than seasonal MA and STAR, but not significantly. As LSTARIMA also considers the local spatial and time dynamics and still keeps the MA component, the prediction accuracy was always better than the LSTAR model in the simulation results. However, the decrease in LSTAR prediction accuracy was very minor when compared to LSTARIMA, and was not statically significant. Furthermore, the computational complexity of the LSTAR model was also lower than that of the LSTARIMA model. Therefore, there existed a tradeoff between the prediction accuracy and the computational complexity for the two models.

Future studies will be carried out to assess the performance of the LSTAR model with different real-world traffic data and the usage of prediction data for different urban traffic applications. We will also conduct the simulation and performance evaluation of the traffic information collection via the VANET approach.

Acknowledgments: This work is supported by the NSF of China under Grants No. 61772130, No. 61301118, No. 71171045; The International S&T Cooperation Program of Shanghai Science and Technology Commission under Grant No. 15220710600, and the Innovation Program of Shanghai Municipal Education Commission under Grant No. 14YZ130.

Author Contributions: Jianbin Chen and Demin Li conceived and designed the method. Jianbin Chen performed the experiments and analyzed the experimental data. Finally, Jianbin Chen and Xiaolu Zhang wrote the paper with the help of Demin Li and Guanglin Zhang.

Conflicts of Interest: The authors declare no conflict of interest.

References

1. United States Department of Transportation, National Transportation Statistics. Table 1-72: Annual Highway Congestion Cost. 2017. Available online: https://www.rita.dot.gov/bts/sites/rita.dot.gov.bts/files/NTS_Entire_2017Q2.pdf (accessed on 8 January 2018).
2. Alam, M.; Ferreira, J.; Fonseca, J. *Introduction to Intelligent Transportation Systems*; Springer: Cham, Switzerland, 2016; pp. 552–557. [CrossRef]
3. Kong, Q.J.; Xu, Y.; Lin, S.; Wen, D.; Zhu, F.; Liu, Y. UTN-Model-Based Traffic Flow Prediction for Parallel-Transportation Management Systems. *IEEE Trans. Intell. Transp. Syst.* **2013**, *14*, 1541–1547. [CrossRef]

4. Lighthill, M.J.; Whitham, G.B. On kinematic waves II. A therory of traffic flow on long crowded roads. *Proc. R. Soc. A Math. Phys. Eng. Sci.* **1955**, *229*, 317–345. [CrossRef]

5. Richards, P.I. Shock Waves on the Highway. *Op. Res.* **1956**, *4*, 42–51. [CrossRef]

6. Tian, J.F.; Li, G.Y.; Treiber, M.; Jiang, R.; Jia, N.; Ma, S.F. Cellular automaton model simulating spatiotemporal patterns, phase transitions and concave growth pattern of oscillations in traffic flow. *Trans. Res. B Methodol.* **2016**, *93*, 560–575. [CrossRef]

7. Box, G.E.; Jenkins, G.M. *Time Series Analysis: Forecasting and Control*; Holden-Day: Oakland, CA, USA, 1976; Volume 31, p. 303.

8. Williams, B.M.; Hoel, L.A. Modeling and Forecasting Vehicular Traffic Flow as a Seasonal ARIMA Process: Theoretical Basis and Empirical Results. *J. Trans. Eng.* **2003**, *129*, 664–672. [CrossRef]

9. Kamarianakis, Y.; Prastacos, P. Forecasting traffic flow conditions in an urban network—Comparison of multivariate and univariate approaches. *Trans. Res. Rec.* **2003**, 74–84. [CrossRef]

10. Pfeifer, P.E.; Deutsch, S.J. A Three-Stage Iterative Procedure for Space-Time Modeling. *Technometrics* **1980**, *22*, 35–47. [CrossRef]

11. Kamarianakis, Y.; Prastacos, P. Space–time modeling of traffic flow. *Comput. Geosci.* **2005**, *31*, 119–133. [CrossRef]

12. Min, X.; Hu, J.; Chen, Q.; Zhang, T.; Zhang, Y. Short-term traffic flow forecasting of urban network based on dynamic STARIMA model. In Proceedings of the International IEEE Conference on Intelligent Transportation Systems, St. Louis, MO, USA, 4–7 October 2009; pp. 1–6. [CrossRef]

13. Cheng, T.; Wang, J.; Haworth, J.; Heydecker, B.; Chow, A. A Dynamic Spatial Weight Matrix and Localized Space—Time Autoregressive Integrated Moving Average for Network Modeling. *Geogr. Anal.* **2014**, *46*, 75–97. [CrossRef]

14. Wan, Y.; Huang, Y.; Buckles, B. Camera calibration and vehicle tracking: Highway traffic video analytics. *Trans. Res. Part C* **2014**, *44*, 202–213. [CrossRef]

15. Mehta, V.; Chana, I. Urban Traffic State Estimation Techniques Using Probe Vehicles: A Review. In *Computing and Network Sustainability*; Vishwakarma, H., Akashe, S., Eds.; Lecture Notes in Networks and Systems; Springer: Singapore, 2017; Volume 12, pp. 273–281. [CrossRef]

16. Lai, W.-K.; Kuo, T.-H.; Chen, C.-H. Vehicle Speed Estimation and Forecasting Methods Based on Cellular Floating Vehicle Data. *Appl. Sci.* **2016**, *6*, 47. [CrossRef]

17. Zhang, G.; Xu, Y.; Wang, X.; Tian, X.; Liu, J.; Gan, X.; Qian, L. Multicast capacity for VANETs with directional antenna and delay constraint. *IEEE J. Sel. Areas Commun.* **2012**, *30*, 818–833. [CrossRef]

18. Zhang, G.; Liu, J.; Ren, J. Multicast capacity of cache enabled content-centric wireless Ad Hoc networks. *China Commun.* **2017**, *14*, 1–9. [CrossRef]

19. Ren, J.; Zhang, G.; Li, D. Multicast capacity for VANETs with directional antenna and delay constraint under random walk mobility model. *IEEE Access* **2017**, *5*, 3958–3970. [CrossRef]

20. Guo, C.; Li, D.; Zhang, G.; Cui, Z. Data delivery delay reduction for VANETs on bi-directional roadway. *IEEE Access* **2017**, *4*, 8514–8524. [CrossRef]

21. Hussain, R.; Kim, S.; Oh, H. Traffic Information Dissemination System: Extending Cooperative Awareness among Smart Vehicles with Only Single-Hop Beacons in VANET. *Wirel. Pers. Commun.* **2016**, *88*, 151–172. [CrossRef]

22. Li, D.; Li, Q.; Wang, J. Traffic information collecting algorithms for road selection decision support in vehicle ad hoc networks. *Int. J. Simul. Proc. Modell.* **2012**, *7*, 50–56. [CrossRef]

23. Darwish, T.; Bakar, A.K. Traffic density estimation in vehicular ad hoc networks: A review. *IEICE Trans. Inf. Syst.* **2015**, *24*, 337–351. [CrossRef]

24. Guo, J.; Huang, W.; Williams, B.M. Adaptive Kalman filter approach for stochastic short-term traffic flow rate prediction and uncertainty quantification. *Transp. Res. Part C Emerg. Technol.* **2014**, *43*, 50–64. [CrossRef]

25. Abidin, A.F.; Kolberg, M. Towards improved vehicle arrival time prediction in public transportation: integrating SUMO and Kalman filter models. In Proceedings of the 2015 17th UKSim-AMSS International Conference on Modelling and Simulation (UKSim), Cambridge, UK, 25–27 March 2015; pp. 147–152. [CrossRef]

26. Çetiner, B.G.; Sari, M.; Borat, O. A Neural Network Based Traffic-Flow Prediction Model. *Math. Comput. Appl.* **2010**, *15*, 269–278. [CrossRef]

27. Tang, J.; Liu, F.; Zou, Y.; Zhang, W.; Wang, Y. An Improved Fuzzy Neural Network for Traffic Speed Prediction Considering Periodic Characteristic. *IEEE Trans. Intell. Transp. Syst.* **2017**, *18*, 2340–2350. [CrossRef]

28. Ma, Y.; Chowdhury, M.; Sadek, A.; Jeihani, M. Integrated Traffic and Communication Performance Evaluation of an Intelligent Vehicle Infrastructure Integration (VII) System for Online Travel-Time Prediction. *IEEE Trans. Intell. Transp. Syst.* **2012**, *13*, 1369–1382. [CrossRef]

29. Deng, L.; He, Z.; Zhong, R. The Bus Travel Time Prediction Based on Bayesian Networks. In Proceedings of the 2013 International Conference on Information Technology and Applications, Chengdu, China, 16–17 November 2013; pp. 282–285. [CrossRef]

30. Yu, B.; Song, X.L.; Guan, F.; Yang, Z.M.; Yao, B.Z. k-Nearest Neighbor Model for Multiple-Time-Step Prediction of Short-Term Traffic Condition. *J. Transp. Eng.* **2016**, *142*. [CrossRef]

31. Qi, Y.; Ishak, S. A Hidden Markov Model for short term prediction of traffic conditions on freeways. *Transp. Res. Part C Emerg. Technol.* **2014**, *43*, 95–111. [CrossRef]

32. Lv, Y.; Duan, Y.; Kang, W.; Li, Z.; Wang, F.Y. Traffic Flow Prediction With Big Data: A Deep Learning Approach. *IEEE Trans. Intell. Transp. Syst.* **2015**, *16*, 865–873. [CrossRef]

33. Dhivyabharathi, B.; Hima, E.S.; Vanajakshi, L. Stream travel time prediction using particle filtering approach. *Transp. Lett. Int. J. Transp. Res.* **2016**, 1–8. [CrossRef]

34. Martino, L.; Read, J.; Elvira, V.; Louzada, F. Cooperative parallel particle filters for online model selection and applications to urban mobility. *Digit. Signal Proc.* **2017**, *60*, 172–185. [CrossRef]

35. Liebig, T.; Piatkowski, N.; Bockermann, C.; Morik, K. Dynamic route planning with real-time traffic predictions. *Inf. Syst.* **2017**, *64*, 258–265. [CrossRef]

36. Florin, R.; Olariu, S. A survey of vehicular communications for traffic signal optimization. *Veh. Commun.* **2015**, *2*, 70–79. [CrossRef]

37. Bo, W. Estimation of Autoregressive Moving-Average Models via High-Order Autoregressive Approximations. *J. Time* **2010**, *10*, 283–299. [CrossRef]

38. Griffith, D.A.; Heuvelink, G.B.M. Deriving Space-Time Variograms from Space-Time Autoregressive (STAR) Model Specifications. In Proceedings of the StatGIS09: Geo Informatics for Environmental Surveillance, Milos, Greece, 17–19 June 2009; Volume 38, pp. 285–303. [CrossRef]

39. Behrisch, M.; Bieker, L.; Erdmann, J.; Krajzewicz, D. *SUMO—Simulation of Urban Mobility: An Overview*; SIMUL: Barcelona, Spain, 2011; pp. 63–68.

40. Haklay, M.; Weber, P. OpenStreetMap: User-Generated Street Maps. *IEEE Pervasive Comput.* **2008**, *7*, 12–18. [CrossRef]

41. Wang, Y.; Jiang, J.; Mu, T. Context-Aware and Energy-Driven Route Optimization for Fully Electric Vehicles via Crowdsourcing. *IEEE Trans. Intell. Transp. Syst.* **2013**, *14*, 1331–1345. [CrossRef]

42. Griggs, W.M.; Ordóñez-Hurtado, R.H.; Crisostomi, E.; Häusler, F.; Massow, K.; Shorten, R.N. A Large-Scale SUMO-Based Emulation Platform. *IEEE Trans. Intell. Transp. Syst.* **2015**, *16*, 3050–3059. [CrossRef]

43. Diebold, F.X.; Mariano, R.S. Comparing Predictive Accuracy. *J. Bus. Econ. Stat.* **1995**, *20*, 134–144. [CrossRef]

44. Coreteam, R. R: A language and environment for statistical computing. *Computing* **2015**, *1*, 12–21.

applied sciences

MDPI

Article

Cellular Automaton to Study the Impact of Changes in Traffic Rules in a Roundabout: A Preliminary Approach

Krzysztof Małecki * and Jarosław Wątróbski

Department of Computer Science, West Pomeranian University of Technology, 52 Żołnierska Str., 71-210 Szczecin, Poland; jwatrobski@wi.zut.edu.pl
* Correspondence: kmalecki@wi.zut.edu.pl; Tel.: +48-91-449-5661

Academic Editor: Felipe Jimenez
Received: 16 June 2017; Accepted: 18 July 2017; Published: 21 July 2017

Featured Application: The current article proposes traffic modelling in a roundabout on the basis of cellular automata. It also considers roundabout traffic reorganisation in order to increase roundabout capacity. The article also presents an analysis on the impacts of pedestrian traffic and on distances between vehicles. An analysis focussing on multi-lane roundabouts is also provided.

Abstract: The current article presents a roundabout traffic model based on cellular automata for computer simulation. The model takes into account various sizes of roundabouts, as well as various types and maximum speeds of vehicles. A realistic vehicle braking phase is presented which is adjusted to the kind of vehicle and weather conditions. It also analyses roundabout traffic options including where the various rules for entering and exiting a roundabout apply. Traffic rules are contained in respective traffic scenarios. The simulation results indicate that there is significant scope for roundabout traffic reorganisation, with a mind to increasing roundabout capacity.

Keywords: cellular automaton (CA); model of CA; computer simulation; roundabout traffic simulation; roundabout traffic rules

---

1. Introduction

In most urban areas, there is an on-going increase in the number of vehicles on the road. For road authorities, taking measures to improve traffic efficiency and the safety of its participants is essential. Amongst research in this field [1–3], studies on traffic theories [3], models [4], and traffic modelling for one- and two-way roads and also roundabouts appear to be the most promising. Such analyses are significantly facilitated by computer simulations [5], which help to visualise the object of the research.

A roundabout is an intersection where traffic moves in a circle around a central island. According to roundabout traffic rules, vehicles approaching a roundabout must give way to vehicles already moving in the roundabout. Roundabout intersections help to solve number of the problems encountered by the traditional intersection which involves a major and a minor road. Roundabouts reduce the number of vehicle-vehicle and vehicle-pedestrian collision points at an intersection, which in turns helps reduce the number of accidents. Research studies conducted in the US have shown a drop in the number of accidents (by 29%) and injured persons (by 81%) as a result of implementing roundabouts [6]. An important aspect of roundabouts implementation is making drivers slow down when approaching a roundabout, which results from the need to give way to vehicles already moving around the roundabout, and from the appropriate structure of a roundabout. Leaf and Preusser [7] found vehicle velocity has a tremendous effect on the scale of injuries sustained by a pedestrian hit by a vehicle: slowing down from 48 km/h to 32 km/h means the pedestrian's chances of surviving increase ninefold. In a study in Sweden [8], major factors of roundabout safety were specified: the number of lanes in a roundabout, island diameter, and also traffic speed and intensity. The authors of the study argued that single-lane roundabouts are the safest, whereas large-radius roundabouts result in a higher permissible speed, which is positively related to the number of accidents. At the same time, a radius of less than 10 m is also problematic as it usually encourages drivers to accommodate the bend, and consequently drive almost straight on without reducing a reduction in speed.

Computer simulations are applied in roundabout traffic modelling to visualise the effects of layout changes. Simulations by their nature obtain information on a given system during the design stage. Computer simulation can also apply hypothetical conditions to a given system. They can provide an examination of the system operation in conditions which occur rarely and conditions yet to eventuate, such as the simulation of the effect of new traffic regulations). Simulation programs are based on mathematical models (traffic flows, multi-agent models, neural networks and cellular automata), which map the process and enable specification of input parameters that affect the simulation effectiveness. For example, Sisiopiku and Heung-Un Oh have determined, by means of a simulation, that roundabouts are particularly recommended in places where the traffic is evenly spread among the feeder roads, and also in places where there is a substantial amount of left-turn traffic (in the case of right-hand traffic) [9]. The pair [10] established factors that affect the capacity of roundabouts, and found that a three-lane roundabout is not necessarily preferable to a two-lane roundabout.

The mentioned examples of studies show that the focus of the research to date has been to ascertain whether a roundabout is preferable to a conventional intersection, and how many lanes in a roundabout is best to ensure optimal traffic flow. The majority of cities face a problem of chronic jammed streets and roundabouts notwithstanding the studies and road traffic regulations established. There have been numerous examples of drivers who fail to obey rules and cause congestion. The human factor is a significant variable any simulation of traffic.

The purpose of the current article is to determinate, based on computer simulation, whether road traffic reorganisation has an effect on roundabout capacity. The authors have developed a multi-lane roundabout model with multi-lane entrance and exit roads. The novelty of the model is the application of varied traffic principles, which contribute to the analysis of driver behaviour. Comparison of the existing roundabout traffic rules with the proposals presented in the studies reveals that roundabout capacity could be increased by over 15%.

The current publication is organised as follows: the following chapter is a review of previous studies, then there is a short note on the theoretical basis, the research problem and the developed model. Following this, simulation scenarios and experimental studies are discussed, and the article is concludes with a summary.

2. Related Works

2.1. Selected Hardware and Software Solutions

Road traffic analysis and modelling is a challenging task due to the complexity of the situation and its stochastic nature. Attempting to avoid inconveniencing the connected driver with attempts to adjust a specific area (e.g., an intersection or a roundabout), road administrators and researchers resort to simulation to determine best road solutions. The literature describes mathematical models and also software solutions and hardware solutions, as well as solutions combining hardware and software. The most numerous are software solutions, for example: MATSim (Multi-Agent Transport Simulation) [11], VISSIM (Verkehr In Städten-SIMulationsmodell) [12], TRANSIMS (TRansportation ANalysis SIMulation System) [13], MITSIM (Massachusetts Institute of Technology SIMulation) [14], AIMSUN (Advanced Interactive Micro-scopic Simulator for Urban Networks) [15,16], SUMO (Simulation of Urban MObility) [17,18] or CORSIM (CORridor SIMulation) [19,20], extensions of existing solutions [21,22], and multi-agent simulations [23] regarding pollution [24,25] or driver behaviour [26]. Some authors also focus their attention on solutions close to real-time simulations [27].

Hardware solutions are broad. They include numerous scale models of selected road junctions [28–30], hardware-in-the loop solutions [31,32], solutions presenting a scale model of a real object [33] or GPU(Graphics Processing Unit)-based solutions [34].

2.2. Selected Studies and Models of Road Traffic

The car-following theory, "follow-the-leader", was developed in the late 1950s [35]. The theory won many supporters and was applied whole-heartedly [36,37], particularly in relation to road traffic simulation [38–43].

Some of the studies based on car-following theory concentrated on finding a gap in a flow of vehicles moving around a roundabout, which makes it possible for a waiting vehicle to enter the roundabout from a feeder road (the so-called gap-acceptance model) [44–47]. The main limitation of the gap-acceptance theory is an assumption that drivers behave in a predictable way, whereas in reality behaviours are slightly different, e.g., resulting from ignorance or intentional disobeying of the rules [48]. The issue was also addressed by more contemporary researchers [49–51]. Some more recent directions in research studies have been application of cellular automata in the modelling of unloading bays [52] and exploitation of game theory in order to shorten the time autonomous vehicles wait to enter roundabouts [53].

2.3. Selected Basic Traffic Models Based on Cellular Automata

Cellular automata are perfect for traffic flow modelling due to their stochastic nature. The Nagel–Schreckenberg (N–Sch) model is a simple model of cellular automata to simulate car traffic. It was developed in 1992 by Nagel and Schreckenberg [4]. The model describes the single-lane car traffic and it is the basis for testing various traffic scenarios [54]. Schreckenberg created a project to inform drivers from Cologne, Aachen and Bonn about traffic jams on the road. Development and utilization of the model enabled the prediction of traffic congestion and notification of the drivers. Another proven framework is the model developed by Biham [55]. It is a simple cellular automaton model displaying the traffic in two intersecting directions. Each array cell can be occupied with a vehicle travelling in one of the two directions (north or east). The vehicle moves by one cell in the chosen direction, if that the adjacent cell is empty. Otherwise, the vehicle remains in place. The most important model showing traffic within the intersection is the model by Chowdhury and Schadschneider [56]. The authors model the traffic in one-way single-lane roads. The development of the model to the version for two-lane and two-way roads is presented in another study by Małecki and Iwan [21]. The new model expands the original idea of the intersections with the mechanism of induction loops activating traffic light to eliminate congestion, namely to keep smooth traffic flow at the intersection. Phenomena taking place in 4-way intersections were addressed in Ławniczak and Di Stefano [57]. An interesting modification of the N–Sch model is presented by Hartman [58], regarding urban traffic simulation.

Cellular automata are an effective tool for the analysis of actual behaviours [59,60]. Cellular automata have been applied to describe traffic jams [61], queue lengths before intersections and roundabouts [10,62] and to compare the capacities of various road solutions [63], as well as to study drivers' behaviours and to provide references for traffic design and management at roundabouts [64].

2.4. Roundabout Cellular Automaton (CA) Models

The rules of the traffic in a roundabout were dealt with in Belz et al. [65]. The work involved developing a model of a cellular automaton for capture of priority-taking and priority-abstaining behaviours in roundabouts.

A number of publications have addressed single-lane roundabout modelling [66–68]. The topics studied have included study road traffic dynamics in roundabouts [69], and the study of probability of road accidents [70,71]. The current publication focuses on developing a model for multi-lane roundabouts. It adds to previous studies in the field [40,72,73]. However, earlier studies failed to address the impact of traffic reorganization in the way the current article does. The authors of the earlier studies investigated the correlation between roundabout size and its capacity. The current paper is the first to examine the approach consisting in changing the traffic rules at roundabouts of a specified size, and its effect on the roundabout capacity. Roundabout traffic reorganisation is presented in the subsequent simulation scenarios described in the further parts of the article.

3. Proposed Approach

3.1. Theoretical Background

The generalised CA roundabout traffic model for a multi-lane roundabout with multi-lane feeder roads was developed based on the Nagel–Schreckenberg model [4] and its modified version by Hartman [58]. The original traffic model (N–Sch) was developed for the purposes of motorway traffic simulations. The length of the automaton single cell was assumed to correspond to 7.5 m of the road, which represents an average length of a car together with its surrounding space. The velocity unit in the N–Sch model corresponds to the actual speed of 27 km/h. In order to reflect actual urban traffic, using Hartman [58], vehicles were classified according to their length and the number of cells taken up by them (Table 1). The length of the automaton single cell corresponds to 2.5 m of the actual road. This translates into the velocities of moving vehicles (Table 2).

Table 1. Vehicle classification (by length).

Vehicle Type	Vehicle Length [m]	Vehicle Length [cells]
Motorcycle	2.5	1
Car	5	2
Van	7.5	3
Minibus	10	4
Bus, commercial van	12.5	5
Truck	15	6

Table 2. Comparison of velocities in the Nagel–Schreckenberg (N–Sch) model and the presented approach.

Velocity [cells/s]	Velocity in N–Sch Model [km/h]	Velocity in the Presented Approach [km/h]
0	0	0
1	27	9
2	54	18
3	81	27
4	108	36
5	135	45
6	162	54

3.2. Research Problem

Studies have been focused on determining the optimum number of lanes in a roundabout in order to greatest capacity. The authors of the current article approach the issue in a different way, by formulating the following thesis: traffic rule modification may lead to the increase of roundabout capacity without changing the roundabout structure. To prove the thesis, the author has analysed the roundabout traffic, developed a model and run experimental studies.

The roundabout traffic analysis was performed based on several month long observations of traffic at various roundabouts, including large and small, single-lane as well as 2- or 3-lane ones. The observations revealed that excessive congestion was found in the outer lane of a roundabout, while the occupancy of the inner lane (in the case of 2-lane roundabouts) or of the inner and middle lanes (in the case of 3-lane roundabouts) was moderate. Figure 1 presents averaged results of the observations, providing numeric values regarding the mean quantity of vehicles that entered the roundabout directly into the outer lane and those which did not take the first possible exit even though they were moving along the outer lane. The analysis shows that in the case of 2-lane roundabouts as many as 69.3% of the drivers entering the roundabout get into the outer lane and then only 19.2% of them take the first possible exit. In the case of 3-lane roundabouts, the values were 57.8% and 21.9%, respectively. The other drivers (80.8% and 78.1%) continue driving in the outer lane up to the subsequent exit or further.

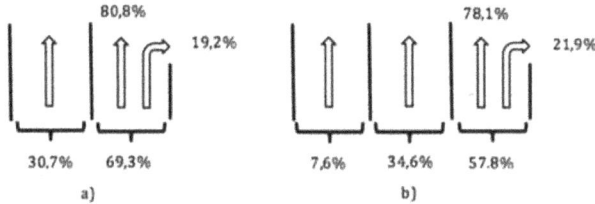

Figure 1. Average tendency to use the individual lanes at (**a**) 2-lane; (**b**) 3-lane roundabouts in Poland. Source: own research.

Pursuant to regulations, such behaviour is legal in most countries. However, the roundabout traffic observations have revealed that such behaviour causes an unnecessary restriction of vehicles queuing in feeder roads to enter the intersection (especially when the distance between the adjacent entrances is small). The above defined thesis may be further specified by stating that a reduction in the occupancy of the outer lane of a roundabout may increase the chances for the queuing vehicles to enter the roundabout, and increase its capacity. Shorter queues of waiting vehicles also provide the benefits of lower emissions and reduced local pollution.

3.3. The Data Set and the Simulation Setup

The data used for calibration and validations of the model were obtained during a comprehensive traffic study in Szczecin, Poland, during which the city area was divided into 255 communication areas. The boundaries of communication divisions were along major roads, rivers and other traffic barriers. Traffic surveys were conducted. Interviews were conducted with truck drivers and traffic volume and traffic pattern were measured. The following information was collected: the generic structure of vehicles (passenger cars, vans, minibuses, trucks without trailers, trucks with trailers and semi-trailers), and number of people in the vehicle. For external areas of the city, the country of origin of the vehicle, the source and destination of the journey, the frequency of the journey and the motives of the journey were also ascertained. The survey was conducted on weekdays from 6 a.m. to 10 p.m.

The field data provided information on the average number of vehicles passing through the roundabout (Table 3).

Table 3. List of roundabouts (Szczecin, Poland) that were under observation.

No.	Name of Roundabout	Number of Lanes	Average Number of Vehicles/day	Average Number of Vehicles/h
1.	Herman Haken Roundabout	3	6346	397
2.	University Roundabout	3	5579	349
3.	Gierosa Roundabout	2	5893	368
4.	Giedroycia Roundabout	2	8174	511
5.	Grunwaldzki Square	3	12,132	758
6.	Odrodzenia Square	2	10,972	686
7.	Szarych Szeregów Square	3	11,526	720
8.	Ułanów Podolskich Roundabout	2	8538	533
9.	Ronald Reagan Roundabout	2	4712	295
10.	Łupaszka Roundabout	1	3171	198

Based on the data obtained, the number of vehicles in the model was set at 500 vehicles per hour (average value resulting from the table is 481.5).

3.4. Assumptions of the Roundabout Traffic Model

The model was prepared so as to enable traffic flow simulation on a multi-lane roundabout with multiple entrance and exit roads. The model describes right-hand traffic.

The lanes on the roundabout are numbered from 0 (inner lane) to $n - 1$, where n corresponds to the total number of lanes on the roundabout. The feeder road consists of the entrance road and the exit road. The right-hand lane of the entrance road and the right-hand lane of the exit road are to first lanes to be numbered starting from zero. The right-hand lane is the lane on the right-hand side from the point of view of a driver travelling on the road in the given direction. Due to different lengths of the roundabout lanes, it is impossible to divide them into the same number of cells. Assuming that the cell length is constant and amounts to 2.5 m, the outer lanes will have more cells than the inner ones, reflecting the actual differences in the lanes lengths. The lane length, taking its inner (shorter) edge, may be determined using the formula:

$$l_{lane} = 2 \times \pi \times (r_{island} + n \times w_{lane}) \qquad (1)$$

where l_{lane} is the lane length, r_{island} is the island radius, n is the subsequent lane number, starting from the inner lane numbered 0, and w_{lane} is the single lane width.

The developed model posits that the width of each lane w_{lane} is the same. By parameterising the model in relation to the actual size of the roundabout, the number of cells in each lane is determined in relation to its length:

$$n_{cells} = \left\lfloor \frac{l_{lane}}{2.5} \right\rfloor \qquad (2)$$

The resulting value is rounded down, as the number of cells making up a lane must be an integral number. Assuming that the radius of a sample roundabout amounts to 28 m, and the lane width is 4.5 m, the number of cells on subsequent lanes will then be, respectively: $n_{cells} = \left\lfloor \frac{2 \times \pi \times (28 \ m + 0 \times 4.5)}{2.5 \ m} \right\rfloor =$ $\lfloor 70.37 \rfloor = 70$ and $n_{cells} = \left\lfloor \frac{2 \times \pi \times (28 \ m + 1 \times 4.5)}{2.5 \ m} \right\rfloor = \lfloor 81.68 \rfloor = 81$. The outer lane will be divided into 81 cells, and the inner lane into 70 cells.

Distance from the exit. If the cells in the roundabout lane are numbered with subsequent values from the range $< 0, \ c_{max} >$, the distance to the nearest exit d_{exit} may be computed using the formulas:

$$d_{exit} = c_{exit} - c_{current} \ dla \ c_{current} \leq c_{exit} \qquad (3)$$

$$d_{exit} = c_{max} + 1 - |c_{exit} - c_{current}| \ dla \ c_{current} > c_{exit} \qquad (4)$$

Distance from the entrance. As the cells in the entrance road are numbered with subsequent values from the range $< 0, \ c_{max} >$, the distance of the simulated vehicle from the entrance into the roundabout $d_{entrance}$ equals the difference between the cell connected with the roundabout $c_{entrance}$ and the cell currently containing the front of the vehicle $c_{current}$.

$$d_{entrance} = c_{entrance} - c_{current} \qquad (5)$$

Calculation of the cell value in the subsequent iteration for vehicles entering and exiting the roundabout is dependent on information on adjacent cells. Adjacent cells are those which link the feeder road with the roundabout internal road. Adjacency does not need to be of the first degree. For example, cells in the inner lane of the roundabout are not directly adjacent to any cell in the feeder road. However, there are cells located between the aforementioned cells which link the two lanes and enable movement of vehicles. When a vehicle exits the roundabout, an adjacent cell is the last one which the vehicle covers on its lane before moving into a cell located in the feeder road lane. Depending on the lane on which the vehicle is located and the lane onto which it is going to move, these will be different cells corresponding to the natural turns taken by vehicles. The values of CA cells from which vehicles may exit the roundabout are determined by the formula:

$$C_{exit} = \left[\frac{n_{cells}}{4} \times r - 2 + 1 \times l_{exit} \right] \tag{6}$$

where c_{exit} is the cell from which a vehicle may exit the roundabout, n_{cells} is the quantity of cells in the roundabout lane in which the vehicle is travelling, r is the multiplier depending on the road, N(North)-1, E(East)-2, S(South)-3, W(West)-4, l_{exit} is the the exit lane, symbol $[\ ]$ denotes rounding to an integral number.

For the vehicles entering the intersection, an adjacent cell is the first one into which the vehicle moves upon entering the roundabout. The interdependence is shown by the formula:

$$C_{entrance} = \left[\frac{n_{cells}}{4} \times r + 2 - 1 \times l_{entrance} \right] \tag{7}$$

where: $c_{entrance}$ is the cell taken up by a vehicle upon entering the roundabout, n_{cells} is the quantity of cells in the roundabout lane in which the vehicle is travelling, r is the multiplier depending on the road, N-1, E-2, S-3, W-4, and $l_{entrance}$ is the entrance lane.

3.5. Simulation Scenarios

Figure 2 shows the possible trajectories of vehicles that enter and exit the roundabout. The observations of roundabouts have revealed that a possibility of taking up any lane in a roundabout causes a risk of collision. When a vehicle entering a roundabout from the right-hand lane of a feeder road takes up the inner lane of the roundabout, it prevents the vehicle on the left-hand side from entering the roundabout. The event leads to temporary blocking of the vehicles located in the left-hand lane. Additionally, when exiting the roundabout, the driver in the outer lane of the roundabout is more privileged. As a result, drivers tend to choose the right-hand lane as "safer", i.e., the lane in which the driver always has right of way (Figure 1), which consequently leads to excessive occupancy of the outer lane of a roundabout while the inner lanes are under occupied.

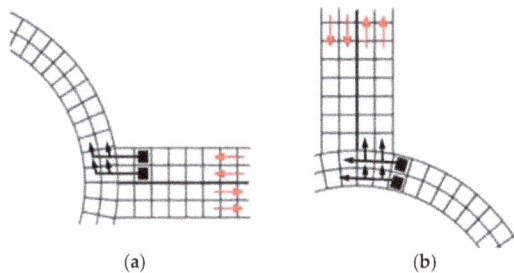

(a)   (b)

Figure 2. Scenario 1: possible directions of vehicle movement in the roundabout (**a**) enter; (**b**) exit.

Scenario 1 (Figure 2) shall be the reference scenario. The subsequent scenarios will refer to Scenario 1 in the comparative aspect.

The proposed traffic scenario (Scenario 2) is as follows: the vehicle in the right-hand lane of the feeder road should take up the outer or middle lane (in the case of 3-lane roundabouts), whereas the vehicle in the left-hand lane of the feeder road should go into the inner or middle lane, giving way to the vehicle on its right-hand side in cases where both of them are trying to get into the middle lane. The tested scenario results in a greater quantity of vehicles simultaneously entering the roundabout (Figure 3).

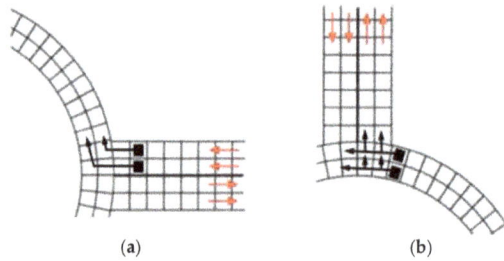

Figure 3. Scenario 2: possible directions of vehicle movement in the roundabout (**a**) enter; (**b**) exit.

Pursuant to the current EU regulations, a vehicle exiting a roundabout from the outer lane may take up any lane of the exit road, whereas the vehicle moving along the inner lane has to yield to it (Figure 4b: vehicle whose trajectory is marked with a dotted line). It may be hard for the driver moving along the inner lane to assess whether the driver on the right-hand side is driving straight on or moving into the outer lane. The proposed traffic scenario (Scenario 3) is as follows: the vehicle in the outer lane may take up the right-hand lane of the exit road, and the vehicle in the inner lane may take up the left-hand lane of the exit road, giving way to the vehicle on the right-hand side which is driving straight.

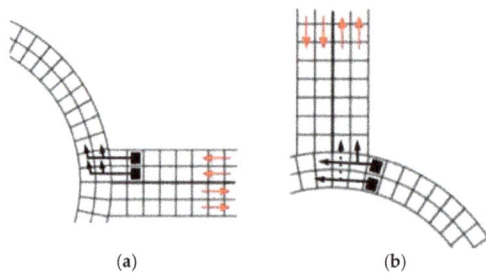

Figure 4. Scenario 3: possible directions of vehicle movement in the roundabout (**a**) enter; (**b**) exit.

The tested scenario is for two cars exiting the roundabout simultaneously, if they are taking the same exit (Figure 4b), provided that the entrance into the roundabout is compliant with Scenario 1.

The fourth analysed traffic scenario (Scenario 4) is a combination of Scenario 2 and Scenario 3. The tested scenario is of two vehicles entering the roundabout simultaneously and exiting the roundabout simultaneously in the same direction (Figure 5).

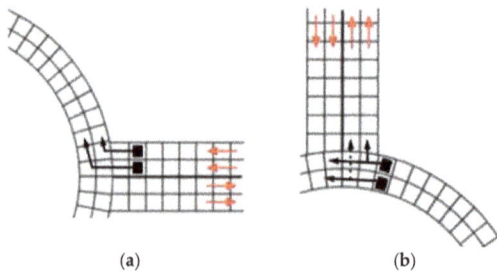

Figure 5. Scenario 4: possible directions of vehicle movement in the roundabout (**a**) enter; (**b**) exit.

The last of the examined scenarios (Scenario 5) regards roundabout traffic reorganisation (Figure 6). The vehicles travelling along the outer lane of the roundabout have to yield to the vehicles driving on the left-hand side, which are exiting the roundabout (Figure 6b: vehicle whose trajectory is marked with a dotted line). The experiment is motivated by the fact that drivers of the vehicles on the right-hand side are better able to see the manoeuvres of the vehicles driving on the left-hand side (in the case of right-hand traffic). Additionally, the scenario also applies the previously proposed changes regarding entering and exiting a roundabout, considered in Scenario 4.

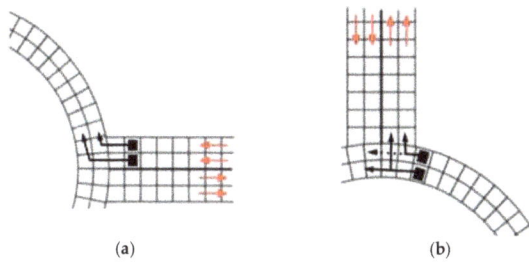

(a)   (b)

Figure 6. Scenario 5: possible directions of vehicle movement in the roundabout (**a**) enter; (**b**) exit.

3.6. The System Developed for the Simulation

The application was developed in JavaScript and it can be operated in web browsers as well as by means of a console. Running the application by means of a console is possible via a Node.js runtime environment based on a V8 engine in the Chrome browser. Running the application in such a way is faster, as the program can operate without the graphics layer. The application makes use of many free tools enabling the programming works. Figure 7 shows the structure of the application together with the tools applied.

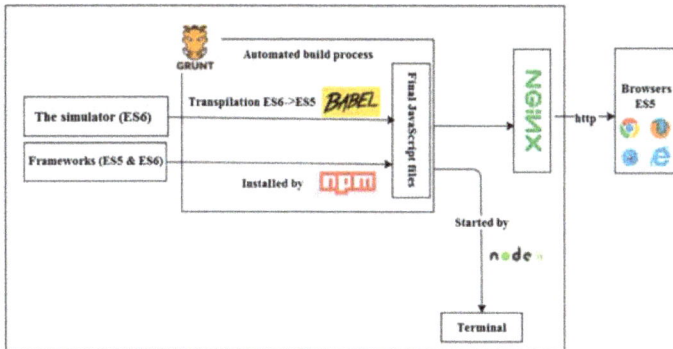

Figure 7. The application structure.

Frequent problems with running the application are due to different runtime environments. The problems are solved by the combination of Vagrant and Ansible tools. Vagrant enables management of virtual machines, offering identical runtime conditions for applications. Ansible is used to ensure that the virtual machine is always equipped with any indispensable libraries. Upon starting, the program compares the current state of the machine with the expected one and carries out any necessary setups. Before that, it is necessary to prepare the configuration files that define the dependencies.

The application is written in accordance with the latest standards of ECMAScript 6. To enable correct operation of the application in web browsers, the Babel transpiler was used to change the code into the one compliant with ECMAScript 5. The process of transpilation and providing the application to the www server was automated by means of the Grunt program.

4. Results of the Experimental Studies

The object of the simulation experiment was to check whether roundabout traffic reorganisation has an effect on the roundabout capacity. A study was performed to find out the impact of an increasing number of trucks on roundabout traffic in order to make a reliable choice of the best configuration of regulations. The study accounted for the sizes of such vehicles and their limited technical and physical capabilities, such as e.g., slow acceleration, or low manoeuvring speed.

The experiments were conducted in a 2-lane roundabout with four feeder roads, for right-hand traffic. Each feeder road consists of the entrance road and the exit road, each with two lanes. The roundabout island diameter amounted to 56 m. The roundabout capacity was examined on the sample of 500 vehicles, by measuring the number of iterations needed for moving all the vehicles through the roundabout, and using the formula:

$$b = \frac{n_{vehicles}}{n_{iterations}} \tag{8}$$

where b is roundabout capacity, $n_{vehicles}$ is the number of vehicles that have driven across the roundabout (500), and $n_{iterations}$ is the number of CA iterations.

The simulation was repeated 1000 times in order to obtain results that are not distorted by random events. The quantity of vehicles in the feeder roads was even. The entrance road for each vehicle was assigned at random with the same probability (25%). The exit road was assigned at random by the programme in the same manner. The entrance and exit roads had two lanes, so the individual lanes were assigned to the vehicles at random. The probability of assigning a given exit lane depended on the examined traffic rules and the roundabout lane on which the vehicle was travelling. In Scenario 1 the vehicle was able to take either of the two lanes in the exit road, if it was moving along the outer lane. In the case the probability of choosing either lane in the exit road was 0.5. If the vehicle was moving along the inner lane of the roundabout, it was able to take the left-hand lane only, and the probability of choosing the left-hand lane was 1. Selection of lanes in the roundabout was made on the same principles, if the principles made it possible to take any lane, the probability was $\frac{1}{n_{roundabout\_lanes}}$, while for a 2-lane roundabout it was 0.5. In the case of random assigning of the entrance lane, the probability was always 0.5 in all the scenarios. The experiments were run assuming mixed traffic, with 90% of cars (including motorcycles), and 10% of trucks. An exception was the examination of the impact of the number of trucks on the roundabout capacity. Then the quantity of trucks varied.

4.1. Examining the Impact of the Proposed Changes on the Roundabout Capacity

Table 4 presents the results of the experiment aimed at examining the impact of different forms of roundabout traffic reorganisation on the roundabout capacity. The mean and median values for the number of iterations have been rounded to integral numbers. The values are similar with each other, which meant that the results were absent outliers.

Table 4. Mean number of iterations and median of the iteration quantity.

Examined Traffic Scenarios	Mean Number of Iterations	Median of Iteration Quantity
Scenario 1	526	524
Scenario 2	494	493
Scenario 3	503	502
Scenario 4	456	456
Scenario 5	483	482

In order to make sure that the analysed results are not distorted, their distribution was visualised in the box plot presented in Figure 8. It shows that the results are scattered in a similar manner, proven by the similar sizes of the boxes and their "whiskers". The distribution of the variable is symmetric, as the horizontal line in the box representing the median is located more or less in the middle. During the testing of the individual scenarios there were some outliers that are marked with circles, however, their quantity compared to the number of cycles accounts for less than 1% of the obtained results. The chart shows that there is a correlation between the examined scenarios and the quantity of CA iterations. The traffic organisation described in Scenario 4 caused vehicles to cross the roundabout in the shortest time. The worst case of the traffic organisation is represented by the reference scenario, i.e., Scenario 1 (526 iterations on the average). That means an improvement of 15%.

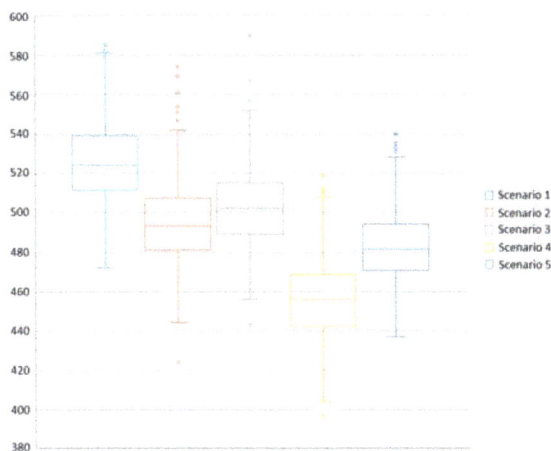

Figure 8. Impact of roundabout traffic organisation on the quantity of cellular automaton (CA) iterations.

The roundabout capacity was calculated (Figure 9) based on each of the obtained results. The guiding principle was the higher the value of the capacity, the better the solution. The measure of roundabout capacity is the quantity of vehicles which exit the roundabout in a time unit which is one iteration of a cellular automaton. The best result was obtained in the case of Scenario 4. The second best result was the roundabout traffic organisation according to Scenario 5.

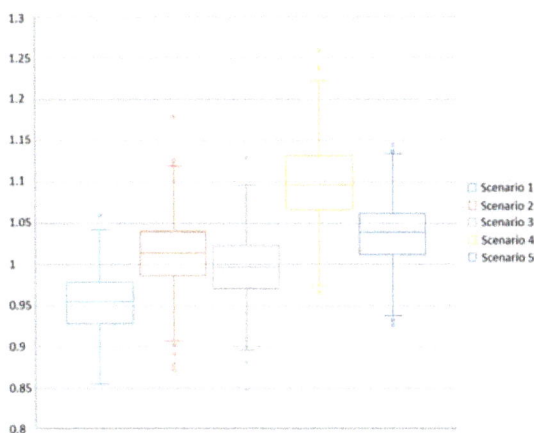

Figure 9. Impact of traffic organisation on the roundabout capacity.

Table 5 shows the percentage differences in relation to Scenario 1. The traffic organisation described in Scenario 4 led to the greatest increase in the roundabout capacity in relation to the current traffic rules (Scenario 1). Scenario 4 is a combination of the rules specified in Scenarios 2 and 3. The changes introduced separately provide an increase of 6.4% and 4.5%, respectively, and cumulatively 10.9%. However, introducing them together increases the roundabout capacity by an extra 4.4%, thus achieving the best result of all the modifications, equalling 15.3%. Such traffic organisation tidies up the road traffic and can be achieved without substantial modifications of the traffic regulations (what it does require is provision of more accurate driver training and at the same time developing appropriate algorithms for autonomous vehicles). The second biggest increment in relation to the reference scenario is demonstrated by Scenario 5. It is an attempt to turn attention to the fact that the traffic regulations in force favour drivers moving along the outer lane of a roundabout. Such vehicles often block roundabouts, while the inner lanes are empty or almost empty. According to the simulations, introducing the changed right of way in roundabouts can cause a result of increase 8.9%, preferable to the status quo (Scenario 1), while at the same time being the second most efficient set of rules.

Table 5. Mean number of iterations and median of the iteration quantity.

Examined Traffic Scenarios	Roundabout Capacity in Relation to Scenario 1 [%]
Scenario 1	0
Scenario 2	6.4
Scenario 3	4.5
Scenario 4	15.3
Scenario 5	8.9

The subsequent studies pertained to the impact of the number of pedestrians at the roundabout on the roundabout capacity for the individual traffic scenarios. Figures 10–13 demonstrate the research results for the specified distances between vehicles 1 and 4 and for the specified maximum velocities of cars and trucks. The traffic scenarios return values that are similar for the distance between vehicles equalling 1 (Figure 10).

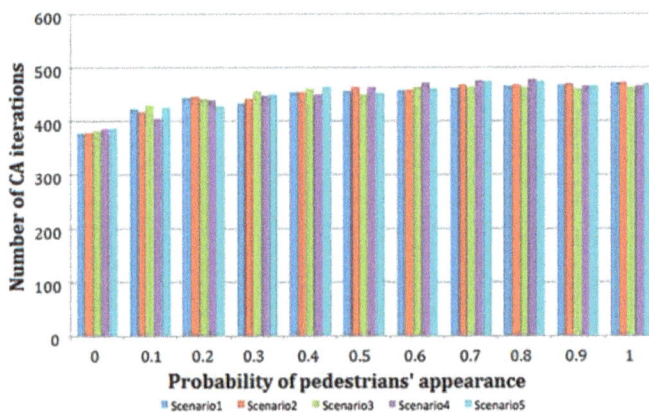

Figure 10. Impact of pedestrian traffic on roundabout capacity as per various traffic scenarios, distance between vehicles = 1. Specified speeds: max car speed = 5, truck speed = 2.

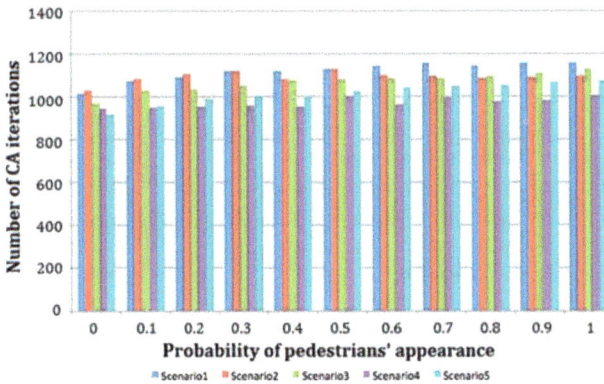

Figure 11. Impact of pedestrian traffic on roundabout capacity as per various traffic scenarios, distance between vehicles = 2. Specified speeds: max car speed = 5, truck speed = 2.

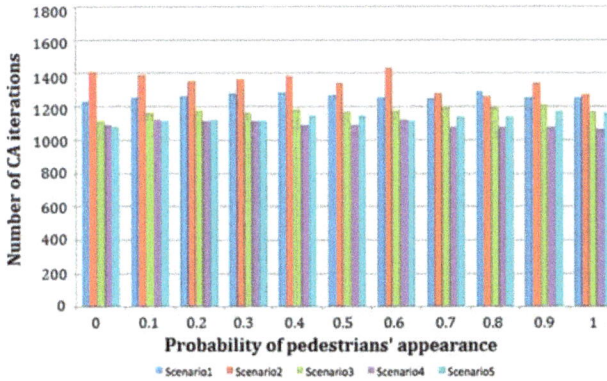

Figure 12. Impact of pedestrian traffic on roundabout capacity as per various traffic scenarios, distance between vehicles = 3. Specified speeds: max car speed = 5, truck speed = 2.

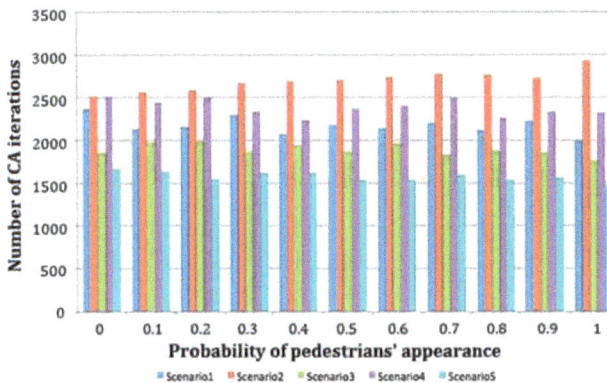

Figure 13. Impact of pedestrian traffic on roundabout capacity as per various traffic scenarios, distance between vehicles = 4. Specified speeds: max car speed = 5, truck speed = 2.

More significant differences are found in the cases shown in Figures 11–13. The figures reveal that road traffic organisation has a considerable impact on roundabout capacity. It was found that the larger the forced distance between vehicles, the smaller the roundabout capacity, due to the fact that vehicles queuing to enter the roundabout must wait longer. However, the distance between moving vehicles is insufficient for the waiting vehicles. In the case presented in Figure 13, the most effective type of traffic organisation is that applied in Scenario 5. It is 20–30% more effective in relation to the reference scenario for various numbers of pedestrians participating in the traffic. For smaller distances between vehicles, the differences are much smaller, ca. 10%. The traffic organisation that proved to be the least susceptible to changing the distances between vehicles was the one applied in Scenario 2, and at the same time it was the least effective.

Figure 14 presents the aggregated results of the research study for the individual scenarios, specified speeds (1–4), probability of pedestrians' appearance at the roundabout (0; 0.5; 1) and different maximum speeds. The observed value is the number of CA iterations.

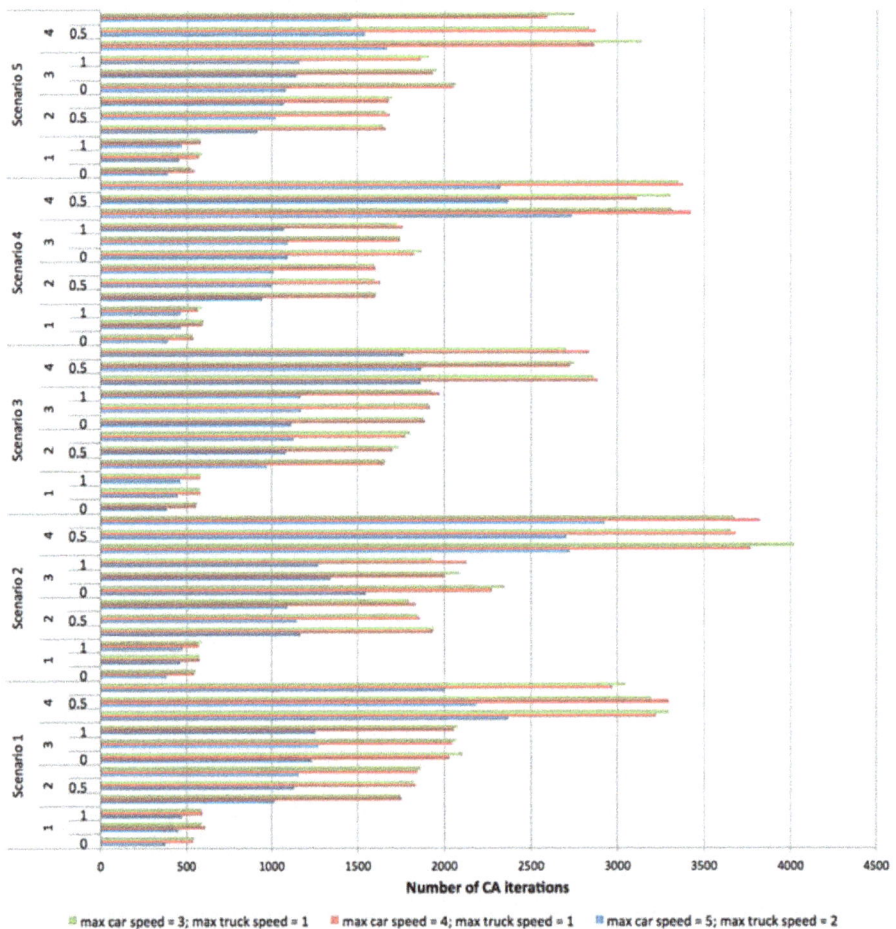

Figure 14. Impact of pedestrian traffic and distances between vehicles on roundabout capacity as per various traffic scenarios.

The subsequent study focused on finding out how roundabout capacity changes in different scenarios of traffic organisation (Scenarios 1–5) in the context of deteriorated adhesion. The developed model was prepared in such a way so that the braking process accounts for delay resulting from lack of tyre adhesion to various surfaces. The adhesion studies were performed for wet (Figure 15) and for snowy (Figure 16) surfaces. In this case, roundabout capacity was studied for 100 vehicles.

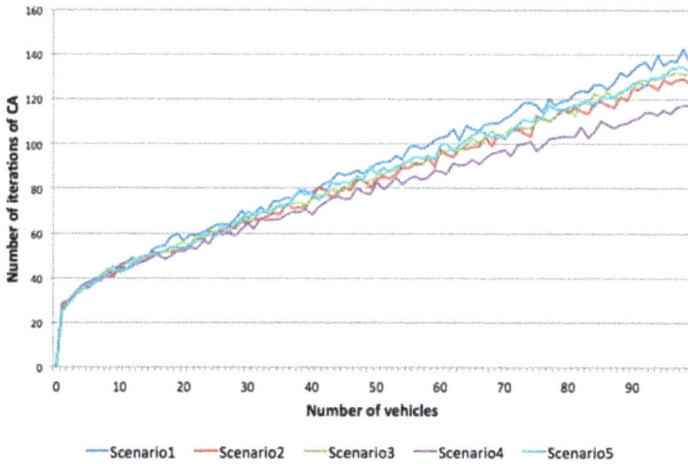

Figure 15. Examination of the impact of the number of vehicles moving on wet surfaces on roundabout capacity, as per various traffic scenarios.

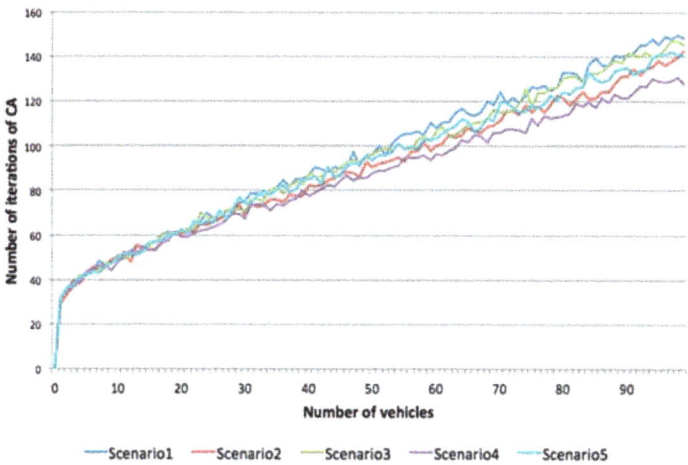

Figure 16. Examining the impact of the number of vehicles moving on snowy surfaces on roundabout capacity, as per various traffic scenarios.

The analysis of the results has shown that again reference Scenario 1 proved to be the least effective form of traffic organisation at the roundabout. Scenarios 3–5 were more effective.

4.2. Examining the Impact of Freight Vehicles on Roundabout Capacity

The aim of the study was to determine whether an increase in the number of trucks significantly affects roundabout capacity. The first stage of the study was examining the roundabout capacity without trucks. In total 500 cars crossed the roundabout on average in 375 iterations of the cellular automaton, applying the most effective traffic organisation—Scenario 4. In the subsequent tests, the percentage of trucks was increased by 1%, which resulted in an increase in the time needed for the constant number of vehicles to cross the roundabout. The distribution of the received results is presented in Figure 17.

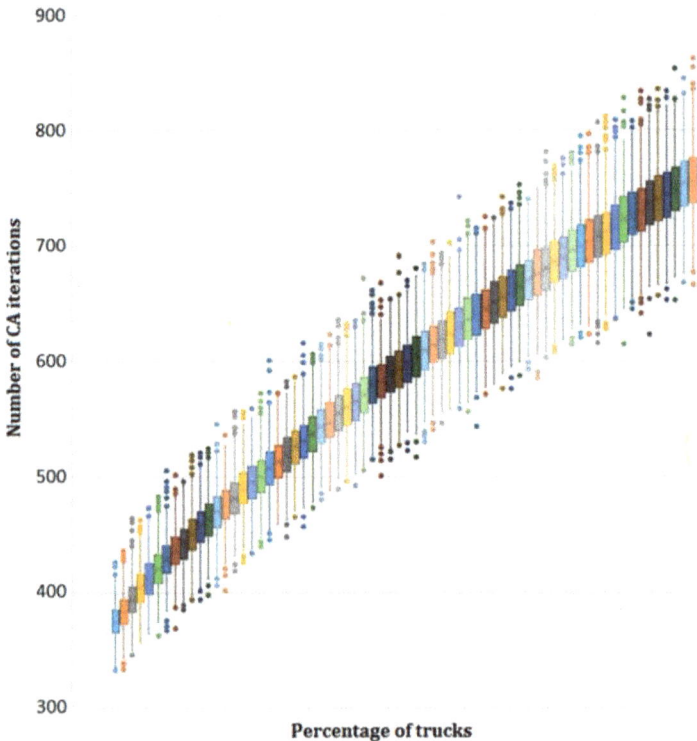

Figure 17. Distribution of the number of CA iterations depending on the percentage of trucks.

It is visible that the more trucks are involved, the longer the average duration of the experiment, which means that the roundabout capacity is getting smaller. A more accurate analysis has shown that the more trucks there are, the more unstable the traffic becomes. The differences between the maximum and minimum values start to increase.

5. Interpretation of the Results and Future Works

The above studies described have shown that it is possible to increase roundabout capacity by changing traffic regulations. Also, trucks have a significant impact on urban traffic fluidity: their growing share in the total number of vehicles (simulated in one of the experiments) led to a significant reduction of the roundabout capacity.

In order to maximise the number of vehicles crossing a roundabout, it should be possible for vehicles to enter and exit the roundabout simultaneously. The multi-lane area of a roundabout can be used most effectively, which results in capacity maximisation. This was demonstrated by Scenario 4. Such traffic organisation may be easily implemented, e.g., by appropriate marking of solid and broken lines within roundabouts. The analysis of urban traffic performed in Poland reveals that more appropriate marking may lead to increasing safety of road traffic participants. An additional criterion to improve traffic safety seems to be driver training. The observations of roundabout traffic have revealed numerous behaviours of drivers that generated unnecessary traffic collisions. Development of Scenario 5 was inspired by actual events observed at roundabouts. Many a time, the vehicles driving in the inner lane (in the case of 2-lane roundabouts) or in the middle lane (in the case of 3-lane roundabouts), when trying to exit the roundabout would cut in on the vehicles moving in the outer lane. It was observed that e.g., less experienced or elderly drivers tended to keep to the outer lane. The traffic organisation described in Scenario 5 could reduce such behaviours, which in turn might lead to increased traffic capacity in cities having a high number of roundabouts.

Another aspect, which should be noted in order to increase traffic capacity, is reduction of freight vehicles on roundabouts affected by heavy traffic. Of course it is impossible to totally eliminate freight traffic in cities. However, it is possible to limit the number of trucks in cities, e.g., by providing ring-roads. Also, a controversial idea seems to be a prohibition to drive trucks in peak hours. When the number of trucks is significant and there are limited other possibilities to increase the traffic capacity, this idea might be taken into account.

It should be emphasised that not each roundabout has the same shape or the number of entrances and exits. Moreover, a computer simulation takes into account merely a fragment of the reality. Nevertheless, the results of the study are promising and this issue deserves further attention. The model does not consider drivers' behaviour, experience, fatigue nor ongoing concentration in navigating through roundabouts. The future plans for this task include an ongoing linkage between the model and metrics form the urban system, which will record quantitative data (although it would certainly be useful to develop a methodology for automatically retrieving qualitative data such as vehicle data). The model also does not specifically look at the specific characteristics of pedestrians walking by the roundabout, treating them only as a parameter that has some influence on the scenarios studied. However, incorporating pedestrians' behaviour into the launched study may point to further important aspects. The developed model does not account for cyclists, either. The aforementioned facts show an area that could contribute to the subsequent studies in this respect. Other aspects that may be of some importance, but were not investigated, are the distances between the individual feeder roads. The developed model could also be extended to include the possibility of turbo roundabout simulations. It would require small modifications of the principles contained in Scenario 5. It also might provide an answer to the question whether in the case of a given roundabout it is better to apply the classical traffic organisation, or a directional traffic organisation.

It must be stressed that the model has been tested in a simulation environment, which can be perceived as an incomplete scientific pathway. Therefore, in the future, some works will be carried out towards validation of the model, using available measures and current real data.

6. Conclusions

The article focuses on the analysis of traffic efficiency at roundabout intersections. The current roundabout traffic rules were presented, and then other forms of roundabout traffic organisation were proposed by means of specifying various traffic scenarios. A generalised model was developed, which was based on cellular automata aimed at examining multi-lane roundabout capacity for different road traffic rules. The developed model was implemented in the form of a simulation system, which enabled numerous simulations. The obtained results proved that modification of the existing traffic regulations has an effect on increasing roundabout capacity. The greatest roundabout capacity is provided by the form of traffic organisation that makes vehicles enter and exit a roundabout in parallel. The implication is that several vehicles may easily enter and exit a roundabout at the same time. The results obtained were better by 15.3% (Scenario 4) and 8.9% (Scenario 5) compared to regulations in force (Scenario 1). Another important result of the study was determination that a reduction of freight traffic has a square function effect on roundabout capacity. It means even a small number of trucks within the total quantity of vehicles may significantly reduce roundabout capacity.

Studies such as the current one cannot possibly be performed in real conditions, mainly due to a possibility of collision, and difficulties in ensuring that all the traffic participants will understand the conditions of the experiment and will behave as prescribed. Computer simulations were applied to bridge this gap. A mathematical model and a simulation system were developed, and the obtained results were presented in the figures and tables. The simulation applied at the microscopic level made it possible to reproduce the phenomena of the real world, such as road traffic, and enabled answering of new questions regarding possible behaviours of drivers.

During the research some possible areas of improvement of the proposed approach have been identified. The most interesting ones seem to be the validation of the model and the inclusion of the human factors to reduce model simplification.

The aim of the study was to demonstrate that a modification of road traffic rules could lead to increasing roundabout capacity. The results described have shown that the aim of the study has been attained, and the results make it possible to take up further studies.

Author Contributions: Krzysztof Małecki wrote the paper, prepared and implemented CA model and performed data processing; Jarosław Wątróbski supervised manuscript and formal analyses and prepared the final amendments. All authors have read and approved the final manuscript.

Conflicts of Interest: The authors declare no conflict of interest.

References

1. Chowdhury, D.; Santen, L.; Schadschneider, A. Statistical physics of vehicular traffic and some related systems. *Phys. Rep.* **2000**, *329*, 199–329. [CrossRef]
2. Helbing, D. *Verkehrsdynamik: Neue Physikalische Modellierungskonzepte*; Springer: Berlin, Germany, 2013.
3. May, A.D. *Traffic Flow Fundamentals*; Prentice Hall: Englewood Cliffs, NJ, USA, 1990.
4. Nagel, K.; Schreckenberg, M. A cellular automaton model for freeway traffic. *J. Phys. I* **1992**, *2*, 2221–2229. [CrossRef]
5. Heermann, D.W.; Burkitt, A.N. Computer Simulation Methods. In *Parallel Algorithms in Computational Science*; Springer: Berlin, Germany, 1991; pp. 5–35.
6. Rodegerdts, L.; Blogg, M.; Wemple, E.; Myers, E.; Kyte, M.; Dixon, M.; List, G.; Flannery, A.; Troutbeck, R.; Briton, W.; et al. *NCHRP Report 572: Roundabouts in the United States*; Transportation Research Board of the National Academies: Washington, DC, USA, 2007.
7. Leaf, W.A.; Preusser, D.F. *Literature Review on Vehicle Travel Speeds and Pedestrian Injuries*; National Highway Traffic Safety Administration: Washington, DC, USA, 1999.
8. Brude, U.; Larsson, J. What roundabout design provides the highest possible safety? *Nord. Road Transp. Res.* **2000**, *12*, 17–21.
9. Sisiopiku, V.P.; Oh, H.U. Evaluation of roundabout performance using SIDRA. *J. Transp. Eng.* **2001**, *127*, 143–150. [CrossRef]
10. Wang, R.; Liu, M. A Realistic Cellular Automata Model to Simulate Traffic Flow at Urban Roundabouts. In Proceedings of the 5th International Conference on Computational Science, Atlanta, GA, USA, 22–25 May 2005; pp. 17–36.
11. Balmer, M.; Rieser, M.; Meister, K.; Charypar, D.; Lefebvre, N.; Nagel, K. MATSim-T: Architecture and simulation times. In *Multi-Agent Systems for Traffic and Transportation Engineering*; IGI Global: Hershey, PA, USA, 2009; pp. 57–78.
12. Fellendorf, M. VISSIM: A microscopic simulation tool to evaluate actuated signal control including bus priority. In Proceedings of the 64th Institute of Transportation Engineers Annual Meeting, Dallas, TX, USA, 16–19 October 1994; Technical Paper. pp. 1–9.
13. Smith, L.; Beckman, R.; Baggerly, K.; Anson, D.; Williams, M. *Overview of TRANSIMS, the TRansportation ANalysis and SIMulation System (Combined One-Pagers)*; Los Alamos National Security: New Mexico, NM, USA, 1995.

14. Yang, Q.; Koutsopoulos, H.N. A microscopic traffic simulator for evaluation of dynamic traffic management systems. *Transp. Res. C Emerg. Technol.* **1996**, *4*, 113–129. [CrossRef]

15. Barceló, J.; Ferrer, J.L. *AIMSUN2: Advanced Interactive Microscopic Simulator for Urban Networks, User's Manual*; Universidad Politécnica de Cataluña: Barcelona, Spain, 1997.

16. Barceló, J.; Codina, E.; Casas, J.; Ferrer, J.L.; García, D. Microscopic traffic simulation: A tool for the design, analysis and evaluation of intelligent transport systems. *J. Intell. Robot. Syst.* **2005**, *41*, 173–203. [CrossRef]

17. Krajzewicz, D.; Hertkorn, G.; Rössel, C.; Wagner, P. SUMO (Simulation of Urban MObility)—An open-source traffic simulation. In Proceedings of the 4th Middle East Symposium on Simulation and Modelling (MESM2002), Berlin-Adlershof, Germany, 1–30 September 2002; pp. 183–187.

18. Krajzewicz, D.; Erdmann, J.; Behrisch, M.; Bieker, L. Recent development and applications of SUMO-Simulation of Urban MObility. *Int. J. Adv. Syst. Meas.* **2012**, *5*, 128–138.

19. Halati, A.; Lieu, H.; Walker, S. CORSIM-corridor traffic simulation model. In Proceedings of the Traffic Congestion and Traffic Safety in the 21st Century: Challenges, Innovations, and Opportunities, Chicago, IL, USA, 8–11 June 1997.

20. Bloomberg, L.; Dale, J. A comparison of the VISSIM and CORSIM traffic simulation models. In Proceedings of the Institute of Transportation Engineers Annual Meeting, Oakland, CA, USA, 6–9 August 2000; Volume 1727, pp. 52–60.

21. Małecki, K.; Iwan, S. Development of Cellular Automata for Simulation of the Crossroads Model with a Traffic Detection System. In Proceedings of the International Conference on Transport Systems Telematics, Katowice-Ustroń, Poland, 10–13 October 2012; pp. 276–283.

22. Li, X.; Li, X.; Xiao, Y.; Jia, B. Modeling mechanical restriction differences between car and heavy truck in two-lane cellular automata traffic flow model. *Phys. A Stat. Mech. Its Appl.* **2016**, *451*, 49–62. [CrossRef]

23. Doniec, A.; Mandiau, R.; Piechowiak, S.; Espié, S. A behavioral multi-agent model for road traffic simulation. *Eng. Appl. Artif. Intell.* **2008**, *21*, 1443–1454. [CrossRef]

24. Jin, T.; Fu, L. Application of GIS to modified models of vehicle emission dispersion. *Atmos. Environ.* **2005**, *39*, 6326–6333. [CrossRef]

25. Krzysztof, M. The Importance of Automatic Traffic Lights time Algorithms to Reduce the Negative Impact of Transport on the Urban Environment. *Transp. Res. Procedia* **2016**, *16*, 329–342. [CrossRef]

26. Wang, Y.; Chen, Y.Y. Modeling the effect of microscopic driving behaviors on Kerner's time-delayed traffic breakdown at traffic signal using cellular automata. *Phys. A Stat. Mech. Its Appl.* **2016**, *463*, 12–24. [CrossRef]

27. Franzese, O.; Joshi, S. Transportation applications of simulation: Traffic simulation application to plan real-time distribution routes. In Proceedings of the 34th Conference on Winter Simulation: Exploring New Frontiers, San Diego, CA, USA, 8–11 December 2002; pp. 1214–1218.

28. Popescu, M.C.; Ranea, C.; Grigoriu, M. Solutions for traffic lights intersections control. In Proceedings of the 10th WSEAS, Prague, Czech Republic, 23–25 March 2010.

29. Han, X.; Sun, H. The Implementation of Traffic signal Light Controlled by PLC. *J. Changchun Inst. Opt. Fine Mech.* **2003**, *4*, 29.

30. Kwon, E.; Kim, S.; Kwon, T.M. Pseudo real-time evaluation of adaptive traffic control strategies using hardware-in-loop simulation. In Proceedings of the 27th Annual Conference of the IEEE on Industrial Electronics Society (IECON'01), Denver, CO, USA, 29 November–2 December 2001; Volume 3, pp. 1910–1914.

31. Bullock, D.; Urbanik, T. Hardware-in-the-loop evaluation of traffic signal systems. In Proceedings of the 10th International Conference on Road Transport Information and Control, London, UK, 4–6 April 2000; pp. 177–181.

32. Wells, R.B.; Fisher, J.; Zhou, Y.; Johnson, B.K.; Kyte, M. Hardware and software considerations for implementing hardware-in-the loop traffic simulation. In Proceedings of the 27th Annual Conference of the IEEE on Industrial Electronics Society (IECON'01), Denver, CO, USA, 29 November–2 December 2001; Volume 3, pp. 1915–1919.

33. Jaszczak, S.; Małecki, K. Hardware and software synthesis of exemplary crossroads in a modular programmable controller. *Przegląd Elektrotechniczny* **2013**, *89*, 121–124.

34. Shen, Z.; Wang, K.; Zhu, F. Agent-based traffic simulation and traffic signal timing optimization with GPU. In Proceedings of the 14th International IEEE Conference on Intelligent Transportation Systems (ITSC), Washington, DC, USA, 5–7 October 2011; pp. 145–150.

35. Herman, R.; Montroll, E.W.; Potts, R.B.; Rothery, R.W. Traffic dynamics: analysis of stability in car following. *Oper. Res.* **1959**, *7*, 86–106. [CrossRef]

36. Gazis, D.C.; Herman, R.; Rothery, R.W. Nonlinear follow-the-leader models of traffic flow. *Oper. Res.* **1961**, *9*, 545–567. [CrossRef]

37. Farzaneh, M.; Rakha, H. Impact of differences in driver-desired speed on steady-state traffic stream behavior. *Transp. Res. Rec. J. Transp. Res. Board* **2006**, *1965*, 142–151. [CrossRef]

38. Dowling, R.; Skabardonis, A.; Halkias, J.; McHale, G.; Zammit, G. Guidelines for calibration of microsimulation models: Framework and applications. *Transp. Res. Rec. J. Transp. Res. Board* **2004**, *1876*, 1–9. [CrossRef]

39. Vision PTV. *VISSIM 5.30-05 User Manual*; Planung Transport Verkehr AG: Braunschweig, Germany, 2011.

40. Schroeder, B.; Rouphail, N.; Salamati, K.; Bugg, Z. Effect of pedestrian impedance on vehicular capacity at multilane roundabouts with consideration of crossing treatments. *Transp. Res. Rec. J. Transp. Res. Board* **2012**, *2312*, 14–24. [CrossRef]

41. Vaiana, R.; Gallelli, V.; Iuele, T. Analysis of roundabout stop-line delays: Effects of kinematical and behavioural parameters in the simulation process of observed traffic conditions. In Proceedings of the 91th TRB Annual Meeting, Washington, DC, USA, 22–26 January 2012.

42. Vaiana, R.; Gallelli, V.; Iuele, T. Experimental analysis and methodological approach by micro-simulation of crossing speed distribution on roundabouts of small and large diameter. In Proceedings of the 92nd Annual Meeting of the Transportation Research Board, Washington, DC, USA, 13–17 January 2013.

43. Wei, T.; Shah, H.R.; Ambadipudi, R.P. VISSIM Calibration for Modeling Single-Lane Roundabouts: Capacity-Based Strategies. In Proceedings of the Transportation Research Board 91st Annual Meeting, Washington, DC, USA, 22–26 January 2012.

44. Hagring, O. Estimation of critical gaps in two major streams. *Transp. Res. B Methodol.* **2000**, *34*, 293–313. [CrossRef]

45. Tian, Z.Z.; Troutbeck, R.; Kyte, M.; Brilon, W.; Vandehey, M.A.R.K.; Kittelson, W.; Robinson, B. A further investigation on critical gap and follow-up time. In Proceedings of the 4th International Symposium on Highway Capacity, Maui, Hawaii, 27 June–1 July 2000; pp. 409–421.

46. Kay, N.; Ahuja, S.; Cheng, T.N.; Van Vuren, T.; MacDonald, M. *Estimation and Simulation Gap Acceptance Behaviour at Congested Roundabouts*; Association for European Transport and Contributors: Strasbourg, France, 2006; pp. 1–16.

47. Alexander, L.; Cheng, P.; Donath, M.; Gorjestani, A.; Menon, A.; Newstrom, B.; Starr, R. Lag acceptance analysis for a rural unsignalized intersection. In Proceedings of the 86th Annual Meeting of the TRB, Washington, DC, USA, 21–25 January 2007.

48. Pollatschek, M.A.; Polus, A.; Livneh, M. A decision model for gap acceptance and capacity at intersections. *Transp. Res. B Methodol.* **2002**, *36*, 649–663. [CrossRef]

49. Macioszek, E.; Sierpiński, G.; Czapkowski, L. Problems and issues with running the cycle traffic through the roundabouts. In Proceedings of the Transport Systems Telematics, Katowice-Ustron, Poland, 20–23 October 2010; pp. 107–114.

50. Macioszek, E. Analysis of the Effect of Congestion in the Lanes at the Inlet to the Two-Lane Roundabout on Traffic Capacity of the Inlet. In Proceedings of the Transport Systems Telematics, Katowice-Ustron, Poland, 23–26 October 2013; pp. 97–104.

51. Macioszek, E. Analysis of Significance of Differences Between Psychotechnical Parameters for Drivers at the Entries to One-Lane and Turbo Roundabouts in Poland. In *Intelligent Transport Systems and Travel Behaviour, Proceedings of the International Conference Transport System Theory and Practice, Katowice, Poland, 18–20 September 2017*; Springer: Berlin/Heidelberg, Germany, 2017; pp. 149–161.

52. Iwan, S.; Kijewska, K.; Johansen, B.G.; Eidhammer, O.; Małecki, K.; Konicki, W.; Thompson, R.G. Analysis of the environmental impacts of unloading bays based on cellular automata simulation. *Transp. Res. D Transp. Environ.* **2017**. [CrossRef]

53. Banjanovic-Mehmedovic, L.; Halilovic, E.; Bosankic, I.; Kantardzic, M.; Kasapovic, S. Autonomous Vehicle-to-Vehicle (V2V) Decision Making in Roundabout using Game Theory. *Int. J. Adv. Comput. Sci. Appl.* **2016**, *7*, 292–298. [CrossRef]

54. Nagel, K.; Wolf, D.E.; Wagner, P.; Simon, P. Two-lane traffic rules for cellular automata: A systematic approach. *Phys. Rev. E* **1998**, *58*, 1425–1437. [CrossRef]

55. Biham, O.; Middleton, A.A.; Levine, D. Self-organization and a dynamical transition in traffic-flow models. *Phys. Rev. A* **1992**, *46*. [CrossRef]

56. Chowdhury, D.; Schadschneider, A. Self-organization of traffic jams in cities: Effects of stochastic dynamics and signal periods. *Phys. Rev. E* **1999**, *59*. [CrossRef]

57. Lawniczak, A.T.; Di Stefano, B.N. Development of multi CA model of 4-way road intersection. In Proceedings of the Electrical and Computer Engineering, St. John's, NL, Canada, 3–6 May 2009.

58. Hartman, D. Head Leading Algorithm for Urban Traffic Modeling. *Positions* **2004**, *2*, 1–6.

59. Daganzo, C.F. In traffic flow, cellular automata = kinematic waves. *Transp. Res. B Methodol.* **2006**, *40*, 396–403. [CrossRef]

60. Daganzo, C.F.; Gayah, V.V.; Gonzales, E.J. The potential of parsimonious models for understanding large scale transportation systems and answering big picture questions. *EURO J. Transp. Logist.* **2012**, *1*, 47–65. [CrossRef]

61. Wagner, P.; Nagel, K.; Wolf, D.E. Realistic multi-lane traffic rules for cellular automata. *Phys. A Stat. Mech. Its Appl.* **1997**, *234*, 687–698. [CrossRef]

62. Feng, Y.; Liu, Y.; Deo, P.; Ruskin, H.J. Heterogeneous traffic flow model for a two-lane roundabout and controlled intersection. *Int. J. Mod. Phys. C* **2007**, *18*, 107–117. [CrossRef]

63. Fouladvand, M.E.; Sadjadi, Z.; Shaebani, M.R. Characteristics of vehicular traffic flow at a roundabout. *Phys. Rev. E* **2004**, *70*. [CrossRef]

64. Ming-Zhe, L.; Shi-Bo, Z.; Rui-Li, W. A Cellular Automaton Model for Heterogeneous and Incosistent Driver Behavior in Urban Traffic. *Commun. Theor. Phys.* **2012**, *58*, 744.

65. Belz, N.P.; Aultman-Hall, L.; Montague, J. Influence of priority taking and abstaining at single-lane roundabouts using cellular automata. *Transp. Res. C Emerg. Technol.* **2016**, *69*, 134–149. [CrossRef]

66. Wang, R.; Ruskin, H.J. Modeling traffic flow at a single-lane urban roundabout. *Comput. Phys. Commun.* **2002**, *147*, 570–576. [CrossRef]

67. Huang, D.W. Phase diagram of a traffic roundabout. *Phys. A Stat. Mech. Its Appl.* **2007**, *383*, 603–612. [CrossRef]

68. Lakouari, N.; Ez-Zahraouy, H.; Benyoussef, A. Traffic flow behavior at a single lane roundabout as compared to traffic circle. *Phys. Lett. A* **2014**, *378*, 3169–3176. [CrossRef]

69. Echab, H.; Lakouari, N.; Ez-Zahraouy, H.; Benyoussef, A. Phase diagram of a single lane roundabout. *Phys. Lett. A* **2016**, *380*, 992–997. [CrossRef]

70. Echab, H.; Lakouari, N.; Ez-Zahraouy, H.; Benyoussef, A. Simulation study of traffic car accidents at a single lane roundabout. *Int. J. Mod. Phys. C* **2016**, *27*. [CrossRef]

71. Echab, H.; Ez-Zahraouy, H.; Lakouari, N. Simulation study of interference of crossings pedestrian and vehicle traffic at a single lane roundabout. *Phys. A Stat. Mech. Its Appl.* **2016**, *461*, 854–864. [CrossRef]

72. Wang, R.; Ruskin, H. Modelling traffic flow at a multilane intersection. In *Computational Science and Its Applications—ICCSA*; Springer: Berlin, Germany, 2003; p. 964.

73. Wang, R.; Ruskin, H.J. Modelling traffic flow at multi-lane urban roundabouts. *Int. J. Mod. Phys. C* **2006**, *17*, 693–710. [CrossRef]

applied
sciences

MDPI

Article

Adaptive Global Fast Sliding Mode Control for Steer-by-Wire System Road Vehicles

Junaid Iqbal [1] [ORCID]**, Khalil Muhammad Zuhaib [1], Changsoo Han [2],*, Abdul Manan Khan [2] and Mian Ashfaq Ali [3]**

[1] Department of Mechatronics Engineering, Hanyang University ERICA Campus, Ansan 15588, Korea; jibssp@gmail.com (J.I.); kmzuhaib@gmail.com (K.M.Z.)
[2] Department of Robot Engineering, Hanyang University ERICA Campus, Ansan 15588, Korea; kam@hanyang.ac.kr
[3] School of Mechanical and Manufacturing Engineering (SMME), National University of Science and Technology (NUST), Islamabad 44000, Pakistan; ishfaqaries@gamil.com
*   Correspondence: cshan@hanyang.ac.kr; Tel.: +82-31-400-4062

Academic Editor: Felipe Jimenez
Received: 21 June 2017; Accepted: 13 July 2017; Published: 19 July 2017

Abstract: A steer-by-wire (SbW) system, also known as a next-generation steering system, is one of the core elements of autonomous driving technology. Navigating a SbW system road vehicle in varying driving conditions requires an adaptive and robust control scheme to effectively compensate for the uncertain parameter variations and external disturbances. Therefore, this article proposed an adaptive global fast sliding mode control (AGFSMC) for SbW system vehicles with unknown steering parameters. First, the cooperative adaptive sliding mode observer (ASMO) and Kalman filter (KF) are established to simultaneously estimate the vehicle states and cornering stiffness coefficients. Second, based on the best set of estimated dynamics, the AGFSMC is designed to stabilize the impact of nonlinear tire-road disturbance forces and at the same time to estimate the uncertain SbW system parameters. Due to the robust nature of the proposed scheme, it can not only handle the tire–road variation, but also intelligently adapts to the different driving conditions and ensures that the tracking error and the sliding surface converge asymptotically to zero in a finite time. Finally, simulation results and comparative study with other control techniques validate the excellent performance of the proposed scheme.

Keywords: adaptive global fast sliding mode (AGFSM); adaptive sliding mode observer (ASMO); Kalman filter (KF); Steer-by-Wire (SbW)

1. Introduction

The automobile industry is immensely working to transform conventional road vehicles into partial/full autonomous vehicles. SAE International and NHTSA have classified six levels of driving autonomy from "no automation" to "full automation" [1,2]. In particular, from the lane-keeping assistance system [3] to fully automated maneuvering [4–6], Steer-by-Wire (SbW) technology is playing a fundamental role in advanced driving assistance systems [7]. Nissan introduced the first commercialized SbW system in 2013 with the Infiniti Q50 vehicle [8,9]. The SbW system delivers better overall steering performance with comfort, reduces power consumption, provides active steering control, and significantly improves the passenger safety. Compared with a conventional steering system, the SbW system has replaced the mechanical shaft between the steering wheel and front wheels with two actuators, controllers, and sensors. The first actuator steers the front wheels and the second actuator provides steering feel feedback to the driver, obtained from the road and tire dynamics.

Over the last decade, many researchers have proposed a number of control techniques to compensate for the system parameter variation, change in road conditions, and external disturbances for obtaining the robust performance of the SbW system. In [10,11], sliding mode based control schemes are proposed for a partially known SbW system with unknown lumped uncertainties to track the reference signal. However, it is hard to classify the wide range of nominal parameters under the sideslip, and the robust performance may not be guaranteed over different road conditions. In [12–16], the upper bound sliding mode control (SMC) technique is proposed for the bounded unknown SbW system parameters and uncertain dynamics. However, the process of obtaining these proper bounds is not evident. In [17–20] proportional-derivative (PD) control is proposed to follow the driver's steering wheel signal closely. However, under uncertain dynamics, it is difficult to achieve satisfactory performance with a conventional control scheme. In [21] cornering stiffness and chassis side slip angle are estimated to calculate the self-aligning torque. The authors used the proportion of estimated torque as a feedback to the driver for artificial steering feel. In [22] three suboptimal sliding mode techniques are evaluated for yaw-rate tracking problem in over-actuated vehicles. In [23,24], adaptive control is implemented for path tracking via SbW system and the authors estimated the sliding gains by considering the known steering parameters and cornering coefficients. However, they did not use any mechanism to stop the estimation. Consequently, the controller could lead to saturation by estimating too large a sliding gain. In [25,26] the authors proposed a hyperbolic tangent function with adaptive SMC based schemes to counter the effect of self-aligning torque. In [27] the frictional torque and self-aligning torque are replaced by a second-order polynomial function that acts as an external disturbance over the SbW system. The authors proposed an adaptive terminal SMC (ATSMC) to estimate the upper bounds of parameters and disturbance.

Apart from the control design, a robust estimation methodology is also needed for the SbW system to estimate the vehicle states, uncertain parameters, and tire–road conditions for eliminating the effect of external disturbances from the controller. For instance, in recent years Kalman filter (KF) and nonlinear observers have gained much more attention from researchers; for example, in [28] a dual extended KF is used to estimate vehicle states and road friction. In [29] the authors estimated five DOF vehicle states and inertial parameters, such as overloaded vehicle's additional mass, respective yaw moment of inertia, and its longitudinal position using the dual unscented KF by considering the constant road–tire friction over a flat road. In [30–32] a fixed gain based full-state nonlinear observer is designed to estimate the longitudinal, lateral, and yaw velocities of the vehicle. However, for good estimation performance the observer gains must be tuned to a wide range of driving conditions. In order to reduce the burden of gain tuning from a-nonlinear observer [33], employed linear matrix inequality based convex optimization to obtain the gains of reduced order observer for estimating vehicle velocities. In [34] the authors implemented an adaptive gain based sliding mode observer to estimate the battery's charging level and health in electric vehicles.

In this paper first, we have established the adaptive sliding mode observer (ASMO) and the Kalman filter (KF) to simultaneously estimate the vehicle states and cornering stiffness coefficients by using the yaw rate and the strap down [35] lateral acceleration signals. Then, based on the simultaneously estimated dynamics, the two-fold adaptive global fast sliding mode control (AGFSMC) is designed for SbW system vehicles, considering that the steering parameters are unknown. In the first fold, estimated dynamics-based control (EDC) is utilized to stabilize the impact of self-aligning torque and frictional torque. In the second fold, the AGFSMC is developed to estimate the uncertain SbW parameters and eliminate the effect of residual disturbance left out from the EDC. The adaptation capability of the proposed scheme not only intelligently handles the tire–road environmental changes, but also adapts the system parameters and sliding gains according to the different driving conditions.

Finally, for avoiding overestimations of parameters and gains, discontinuous projection mapping [36] is incorporated to stop the estimation and adaptive mechanism as the tracking error converges to the designed dead zone bounds [37]. In the simulated results section, the comparative study will show the effectiveness of the proposed AGFSMC scheme, which ensures that the tracking error and sliding surface converge asymptotically to zero in a finite time.

The rest of the paper is structured as follows: In Sections 2 and 3, vehicle dynamics modeling and SbW system modeling with external disturbance are discussed. In Section 4, the ASMO and KF are established to estimate the vehicle states and parameters. In Section 5, the AGFSMC scheme is developed for the SbW system and the convergence analysis with bounded conditions is discussed in detail. Section 6 describes the simulation results and findings to validate the proposed scheme, followed by the last section that concludes the paper.

2. Vehicle Dynamics Modeling

Figure 1 illustrates the simplest bicycle model of a vehicle, which has a central front wheel and a central rear wheel, in place of two front and two rear wheels. The vehicle has two degrees of freedom, represented by the lateral motion y and the yaw angle ψ. According to Figure 1, the dynamics along the y axis and yaw axis are described as [38,39]:

$$m\left(\ddot{y} + V_x\dot{\psi}\right) = F_{yf}\cos\delta_{fw} + F_{xf}\sin\delta_{fw} + F_{yr} \tag{1}$$

$$I_z\ddot{\psi} = l_f\left(F_{yf}\cos\delta_{fw} + F_{xf}\sin\delta_{fw}\right) - l_rF_{yr}, \tag{2}$$

where $\ddot{y}$, $\dot{\psi}$, and $\ddot{\psi}$. are the acceleration with respect to the y axis motion, yaw rate, and yaw acceleration, respectively. l_f and l_r represent the distance of front and rear axles from the center of gravity, respectively. m and I_z are the mass of vehicle and the moment of inertia along the yaw axis, respectively. V_x denotes the longitudinal vehicle velocity at the center of gravity. F_{xf} and F_{xr} are the longitudinal forces of the front and rear wheels, respectively. F_{yf} and F_{yr} are the lateral frictional forces of front and rear wheels, respectively, as shown in Figure 1.

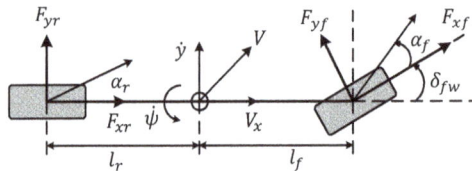

Figure 1. Bicycle model of vehicle.

In order to simplify the model, it is assumed that longitudinal forces Fx_f, Fx_r are equal to zero and by using the small angle approximation, i.e., $\cos\delta_{fw} \approx 1$, the simplified dynamics can be modeled as follows:

$$m\left(\ddot{y} + V_x\dot{\psi}\right) = F_{yf} + F_{yr} \tag{3}$$

$$I_z\ddot{\psi} = l_fF_{yf} - l_rF_{yr}. \tag{4}$$

For small slip angles, the lateral frictional forces are proportional to slip-angle α_f and α_r at the front and rear wheels, respectively. Therefore, lateral forces are defined as:

$$F_{yf} = 2C_f.\alpha_f \tag{5}$$

$$F_{yr} = 2C_r.\alpha_r, \tag{6}$$

where

$$\alpha_f = \delta_{fw} - \frac{V_y + l_f\dot{\psi}}{V_x} \tag{7}$$

$$\alpha_r = -\frac{V_y - l_f \dot{\psi}}{V_x}, \tag{8}$$

where C_f and C_r are the front and rear tires' cornering stiffness coefficients. δ_{fw} denotes the steering angle of front wheels, which is considered the same for both front wheels, and factor 2 accounts for two front and two rear wheels, respectively.

By using the small angle approximation $V_y = \dot{y}$ [38], Equations (7) and (8) can be written as:

$$\alpha_f = \delta_{fw} - \frac{\dot{y} + l_f \dot{\psi}}{V_x} \tag{9}$$

$$\alpha_r = -\frac{\dot{y} - l_f \dot{\psi}}{V_x}. \tag{10}$$

Substituting Equations (5), (6), (9) and (10) into Equations (3) and (4), the state space model is represented as:

$$\dot{x} = Ax + B\delta_{fw}, \tag{11}$$

where

$$x = \begin{bmatrix} \dot{y} \dot{\psi} \end{bmatrix}^T$$

$$A = \begin{bmatrix} -2\left(\frac{C_f+C_r}{mV_x}\right) & -\left(V_x + 2\left(\frac{l_fC_f-l_rC_r}{mV_x}\right)\right) \\ -2\left(\frac{l_fC_f-l_rC_r}{I_zV_x}\right) & -2\left(\frac{l_f^2C_f+l_r^2C_r}{I_zV_x}\right) \end{bmatrix}, B = \begin{bmatrix} \frac{2C_f}{m} \\ \frac{2l_fC_f}{I_z} \end{bmatrix}. \tag{12}$$

3. Steer-by-Wire System Modeling

Figure 2 depicts the standard model of SbW system for road vehicles. As shown, the steering wheel angle sensor is used to detect the driver's reference angle and the feedback motor is used to provide the artificial steering feel.

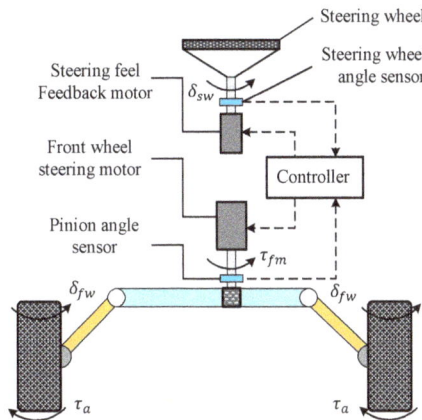

Figure 2. Steer-by-wire model.

Similarly, the front wheel angle is detected by the pinion angle sensor. Based on the error between the reference angle and the front wheel angle, the control signal is provided to the front wheel steering motor to closely steer the front wheels according to the driver's reference angle.

The equivalent second-order dynamics of the front wheels' steering motor is expressed as follows [10,15]:

$$J_{eq}\ddot{\delta}_{fw} + B_{eq}\dot{\delta}_{fw} + \tau_F + \tau_a = ku, \tag{13}$$

where J_{eq} and B_{eq} are the equivalent moment of inertia and the equivalent damping of the SbW system, respectively. u is the front wheels' steering motor control input and k is the steering ratio between the steering wheel angle and the front wheels' angle, given by $\delta_{fw} = \delta_{sw}/k$.

It is known that many modern road vehicles use the variable steering ratio. Therefore, dividing both sides of Equation (13) by k eliminates the impact of the variable steering ratio from the proposed control scheme without compromising the steering performance. Thus, the dynamics of SbW system can be written as:

$$J_{ek}\ddot{\delta}_{fw} + B_{ek}\dot{\delta}_{fw} + \tau_{Fk} + \tau_{ak} = u, \tag{14}$$

where

$$J_{ek} = \frac{J_{eq}}{k} = \frac{J_{fw}}{k} + J_{fm}k \tag{15}$$

$$B_{ek} = \frac{B_{eq}}{k} = \frac{B_{fw}}{k} + B_{fm}k \tag{16}$$

$$\tau_{Fk} = \frac{\tau_F}{k} \tag{17}$$

$$\tau_{ak} = \frac{\tau_a}{k}, \tag{18}$$

where J_{fw} and J_{fm} are the moment of inertia of the front wheels and the front wheel steering motor, respectively. B_{fw} and B_{sm} are the damping factors of the front wheels and the front wheel steering motor, respectively.

When the vehicle is turning, the steering system experiences torque that tends to resist the attempted turn, known as self-aligning torque τ_a. It can be seen from Figure 3 that the resultant lateral force developed by the tire manifests the self-aligning torque. The lateral force is acting behind the tire center on the ground plane and tries to align the wheel plane with the direction of wheel travel. Therefore, the total self-aligning torque is given by [14]:

$$\tau_a = F_{yf}(t_p + t_m), \tag{19}$$

where t_p is the pneumatic trail (the distance between the application point of lateral force F_{yf} to the center of tire), t_m is the mechanical trail, also known as the caster offset, which is the distance between the tire center and the point where the steering axis intersects with the ground plane.

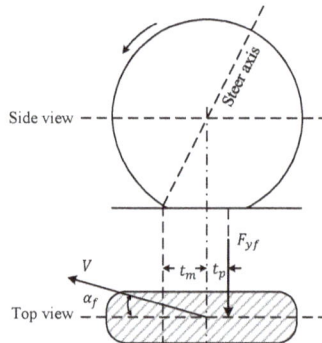

Figure 3. Self-aligning torque on front wheel.

By substituting Equations (5) and (9) into Equation (19), τ_a can be written as:

$$\tau_a = 2C_f\left(\delta_{fw} - \frac{\dot{y} + l_f\dot{\psi}}{V_x}\right)(t_p + t_m). \tag{20}$$

Moreover, τ_F is the coulomb frictional torque acting on SbW system, expressed as [23]:

$$\tau_F = F_{zf}\mu t_p\text{sign}\left(\dot{\delta}_{fw}\right) \tag{21}$$

$$F_{zf} = \frac{mgl_r}{l_f + l_r}, \tag{22}$$

where F_{zf} is the normal load on front axle, μ is the coeffient of friction, mg is the vehicle's weight without any external load, and sign() signum function is used to identify the direction of the frictional torque.

The stability of the SbW system mainly depends on the road and environmental conditions. The uncertain road surface such as dry, wet, or icy can produce considerable variations in the tire cornering stiffness coefficients C_f, C_r, which can adversely affect the controller performance. Therefore, to estimate the vehicle states and the uncertain parameter variation, the cooperative ASMO and KF are designed in the next section.

4. ASMO and KF

In this section, we will first design the ASMO for yaw rate $\dot{\psi}$ and lateral velocity $\dot{y}$, and then use the KF parameter estimator to estimate the tire cornering stiffness coefficients under varying road conditions.

In order to design the observer, a few assumptions are made, such as: the yaw rate $\dot{\psi}$ is directly measurable from yaw rate sensor; and the vehicle's longitudinal velocity is obtained from $V_x = r_e\omega$, where r_e is the effective tire radius and ω is the averaged free wheels angular speed measured from the wheel encoders. Moreover, it is considered that there is no effect of gravitation acceleration g on the lateral acceleration a_y, such that the a_y measurement model is defined as [29]:

$$a_{y,sensor} = \ddot{y} + V_x\dot{\psi}. \tag{23}$$

Therefore, the lateral velocity can be obtained from a strapdown algorithm [35] as follows:

$$\dot{y}(t) = \dot{y}(t-1) + \int (a_{y,sensor} - V_x\dot{\psi})dt, \tag{24}$$

where $\dot{y}(t-1)$ is the prior lateral velocity.

The conventional sliding mode observer (SMO) for vehicle states (Equation (11)) can be designed as:

$$\ddot{\hat{y}} = -A_{11}\dot{\hat{y}} - A_{12}\dot{\hat{\psi}} + B_1\delta_{fw} + L_1\text{sign}\left(\dot{y} - \dot{\hat{y}}\right) \tag{25}$$

$$\ddot{\hat{\psi}} = -A_{21}\dot{\hat{y}} - A_{22}\dot{\hat{\psi}} + B_2\delta_{fw} + L_2\text{sign}\left(\dot{\psi} - \dot{\hat{\psi}}\right), \tag{26}$$

where A_{11}, A_{12}, B_1, and B_2 are the elements of Equation (12) with nominal m_0 and I_{z0}; L_1 and L_2 are the observer gains, which must satisfy the following conditions, such that:

$$L_1 > \max(|A_{11}e_1| + |A_{12}e_2|) \tag{27}$$

$$L_2 > \max(|A_{12}e_1| + |A_{22}e_2|), \tag{28}$$

where $e_1 = \dot{y} - \dot{\hat{y}}$ and $e_2 = \dot{\psi} - \dot{\hat{\psi}}$.

The road surface variation is a critical factor for tuning the observer gains L_1 and L_2 during the design process. Any inappropriate selection of L_1 and L_2 will significantly reduce the SMO performance, resulting in a possible deviation of state estimation from the original trajectory.

Due to the aforementioned fact, an adaptive gain based sliding mode observer [34] is proposed, which improves the estimation performance by adapting the observer gains according to tire road conditions. Therefore, Equations (25) and (26) are changed to new forms, as follows:

$$\dot{\hat{y}} = -A_{11}\hat{y} - A_{12}\dot{\hat{\psi}} + B_1\delta_{fw} + \hat{L}_1(t)\text{sign}(e_1) \tag{29}$$

$$\dot{\hat{\psi}} = -A_{21}\hat{y} - A_{22}\dot{\hat{\psi}} + B_2\delta_{fw} + \hat{L}_2(t)\text{sign}(e_2), \tag{30}$$

where the ASMO gain adaptation law for $i = 1, 2$ is expressed as:

$$\dot{\hat{L}}_i(t) = \begin{cases} \rho_i|e_i|, & |e_i| > \varepsilon_i \\ 0, & \text{otherwise} \end{cases}, \tag{31}$$

where $\hat{L}_i(t) > 0$, is strictly positive time varying adaptive ASMO gain. ρ_i is a positive scalar used to adjust the adaption speed. $\varepsilon_i \ll 1$ are small positive constants used to activate the adaptation mechanism with the condition defined in Equation (31); therefore, as the error converges to the bound $|e_i| \leq \varepsilon_i$ in finite time, $\hat{L}_i(t)$ will stop increasing.

For convergence proof, the Lyapunov function of ASMO for lateral velocity is defined as:

$$V_1 = \frac{1}{2}e_1^2 + \frac{1}{2\rho_1}\tilde{L}_1, \tag{32}$$

where $\tilde{L}_1 = \hat{L}_1 - L_1$ is the adaptive gain convergence error.

The derivative of V_1, with the consideration that $\dot{L}_1 = 0$, is obtained as follows:

$$\begin{aligned} \dot{V}_1 &= e_1\dot{e}_1 + \frac{1}{\rho_1}\tilde{L}_1\dot{\hat{L}}_1 \\ &= e_1[-A_{11}e_1 - A_{12}e_2 - \hat{L}_1\text{sign}(e_1)] + \frac{1}{\rho_1}\tilde{L}_1\dot{\hat{L}}_1 \\ &\leq e_1[-A_{11}e_1 - A_{12}e_2] - \hat{L}_1|e_1| + (\hat{L}_1 - L_1)|e_1| \\ &\leq e_1[-A_{11}e_1 - A_{12}e_2] - L_1|e_1|. \end{aligned} \tag{33}$$

Thus, by considering Equation (27):

$$\dot{V}_1 \leq 0. \tag{34}$$

Similarly, the Lyapunov function V_2 for yaw rate convergence error e_2 and adaptive gain convergence error $\tilde{L}_2 = \hat{L}_2 - L_2$ can be written as:

$$V_2 = \frac{1}{2}e_2^2 + \frac{1}{2\rho_2}\tilde{L}_2. \tag{35}$$

The time derivative of Equation (35) will asymptotically converge to zero, $\dot{V}_2 \leq 0$, by considering $\dot{L}_2 = 0$ and $L_2 > \max(|a_{12}e_1| + |a_{22}e_2|)$.

Remark 1. *In the practical implementation, the direct strapdown of lateral acceleration may incorporate the small continuous noise to the lateral velocity that can diverge the ASMO estimation over time. Therefore, to deal with the issue, a lateral velocity-based damping term [40] is added to cancel out the incremental noise. Now Equation (24) can be written as:*

$$\dot{y}(t) = \dot{y}(t-1)(1-\sigma) + \int(a_{y,sensor} - V_x\dot{\psi})dt, \tag{36}$$

where $\sigma > 0$ is an adjustable small damping parameter.

Remark 2. *The designed ASMO may encounter high-frequency chattering due to the discontinuous signum function sign(); therefore, it is replaced by the continuous function* $e_i/(|e_i| + \varepsilon_i)$, *such that Equations (29) and (30) are rewritten as:*

$$\ddot{\hat{y}} = -A_{11}\dot{\hat{y}} - A_{12}\dot{\hat{\psi}} + B_1\delta_{fw} + \hat{L}_1(t)\frac{e_1}{|e_1| + \varepsilon_1} \tag{37}$$

$$\ddot{\hat{\psi}} = -A_{21}\dot{\hat{y}} - A_{22}\dot{\hat{\psi}} + B_2\delta_{fw} + \hat{L}_2(t)\frac{e_2}{|e_2| + \varepsilon_2}. \tag{38}$$

The estimation performance of the ASMO for lateral velocity and yaw rate primarily depends upon the knowledge of tire cornering stiffness coefficients C_f and C_r, which are unknown in practice and cannot be measured directly from the onboard vehicle sensors. Therefore, a Kalman filter (KF) [41] is proposed in cooperation with ASMO to estimate these stiffness coefficients under different tire–road conditions. Once the KF estimates a sufficient set of tire cornering stiffness coefficients, the parameter estimation can be switched off.

The KF algorithm [41] for tire cornering stiffness estimation is given in Table 1.

Table 1. Kalman filter algorithm.

1. Initialize $\hat{w}_0, P_0$**:**
$\hat{w}_0 = E[w(0)]$ $P_0 = E[(w(0) - \hat{w}_0)(w(0) - \hat{w}_0)^T]$
2. Time Update:
$\hat{w}_t^- = \hat{w}_{t-1}$ $P_t^- = P_{t-1} + Q$
3. Measurement Update:
$K_t = P_t^- H^T (HP_t^- H^T + R)^{-1}$ $\hat{w}_t = \hat{w}_t^- + K_t(z_t - H\hat{w}_t^-)$ $P_t = (I - K_t H)P_t^-$

P denotes the estimate error covariance, Q is the process noise covariance, and $R = r_s^2$ is the measurement noise covariance, whereas r_s represents the sensor's zero-mean white noise.

The tire cornering coefficients vector w and the measurement z, consisting of the lateral acceleration a_y, are defined as:

$$w = \begin{bmatrix} C_f & C_r \end{bmatrix}^T, z = Hw, \tag{39}$$

where

$$z = a_y$$

$$H = \begin{bmatrix} -\frac{2}{m_0}\left(\frac{\dot{y} + l_f\dot{\psi}}{V_x} - \delta_{fw}\right) & -\frac{2}{m_0}\left(\frac{\dot{y} - l_f\dot{\psi}}{V_x}\right) \end{bmatrix}. \tag{40}$$

It is to be noted that the tire cornering stiffness coefficient's vector w is considered as constant, therefore, the time derivative of w is zero, $(\dot{w} = 0)$. Then, w and z can be written in Euler's discretized form as:

$$w(k) = w(k-1) + v(k) \tag{41}$$

$$z(k) = Hw(k) + r(k), \tag{42}$$

where v and r are the zero mean process noise and measurement noise, respectively.

In order to improve the estimation performance and the convergence accuracy of KF, the difference $e_3 = z_t - H\hat{w}_t^-$, known as residual, is utilized to switch off the KF estimator. Therefore, on the basis of

e_3, a bounded condition is selected, such that when e_3 reaches the specified bound $|e_3| \leq \varepsilon_3$, the KF will stop the estimation process and thereafter the estimated parameters will become constant until e_3 exceeds the specified condition. ε_3 ($\varepsilon_3 > 0$) is the small positive constant.

Thus, the estimated tire cornering stiffness-based ASMO for Equations (37) and (38) is revised as:

$$\dot{\hat{y}} = -A_{11}(\hat{w}_{t-1})\dot{\hat{y}} - A_{12}(\hat{w}_{t-1})\dot{\hat{\phi}} + B_1(\hat{w}_{t-1})\delta_{fw} + \hat{L}_1(t)\frac{e_1}{|e_1| + \varepsilon_1} \tag{43}$$

$$\dot{\hat{\phi}} = -A_{21}(\hat{w}_{t-1})\dot{\hat{y}} - A_{22}(\hat{w}_{t-1})\dot{\hat{\phi}} + B_2(\hat{w}_{t-1})\delta_{fw} + \hat{L}_2(t)\frac{e_2}{|e_2| + \varepsilon_2}. \tag{44}$$

5. AGFSMC Control Design

In this section, the estimated dynamics-based adaptive global fast sliding mode control (AGFSMC) is designed in two steps to estimate the uncertain steering parameters and eliminate the effect of varying tire–road disturbance forces, so that the front wheels asymptotically track the driver's reference command in finite time.

The tracking error e_θ between the front wheel angle δ_{fw} and the scaled reference hand wheel angle δ_d is defined as:

$$e_\theta(t) = \delta_{fw}(t) - \frac{\delta_{sw}(t)}{k} = \delta_{fw}(t) - \delta_d(t). \tag{45}$$

The combination of linear sliding surface and the terminal sliding surface is known as the global fast terminal sliding surface, s, which is defined as [42]:

$$s = \dot{e}_\theta + \lambda_1(e_\theta)^{q/p} + \lambda_2 e_\theta, \tag{46}$$

where λ_1 and λ_2 ($\lambda_1, \lambda_2 > 0$), are strictly positive constants, and q and p, are positive odd numbers, such that $q < p$.

Thus, the time derivative of s is obtained as:

$$\dot{s} = \ddot{e}_\theta + \lambda_1 \frac{q}{p}(e_\theta)^{(\frac{q}{p}-1)}\dot{e}_\theta + \lambda_2\dot{e}_\theta. \tag{47}$$

$\dot{s}$ can be written as:

$$\dot{s} = \ddot{\delta}_{fw} - \ddot{\delta}_r, \tag{48}$$

where $\ddot{\delta}_r$ is expressed as:

$$\ddot{\delta}_r = \ddot{\delta}_d - \left(\lambda_1\frac{q}{p}(e_\theta)^{(\frac{q}{p}-1)} + \lambda_2\right)\dot{e}_\theta. \tag{49}$$

Thus, for stabilizing the SbW system (Equation (14)) and exponentially converging the tracking error (Equation (45)) to zero, the two-step closed loop control law u for the SbW system is designed as:

$$u = u_E + u_A, \tag{50}$$

where, in the first step, the estimated dynamics based control (EDC) u_E, is designed to counter the tire–road disturbance acting on the SbW system as follows:

$$u_E = -\text{sign}(s)(|\check{\varsigma}\tau_{ak}| + |\check{\varsigma}\tau_{Fk}|), \tag{51}$$

where $\check{\varsigma}\tau_{ak}$, is the estimated self-aligning torque, which is computed from the best set of estimated vehicle states and front wheel cornering stiffness provided by the ASMO and KF. $\check{\varsigma}\tau_{Fk}$ is the nominal frictional torque obtained from the nominal set of vehicle parameters, such as mass, nominal coefficient of friction, and the geometry of the vehicle.

Both $|\zeta\tau_{ak}|$ and $|\zeta\tau_{Fk}|$ are expressed as:

$$|\zeta\tau_{ak}| = \frac{2\hat{C}_f}{k_o}(t_{p_o} + t_m)\left|\left(\delta_{fw} - \frac{\dot{\hat{y}} + l_f\dot{\hat{\psi}}}{V_x}\right)\right| \tag{52}$$

$$|\zeta\tau_{Fk}| = \frac{m_ogl_r}{(l_f + l_r)k_o}\mu_ot_{p_o}\left|\text{sign}\left(\dot{\delta}_{fw}\right)\right|, \tag{53}$$

where $\dot{\hat{y}}$, $\dot{\hat{\psi}}$, and $\hat{C}_f$ are the observed vehicle states and the front wheel's estimated cornering stiffness, as worked out in the previous section, respectively. μ_o, t_{po}, m_o, and k_o are the nominal system parameters.

Second, to tackle the residual disturbance left by the EDC and estimate the uncertain steering parameters, the adaptive global fast sliding mode control (AGFSMC) u_A is designed as follows:

$$u_A = -\text{sign}(s)\left(|y|\hat{a} + \hat{\mathcal{T}}\left|\frac{\dot{\hat{y}} + l_f\dot{\hat{\psi}}}{V_x}\right| + \hat{\beta}_1|u(t-1)|\right) - \beta_2 s, \tag{54}$$

where $\hat{a}(t)$ is the estimated parameter's vector and $|y|$ is the signal feedback vector; they are defined as follows:

$$\hat{a} = [\hat{J}_{ek}\ \hat{B}_{ek}\ \hat{\mathcal{F}}\ \hat{\mathcal{T}}]^T \tag{55}$$

$$|y| = [|\ddot{\delta}_r||\dot{\delta}_{fw}||\text{sign}(\dot{\delta}_{fw})||\delta_{fw}|]. \tag{56}$$

Moreover, $\hat{\beta}_1$ and β_2, $(\beta_1, \beta_2 > 0)$ are the fixed and adaptive gains used to control the convergence speed of AGFSMC, respectively, and $|u(t-1)|$ is the prior control input obtained at the time step $t-1$. Therefore, the adaptation laws for updating the $\hat{a}(t)$ and $\hat{\beta}_1$ are designed as:

$$\dot{\hat{a}} = \Gamma|y^T|||s| \tag{57}$$

$$\dot{\hat{\beta}}_1 = |s||u(t-1)|, \tag{58}$$

where $\Gamma(\Gamma > 0)$ is the diagonal positive definite gain matrix used to tune the parameter adaptation speed. Figure 4 shows the framework of the proposed AGFSMC scheme.

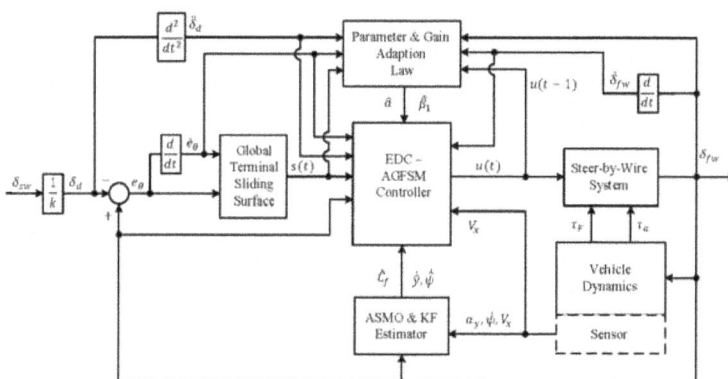

Figure 4. AGFSMC scheme framework.

Convergence Proof

The Lyapunov function candidate is defined as:

$$V_3 = \frac{1}{2}J_{ek}s^2 + \frac{1}{2}\tilde{a}^T\Gamma^{-1}\tilde{a} + \frac{1}{2}\tilde{\beta}_1^{\ 2}, \tag{59}$$

where $\tilde{a}(t) = \hat{a}(t) - a$ is the parameter estimation error, $\tilde{\beta}_1(t) = \hat{\beta}_1(t) - \beta_1$ is the adaptive gain convergence error, and Γ^{-1} is the inverse of gain matrix.

The time derivative of Lyapunov function V_3 in terms of the SbW system (Equation (14)) and the control input (Equation (50)), with the considerations that $\dot{a} = 0$, $\dot{\beta}_1 = 0$, are obtained as follows:

$$
\begin{aligned}
\dot{V}_3 &= sJ_{ek}\dot{s} + \dot{\hat{a}}^T\Gamma^{-1}\tilde{a} + \tilde{\beta}_1\dot{\hat{\beta}}_1 \\
&= s[-J_{ek}\ddot{\delta}_r - B_{ek}\dot{\delta}_{fw} - \tau_{Fk} - \tau_{ak} + u] + \dot{\hat{a}}^T\Gamma^{-1}\tilde{a} + \tilde{\beta}_1\dot{\hat{\beta}}_1 \\
&= s[-J_{ek}\ddot{\delta}_r - B_{ek}\dot{\delta}_{fw} - \tau_{Fk} - \tau_{ak} - \text{sign}(s)(|\xi\tau_{Fk}| \\
&\quad + |\xi\tau_{ak}|) + u_A] + \dot{\hat{a}}^T\Gamma^{-1}\tilde{a} + \tilde{\beta}_1\dot{\hat{\beta}}_1 \\
&= s[-J_{ek}\ddot{\delta}_r - B_{ek}\dot{\delta}_{fw} + u_A] - (s\tau_{Fk} + |s||\xi\tau_{Fk}|) \\
&\quad - (s\tau_{Fk} + |s||\xi\tau_{ak}|) + \dot{\hat{a}}^T\Gamma^{-1}\tilde{a} + \tilde{\beta}_1\dot{\hat{\beta}}_1 \\
&\leq s[-J_{ek}\ddot{\delta}_r - B_{ek}\dot{\delta}_{fw} + u_A] - |s|(|\xi\tau_{Fk}| - |\tau_{Fk}|) \\
&\quad - |s|(|\xi\tau_{ak}| - |\tau_{ak}|) + \dot{\hat{a}}^T\Gamma^{-1}\tilde{a} + \tilde{\beta}_1\dot{\hat{\beta}}_1.
\end{aligned}
\tag{60}
$$

It is considered that $|\xi\tau_{Fk}| < \tau_{Fk}$ and $|\xi\tau_{aFk}| < \tau_{ak}$, such that:

$$|\xi\tau_{Fk}| - |\tau_{Fk}| = -\mathcal{F}|\text{sign}(\dot{\delta}_{fw})| \tag{61}$$

$$|\xi\tau_{ak}| - |\tau_{ak}| = -\mathcal{T}\left(|\delta_{fw}| + \left|\frac{\dot{y} + l_f\dot{\psi}}{V_x}\right|\right), \tag{62}$$

where $\mathcal{F}$ and $\mathcal{T}$ are the uncertain residual parameters of frictional torque and self-aligning torque, respectively.

Substituting Equations (61), (63) and AGFSMC u_A (Equation (54)) into (Equation (60)), then the inequality is written as:

$$
\begin{aligned}
\dot{V}_3 \leq\ & s[-J_{ek}\ddot{\delta}_r - B_{ek}\dot{\delta}_{fw} - \text{sign}(s)\{\hat{J}_{ek}|\ddot{\delta}_r| + \hat{B}_{ek}|\dot{\delta}_{fw}| \\
& + \hat{\mathcal{T}}|\delta_{fw}| + \hat{\mathcal{T}}\left|\frac{\dot{y}+l_f\dot{\psi}}{V_x}\right| + \hat{\beta}_1|u(t-1)|\} - \beta_2 s] \\
& + |s|\mathcal{F}|\text{sign}(\dot{\delta}_{fw})| + |s|\mathcal{T}|\delta_{fw}| + |s|\mathcal{T}\left|\frac{\dot{y}+l_f\dot{\psi}}{V_x}\right| \\
& + \dot{\hat{a}}^T\Gamma^{-1}\tilde{a} + \tilde{\beta}_1\dot{\hat{\beta}}_1 \\
=\ & -(|s|\hat{J}_{ek}|\ddot{\delta}_r| + sJ_{ek}\ddot{\delta}_r) - (|s|\hat{B}_{ek}|\dot{\delta}_{fw}| + sB_{ek}\dot{\delta}_{fw}) \\
& - (|s|\hat{\mathcal{F}}|\text{sign}(\dot{\delta}_{fw})| - s|\mathcal{F}|\text{sign}(\dot{\delta}_{fw})|) \\
& - (|s|\hat{\mathcal{T}}|\delta_{fw}| - |s|\mathcal{T}|\delta_{fw}|) + |s|\mathcal{T}\left|\frac{\dot{y}+l_f\dot{\psi}}{V_x}\right| - |s|\hat{\mathcal{T}}\left|\frac{\dot{y}+l_f\dot{\psi}}{V_x}\right| \\
& - |s|\hat{\beta}_1|u(t-1)| - \beta_2 s^2 + \dot{\hat{a}}^T\Gamma^{-1}\tilde{a} + \tilde{\beta}_1\dot{\hat{\beta}}_1 \\
=\ & -|s||\ddot{\delta}_r|\tilde{J}_{ek} - |s||\dot{\delta}_{fw}|\tilde{B}_{ek} - |s||\text{sign}(\dot{\delta}_{fw})|\tilde{\mathcal{F}} - |s||\delta_{fw}|\tilde{\mathcal{T}} \\
& - |s|\left(\hat{\mathcal{T}}\left|\frac{\dot{y}+l_f\dot{\psi}}{V_x}\right| - \mathcal{T}\left|\frac{\dot{y}+l_f\dot{\psi}}{V_x}\right|\right) - |s|\hat{\beta}_1|u(t-1)| - \beta_2 s^2 \\
& + \dot{\hat{a}}^T\Gamma^{-1}\tilde{a} + \tilde{\beta}_1\dot{\hat{\beta}}_1 \\
=\ & -|s||y|\tilde{a} + \dot{\hat{a}}^T\Gamma^{-1}\tilde{a} - |s|\left(\hat{\mathcal{T}}\left|\frac{\dot{y}+l_f\dot{\psi}}{V_x}\right| - \mathcal{T}\left|\frac{\dot{y}+l_f\dot{\psi}}{V_x}\right|\right) \\
& - |s|\hat{\beta}_1|u(t-1)| + (\hat{\beta}_1 - \beta_1)\dot{\hat{\beta}}_1 - \beta_2 s^2.
\end{aligned}
\tag{63}
$$

With the adaptation laws of $\dot{\hat{a}}$ (Equation (57)) and $\dot{\hat{\beta}}_1$ (Equation (58)), substituting into Equation (63) satisfies:

$$\dot{V}_3 \leq -|s| \left(\hat{\mathcal{T}} \left| \frac{\dot{\hat{y}} + l_f \dot{\hat{\psi}}}{V_x} \right| - \mathcal{T} \left| \frac{\dot{y} + l_f \dot{\psi}}{V_x} \right| \right) - |s| \beta_1 |u(t-1)| - \beta_2 s^2 \tag{64}$$

The convergence proof shows that the proposed AGFSMC is stable and the inequality (Equation (64)) ensures that the global fast terminal sliding surface variable exponentially converges to zero ($s = 0$) in the finite time.

Remark 3. *The signum function* sign(s) *incorporates the chattering and discontinuity in the proposed controller. Therefore, to eliminate the chattering phenomenon the signum function is replaced by the boundary layer saturation function* sat($\cdot$) *such that Equations (51) and (53) are re-written as:*

$$u_E = -\text{sat}(s)(|\xi \tau_{ak}| + |\xi \tau_{Fk}|) \tag{65}$$

$$u_A = -\text{sat}(s) \left(|\mathbf{y}| \hat{a} + \hat{\mathcal{T}} \left| \frac{\dot{\hat{y}} + l_f \dot{\hat{\psi}}}{V_x} \right| + \hat{\beta}_1 |u(t-1)| \right) - \beta_2 s. \tag{66}$$

The boundary layer saturation function is defined as:

$$\text{sat}(s) = \begin{cases} \frac{s}{\phi} & |s| < \phi \\ \text{sign}(s) & \text{otherwise} \end{cases}, \tag{67}$$

where $\phi > 0$ *represents the boundary layer thickness. Due to the boundary layer, the closed-loop error cannot converge to zero. However, a carefully selected value of* ϕ *would lead the error to a user-specified bounded region.*

Remark 4. *In order to avoid overestimation of* $\hat{a}$ *and* $\hat{\beta}_1$, *which can lead the control input* $u(t)$ *to saturation, Equations (57) and (58) can be re-written for the permissible bounds of* e_θ *using the discontinuous projection mapping [36] as follows:*

$$\dot{\hat{a}} = \begin{cases} 0 & \text{if } |e_\theta| \leq \varepsilon_4 \\ \Gamma |\mathbf{y}|^T |s| & \text{otherwise} \end{cases} \tag{68}$$

$$\dot{\hat{\beta}}_1 = \begin{cases} 0 & \text{if } |e_\theta| \leq \varepsilon_5 \\ |s| |u(t-1)| & \text{otherwise} \end{cases}, \tag{69}$$

where ε_4 *and* ε_5 *are defined as dead zone bounds [27] in terms of tracking error. Therefore, when the tracking error converges to the respective dead zone bound, the adaption mechanism will be switched off and after that* $\hat{a}$ *and* $\hat{\beta}_1$ *become constant.*

6. Simulation Results

In this section, the estimation accuracy of vehicle states and cornering stiffness coefficients, and the control input performance of the proposed AGFSMC scheme for SbW system road vehicles, are validated over the three different maneuvering tests, in compression with adaptive sliding mode control (ASMC) and adaptive fast sliding mode control (ATSMC).

The first test (test 1) is sinusoidal maneuvering with varying tire–road conditions—snowy for the first 30 s and a dry asphalt road for the next 30 s—with the selected coefficient of friction as $\mu_{t<30} = 0.45$, $\mu_{t\geq30} = 0.85$ and the tire cornering stiffness coefficients for the front and rear wheels as $C_{f(t<30)} = 4000$, $C_{f(t\geq30)} = 8000$, $C_{r(t<30)} = 5000$, $C_{r(t\geq30)} = 10,000$, respectively. The second test (test 2) is known as circular maneuvering, conducted over a dry asphalt road. Moreover, a high speed cornering test (test 3) is also introduced to further evaluate the robustness of the proposed scheme.

It is worth noting that the first two tests are carried out at longitudinal speed $V_x = 10$ m/s and the third test at $V_x = 20$ m/s with the same sampling rate of $\Delta T = 0.001$ s. Furthermore, the vehicle and SbW system parameters are listed in Table 2.

Table 2. Vehicle and SbW system parameters.

Parameter	Value (s)
m (kg)	1270
I_z (kg·m^2)	1537
l_f, l_r (m)	1.015, 1.895
J_{ek}	0.28
B_{ek}	0.88
k	18
t_m, t_p (m)	0.023, 0.016

The parameters for the proposed cooperative ASMO and KF estimator with the termination bounds are selected as: $\hat{L}_1(0) = \hat{L}_2(0) = 8, \rho_1 = \rho_2 = 10, \sigma = 0.001, \varepsilon_1 = \varepsilon_2 = 0.005, \varepsilon_3 = 0.01, m_0 = 1150$ kg, $I_{z_0} = 1430$ kg·m^2, $\hat{w}_0 = [100\ 100]^T$, $P_0 = 10000 \times I_{2x2}, Q = (1 \times 10^{-6})I_{2 \times 2}$, and $r_s = 0.001$.

In addition, the parameters for the designed AGFSMC scheme with dead zone bounds are chosen as: $\lambda_1 = \lambda_2 = 12, p = 7, q = 5, \phi = 0.8, \beta_2 = 4, t_{p_0} = t_m = 0.016$ m, $\mu_o = 0.6, k_o = 16, \Gamma = I_{4 \times 4}$, $\varepsilon_4 = \varepsilon_5 = 0.002$, and the initial conditions are considered as $\hat{a}(0) = \hat{B}_1(0) = 0$.

To compare the performance of the proposed AGFSMC scheme with the adaptive sliding mode control (ASMC), as designed in [25], we used the following equations:

$$u = \tfrac{1}{k}(J_{e0}(\lambda \dot{e} + \ddot{\delta}_d) + B_{e0}\dot{\delta}_{fw} + \xi_{f0}\text{sign}(\dot{\delta}_{fw}) + \omega s + K\text{sat}(s) + \hat{\rho}_\tau \tanh(\delta_{fw}))$$

$$K = 0.1(J_{e0}(\lambda|\dot{e}| + |\ddot{\delta}_d|) + B_{e0}|\dot{\delta}_{fw}| + \xi_{f0}) \tag{70}$$

$$\dot{\hat{\rho}}_\tau = \mu \tfrac{\omega}{J_{e0}} + \mu \dot{s}\tanh(\delta_{fw}),$$

where the tracking error $e = \delta_d - \delta_{fw}$ and the sliding surface $s = \dot{e} + \lambda e$ with $\dot{s} = (s_k - s_{k-1})/\Delta t$ are defined in Equation (71). The saturation function sat($\cdot$) is also taken to be the same as Equation (67) with boundary layer thickness $\phi = 0.8$. Moreover, the nominal SbW system parameters $J_{e0} = 3$, $B_{e0k} = 12, \xi_{f0} = 100, k = 18$, and the control parameters $\lambda = 12, \omega = 72, \mu = 450$ are selected according to the methodology defined in [25].

For performance comparison with ATSMC, as designed as [27], the calculations are given as follows:

$$u = -\text{sat}(s)\left[\hat{a}_1\left(\ddot{\delta}_d\right) + \hat{b}_1|\dot{\delta}_{fw}| + \hat{c}_0 + \hat{c}_1|\delta_{fw}| + \hat{c}_2|\dot{\delta}_{fw}| + \lambda\hat{a}\tfrac{q}{p}(e)^{(\frac{q}{p}-1)}|\dot{e}|\right] - \tfrac{\hat{\rho}}{2}s$$

$$-k_1\text{sign}(s) - k_2 s$$

$$\dot{\hat{c}}_0 = \eta_1|s|(1 - \sigma\hat{c}_0)$$

$$\dot{\hat{c}}_1 = \eta_2|s||\delta_{fw}|(1 - \sigma\hat{c}_1) \tag{71}$$

$$\dot{\hat{c}}_2 = \eta_3|s||\dot{\delta}_{fw}|(1 - \sigma\hat{c}_2)$$

$$\dot{\hat{a}}_1 = \left(\eta_4|s|\left(\ddot{\delta}_d\right) + \eta_4|s|\lambda\tfrac{q}{p}(e)^{(\frac{q}{p}-1)}|\dot{e}|\right)(1 - \sigma\hat{a}_1)$$

$$\dot{\hat{b}}_1 = \eta_5|s||\dot{\delta}_{fw}|\left(1 - \sigma\hat{b}_1\right), \dot{\hat{\rho}} = \eta_6\tfrac{s^2}{2}(1 - \sigma\hat{\rho}),$$

where $\lambda, p, q, \text{sat}(s)$, and ϕ have the same values as those defined in AGFSMC. Moreover, the control parameters and adjustable parameters for adaptive laws are selected according to [27] as follows: $\eta_1 = 4, \eta_2 = \eta_3 = \eta_4 = \eta_5 = \eta_6 = 2, k_1 = 0.001, k_2 = 4, \ddot{\delta}_d = 2$ and $\rho = 0.001$, resptectively.

6.1. Sinusoidal Maneuvering Test (Test 1)

The reference steering wheel angle is generated by:

$$\delta_d = 0.4\sin(0.5\pi t)\text{rad}. \tag{72}$$

Figure 5 shows the simultaneously estimated lateral velocity, yaw rate, and cornering stiffness coefficients. It is observed that the cooperative ASMO and KF scheme intelligently cope with the tire–road variations and estimate the vehicle states and cornering stiffness coefficients by self-tuning the gains according to the driving environment. Figure 5c shows that the estimated $\hat{C}_f, \hat{C}_r$ have not only converged to the neighborhood of the actual values in both dry and snowy conditions, but also become constant after the condition e_3 reached a specified termination bound.

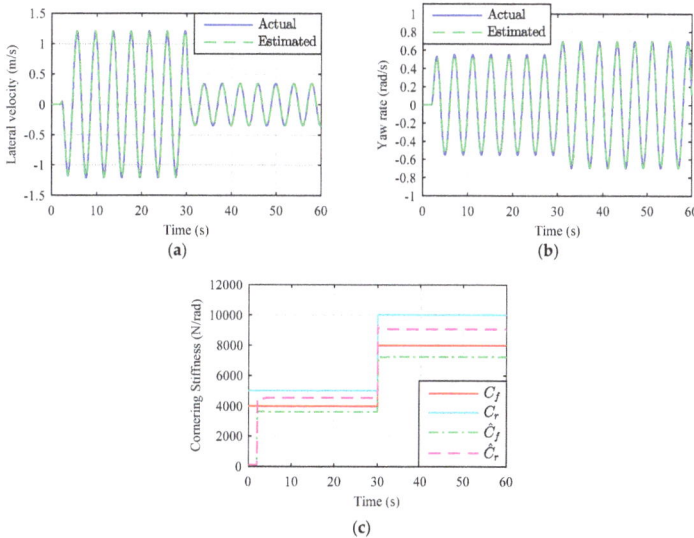

Figure 5. Estimation results of vehicle states and cornering stiffness coefficients in test 1: (**a**) Estimated lateral velocity; (**b**) Estimated yaw rate; (**c**) Estimated cornering stiffness coefficients.

Figure 6 represents the tracking response and the control input performance of the AGFSMC scheme against the varying tire–road disturbance forces. We can see from Figure 6b that the proposed methodology effectively eliminates the impact of self-aligning torque (Equation (20)) and Coulomb frictional torque (Equation (21)) from the SbW system and ensures that the front wheels are precisely tracking the reference steering angle with a steady state tracking error of 0.002 rad. It is noted that at the beginning of sinusoidal maneuvering, after 3 s, the tracking error reached the peak value of 0.01 rad. This is because we started all the parameter estimations from very low values, such as $\hat{w}_0 = [100\ 100]^T, \hat{a}(0) = \hat{B}_1(0) = 0$. Therefore, right after the peak error, all the estimated parameters converged to the sufficient estimation set. As a result, the peak tracking error also converged to the steady-state dead zone region.

Moreover, Figure 7 shows the estimated SbW system parameters and the sliding gain adaptation profile. It is observed that the estimated SbW system parameters did not converge to the listed actual constants, but due to the adaptive capability of the proposed control scheme, all parameters as well as the sliding gain are adaptively adjusted in time for both driving conditions, which ensure the closed-loop stability of the SbW system. Hence, the outstanding steering performance of the SbW system vehicle is achieved against the nonlinear tire–road disturbance forces and the uncertain SbW system parameters.

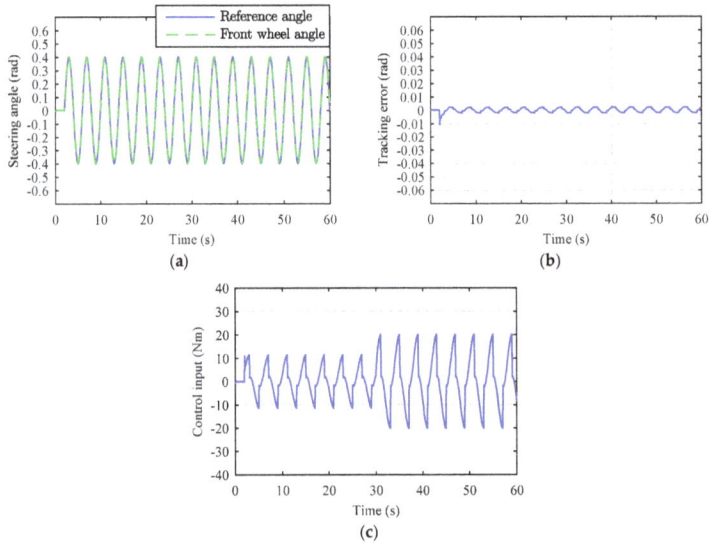

Figure 6. Control performance of the proposed AGFSMC scheme in test 1: (**a**) Tracking performance; (**b**) Tracking error; (**c**) Control input torque.

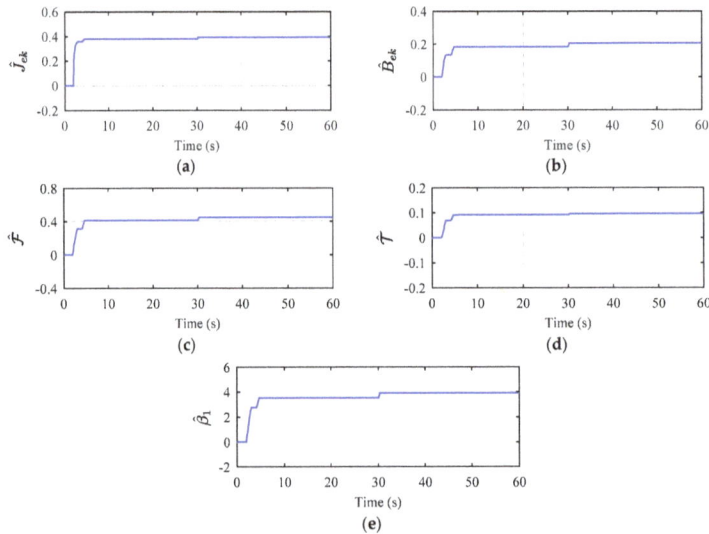

Figure 7. Estimated SbW system parameters and sliding gain with AGFSMC scheme in test 1: (**a–d**) Estimated SbW system parameters; (**e**) Estimated sliding gain.

Figure 8 demonstrates that the steering performance of the ASMC scheme is not as good as that of the proposed AGFSMC scheme. This is because the hyperbolic tangent function used in ASMC is unable to replicate the actual self-aligning torque acting on the SbW system. Also, the adaptation law cannot estimate the appropriate equivalent coefficient of self-aligning torque to compensate for the varying tire–road conditions. Consequently, the overall tracking error is much higher, particularly in the dry asphalt road condition: the tracking error peaks to the steady-state value of 0.06 rad, which is almost 30 times higher than in the proposed scheme. Although the ASMC scheme has the information of nominal parameters and utilized the saturation function, it incorporates high-frequency chattering during the first 3 s of the simulation, where the reference angle is set to zero.

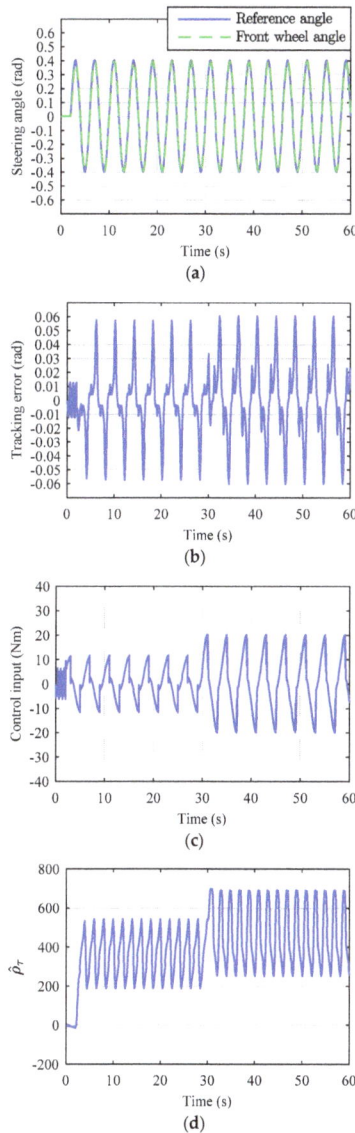

Figure 8. Control performance of adaptive sliding mode controller in test 1: (**a**) Tracking performance; (**b**) Tracking error; (**c**) Control input torque; (**d**) Estimated equivalent coefficient of self-aligning torque.

Figure 9 shows that the overall tracking response of ATSMC is better than the ASMC under the varying driving conditions, while both schemes cannot outperform the proposed AGFSMC. It can be seen that the control input overshoots the allowable control limit, which causes irregular spikes in the tracking error. We noticed two reasons for that: (1) The designed adaptation law for estimating the control parameter $\hat{a}_1$ does not include a provision to maintain the positive estimation; and (2) the ATSMC does not possess any mechanism to bound or stop the parameter adaptation process for avoiding overestimations, as compared to the one proposed in AGFSMC. Therefore, the tracking error is consistently converging to a smaller region with spikes due to the large and continuous parameter estimation, which may lead the controller to saturation state.

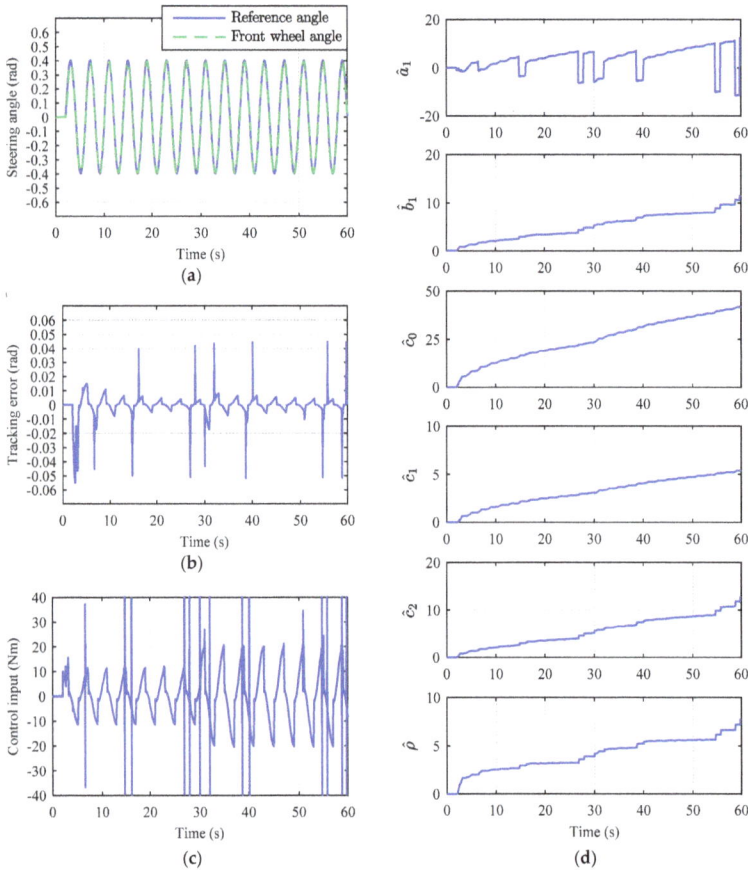

Figure 9. Control performance of adaptive terminal sliding mode control in test 1: (a) Tracking performance; (b) Tracking error; (c) Control input torque; (d) Estimated parameters.

6.2. Circular Maneuvering Test (Test 2)

The circular maneuvering test is carried out over the dry asphalt road for 25 s with these selected tire–road parameters: $C_f = 8000$, $C_r = 10000$, and $u = 0.85$.

Figures 10–12 portray the promising results of the proposed AGFSMC scheme in all aspects during test 2. We can see the fine estimation of vehicle states and cornering coefficients in Figure 10. The estimated cornering coefficients takes less than a second to converge to the sufficient estimation set over the dry asphalt, such as, $\hat{C}_f \cong 7250$, $\hat{C}_r \cong 9050$, and becomes constant after e_3 satisfies the selected ε_3 bound. Thus, Figure 11 exhibits the excellent tracking response of the front wheels with an observed peak tracking error of 0.008 rad, which eventually converged to the ε_4 bound after the rapid adjustment of all adaptive parameters $\hat{J}_{ek}$, $\hat{B}_{ek}$, $\hat{\mathcal{F}}$, $\hat{\mathcal{T}}$, and $\hat{\beta}_1$ to certain constants, as shown in Figure 12.

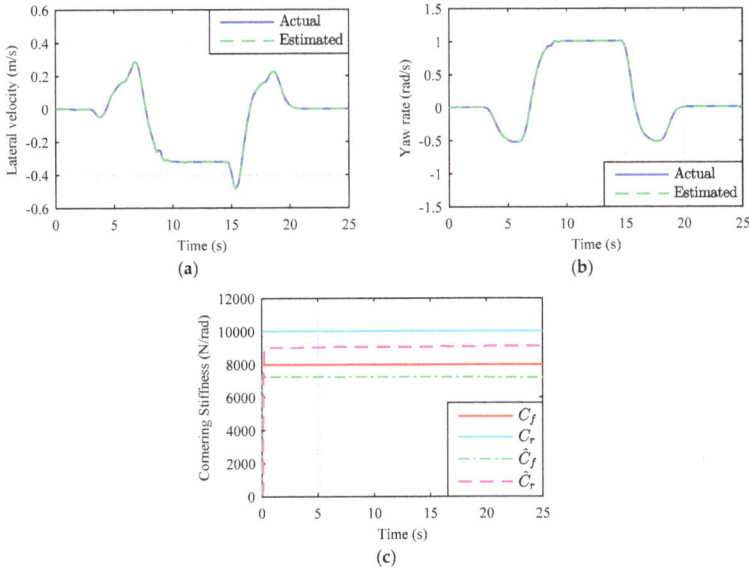

Figure 10. Estimation results of vehicle states and cornering stiffness coefficients in test 2: (**a**) Estimated lateral velocity; (**b**) Estimated yaw rate; (**c**) Estimated cornering stiffness coefficients.

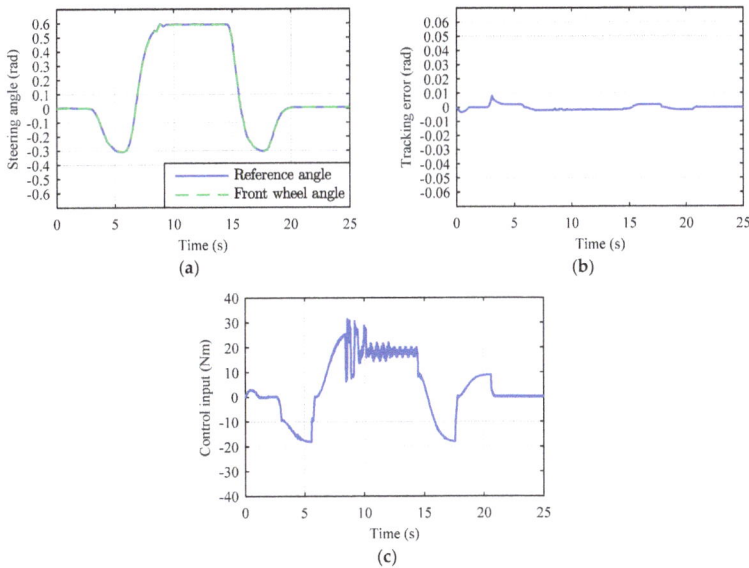

Figure 11. Control performance of the proposed AGFSMC scheme in test 2: (**a**) Tracking performance; (**b**) Tracking error; (**c**) Control input torque.

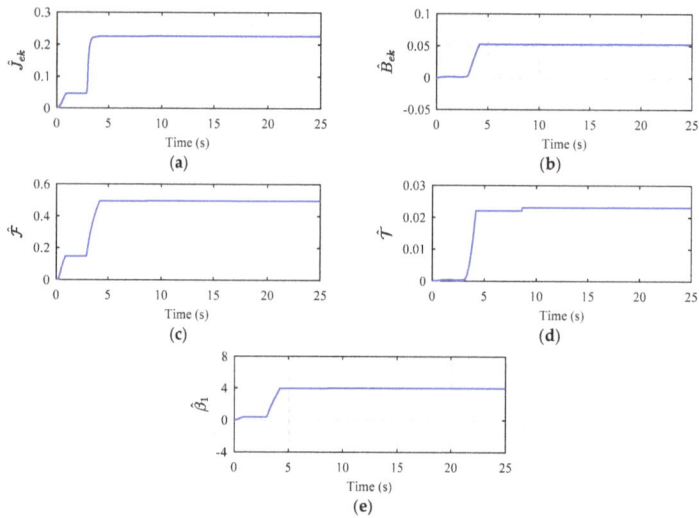

Figure 12. Estimated SbW system parameters and sliding gain with AGFSMC scheme in test 2: (**a–d**) Estimated SbW system parameters; (**e**) Estimated sliding gain.

In contrast to the proposed scheme, the ASMC shows the worst tracking performance throughout test 2. It can be seen from Figure 13 that the tracking error is unable to obtain any steady state bound and reached a peak value of 0.076 rad, which is almost 9.5 times higher than in the proposed AGFSMC scheme. Moreover, the adaptation law also shows inconsistent behavior in the last 7 s of this test, where the estimated coefficient of self-aligning torque rapidly drops to a highly negative value. As a result, neither tracking error nor sliding surface converged to the steady state boundary at a finite time in the Lyapunov's sense.

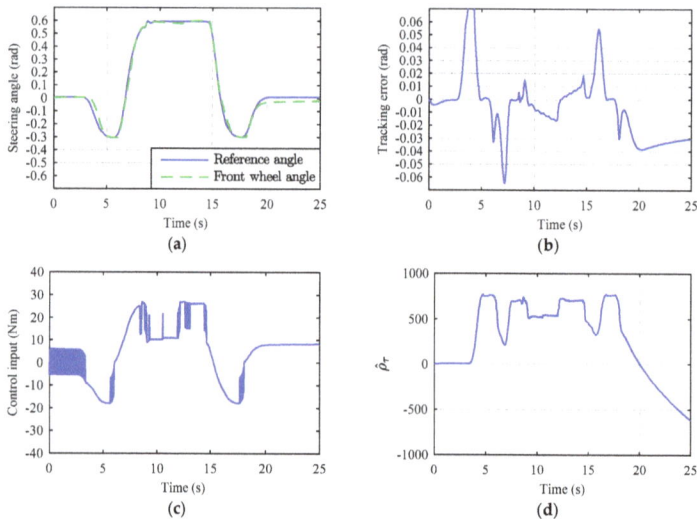

Figure 13. Control performance of adaptive sliding mode controller in test 2: (**a**) Tracking performance; (**b**) Tracking error; (**c**) Control input torque; (**d**) Estimated equivalent coefficient of self-aligning torque.

On the other hand, the ATSMC performed slightly better than the ASMC in terms of tracking response and also managed to converge the tracking error to the steady state bound during test 2. The peak tracking error observed under the ATSMC scheme is 0.067 rad as shown in Figure 14, which is marginally less than the ASMC but almost 8.35 times higher than the proposed scheme. Moreover, the abrupt shift in $\hat{a}_1$ parameter estimation and the multiple control input overshoots are again observed in this test.

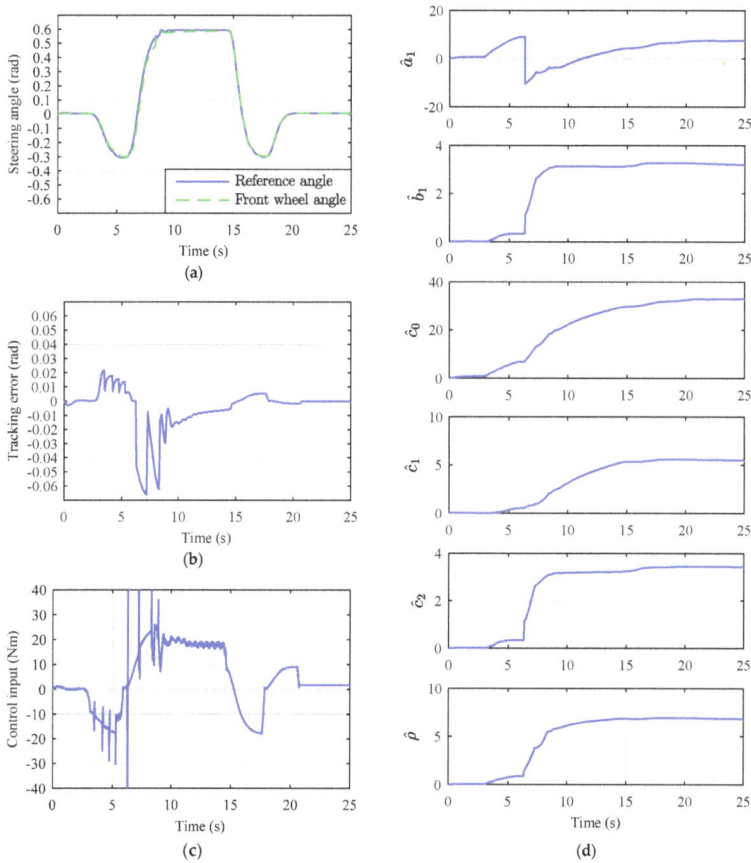

Figure 14. Control performance of adaptive terminal sliding mode control in test 2: (a) Tracking performance; (b) Tracking error; (c) Control input torque; (d) Estimated parameters.

6.3. High Speed Cornering Test (Test 3)

In order to further evaluate the estimation accuracy, tracking response, and control input performance of the proposed AGFSMC scheme, a high-speed cornering test is performed on a dry asphalt road for 45 s.

As expected, Figures 15–17 clearly indicate the remarkable performance of the proposed scheme against the parametric uncertainties and tire–road disturbance. The cooperative ASMO and KF also maintain the robustness and provide adequate estimated dynamics to stabilize the effect of self-aligning torque and frictional torque at high speed. The peak tracking error observed during test 3 under the AGFSMC scheme is 0.0095 rad, which is almost eight times lower than ASMC (0.076 rad), and four times lower than ATSMC (0.04 rad). Compared to other control schemes, Figure 18 shows that the tracking error under ASMC was again unable to attain any steady state bound and also incorporates high-frequency chattering at constant steering angle inputs. The ATSMC shows a decent performance regarding the tracking error convergence as compared to ASMC. However, the sudden parameter estimation shift with control overshoot still exists in this test, as shown in Figure 19.

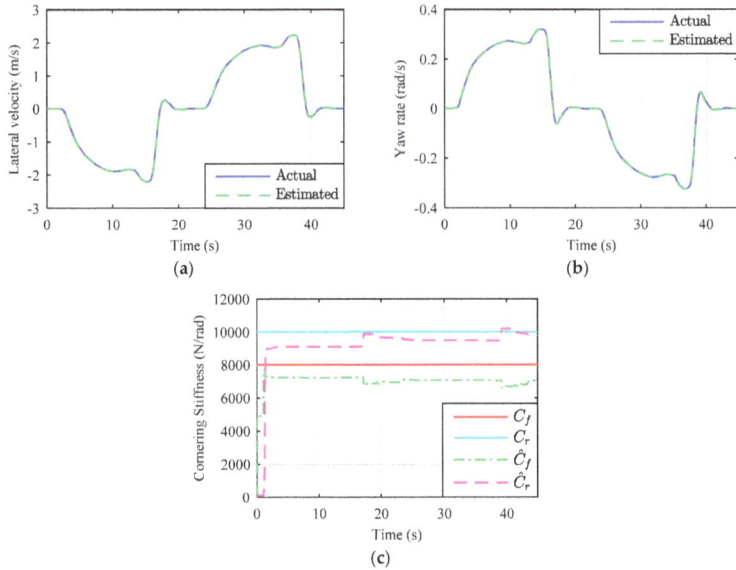

Figure 15. Estimation results of vehicle states and cornering stiffness coefficients in test 3: (**a**) Estimated lateral velocity; (**b**) Estimated yaw rate; (**c**) Estimated cornering stiffness coefficients.

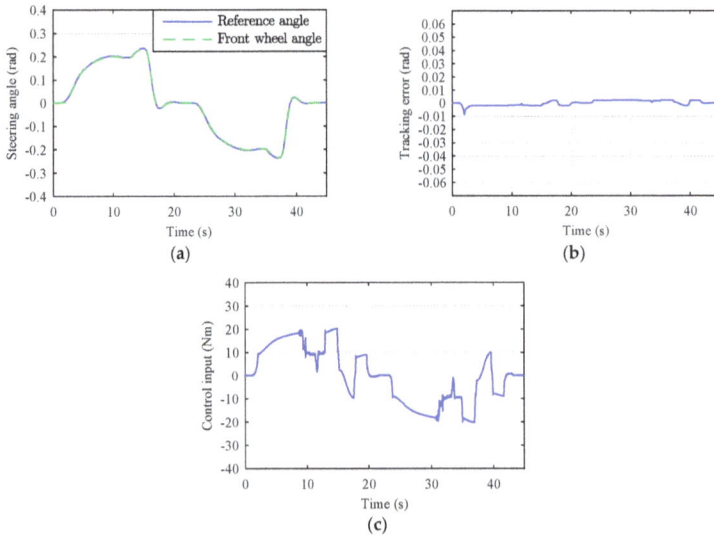

Figure 16. Control performance of the proposed AGFSMC scheme in test 3: (**a**) Tracking performance; (**b**) Tracking error; (**c**) Control input torque.

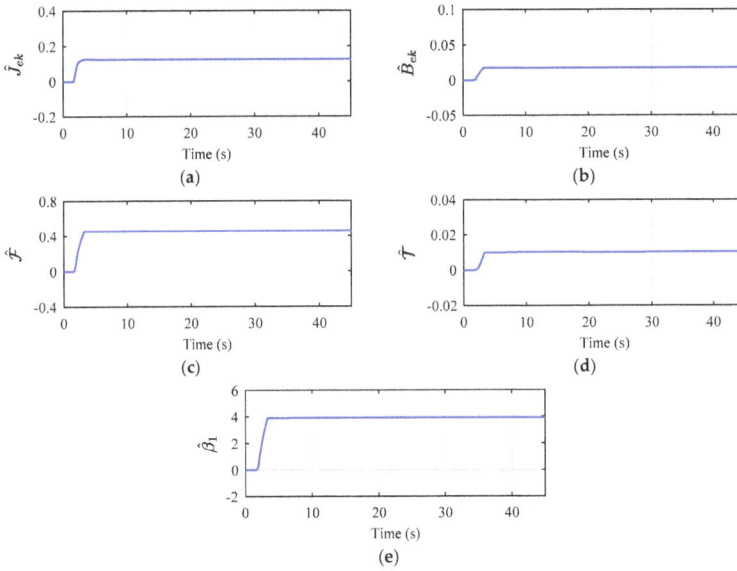

Figure 17. Estimated SbW system parameters and sliding gain with AGFSMC scheme in test 3: (**a–d**) Estimated SbW system parameters; (**e**) Estimated sliding gain.

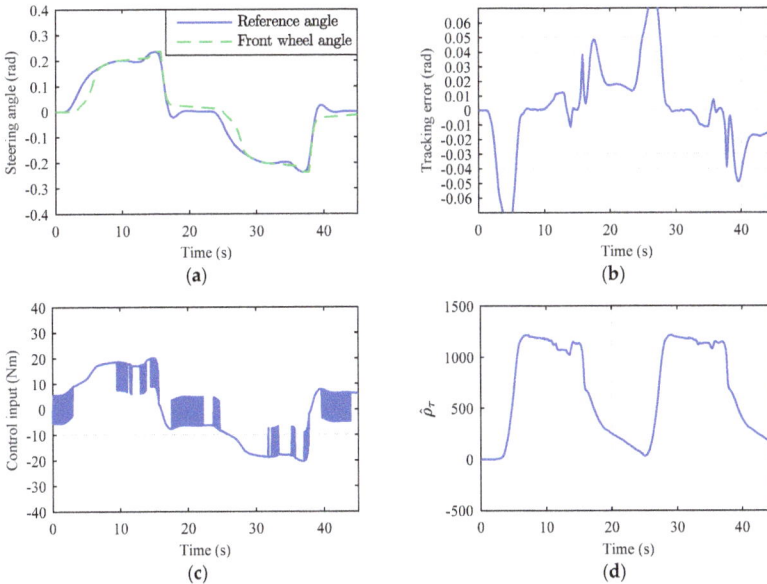

Figure 18. Control performance of adaptive sliding mode controller in test 3: (**a**) Tracking performance; (**b**) Tracking error; (**c**) Control input torque; (**d**) Estimated equivalent coefficient of self-aligning torque.

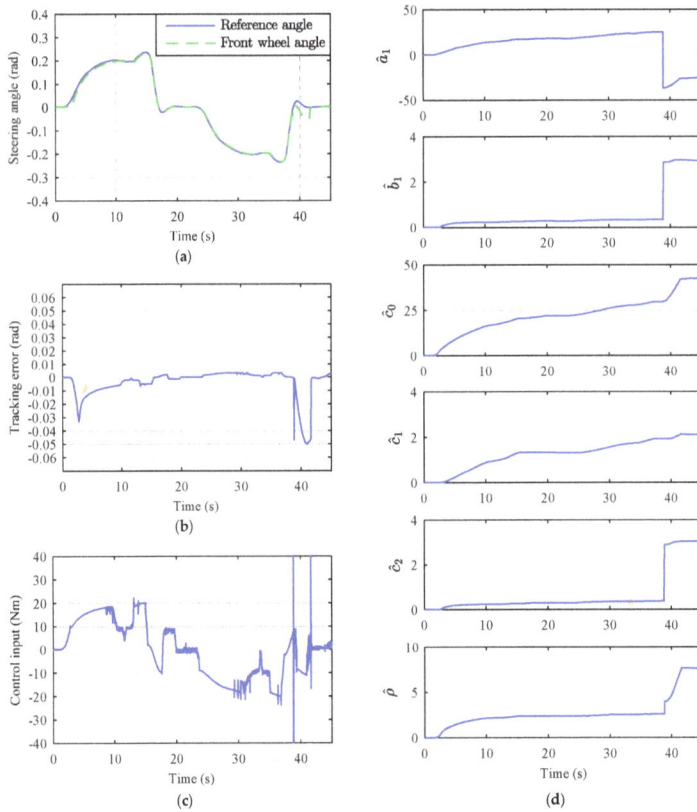

Figure 19. Control performance of adaptive terminal sliding mode control in test 3: (**a**) Tracking performance; (**b**) Tracking error; (**c**) Control input torque; (**d**) Estimated parameters.

7. Conclusions

In this paper, we have developed an AGFSMC scheme for SbW system road vehicles with unknown steering parameters. It has been demonstrated that the cooperative ASMO and KF intelligently cope with tire–road variations and simultaneously provide the best set of estimated vehicle states and cornering stiffness coefficients. Thereafter, the estimated dynamics-based AGFSMC is designed to adapt to the unknown SbW system parameters and eliminate the effect of tire–road disturbance forces. The proposed global fast terminal sliding surface guarantees the precise tracking of front wheels and ensures the asymptotic convergence of tracking error. Finally, the comparative study results are analyzed as follows:

- The adaptive features of cooperative ASMO and KF showed strong robustness against varying driving conditions in all three maneuvering tests, and estimated the sufficient dynamics to stabilize the impact of tire–road frictional torque and self-aligning torque.
- The proposed AGFSMC is proven to attain a smaller peak tracking error and faster convergence of steady-state error to smaller bounds in comparison with ASMC and ATSMC over all three maneuvers.
- The discontinuous projection mapping along with designed dead zones, also effectively managed to restrict the estimation drift problem.

Thus, the comparative study validates the remarkable steering performance of the proposed AGFSMC scheme, carried out over three different driving maneuvers. For our forthcoming work, we will investigate the dynamic behavior of nonlinear self-aligning torque and Coulomb frictional torque in extreme maneuvering conditions. In addition, we are also investigating the adaptive second-order sliding mode control for SbW system road vehicles to improve the estimation accuracy of dynamic tire–road disturbance forces under varying driving conditions.

Acknowledgments: This research was partially supported by the Higher Education Commission of Pakistan by the award letter No. HRDI-UESTPs/Batch-II/South Korea/2012.

Author Contributions: Junaid Iqbal proposed the idea, implemented it by simulations and wrote the manuscript. Mian Ashfaq Ali contributed to comparative study analysis. Khalil Muhammad Zuhaib and Abdul Manan Khan made suggestions in the manuscript correction and improvements. Changsoo Han supervised the study and manuscript writing process.

Conflicts of Interest: The authors declare no conflict of interest.

Abbreviations

The following abbreviations are used in this article:

$V_x, V_y, \dot{\psi}$	Vehicle's velocities and yaw rate
$\ddot{y}, \ddot{\psi}$	Vehicle's lateral and yaw acceleration
m, I_z	Vehicle's mass and mass moment of inertia
F_{yf}, F_{yr}	Lateral force at front and rear wheel
F_{xf}, F_{xr}	Longitudinal force at front and rear wheel
C_f, C_r	Cornering coefficient of front and rear wheel
α_f, α_r	Sideslip angle of front and rear wheel
J_{fw}, J_{fm}, J_{eq}	Moment of inertia of front wheels, actuator, and equivalent SbW system
B_{fw}, B_{fm}, B_{eq}	Viscous damping of front wheels, actuator, and equivalent SbW system
l_f, l_r	Distance of front and rear axles from center of gravity
t_p, t_m	Pneumatic and mechanical trail
τ_F, τ_a	Tire–road frictional torque and self-aligning torque
μ	Tire–road coefficient of dry friction
F_{zf}	Vertical load on front axle
k	Steering ratio
$\dot{\hat{y}}, \dot{\hat{\psi}}$	ASMO estimated vehicle states
$\hat{L}_1, \hat{L}_2$	ASMO adaptive gains
ρ_1, ρ_2	ASMO adaptation law speed adjustment parameters
$\hat{C}_f, \hat{C}_r$	KF estimated cornering coefficients
e_1, e_2, e_3	ASMO and KF estimation errors
$\varepsilon_1, \varepsilon_2, \varepsilon_3$	ASMO and KF termination bounds
δ_{fw}	Front wheel angle
δ_d	Reference angle
e_θ	Tracking error
s	Sliding surface
$\lambda_1, \lambda_2, q, p$	GFTSM surface parameters
$\hat{J}_{ek}, \hat{B}_{ek}, \hat{\mathcal{F}}, \hat{\mathcal{T}}, \hat{\beta}_1, \hat{\beta}_2$	AGFSM estimated and fast convergence parameters
$\varepsilon_4, \varepsilon_5$	AGFSM adaptation law dead zone bounds
u	AGFSM control input

References

1. SAE International Technical Standard Provides Terminology for Motor Vehicle Automated Driving Systems. Available online: http://www.sae.org/autodrive (accessed on 10 December 2016).
2. Federal Automated Vehicles Policy. Available online: https://www.transportation.gov/AV (accessed on 23 January 2017).

3. Ishida, S.; Gayko, J.E. Development, evaluation and introduction of a lane keeping assistance system. In Proceedings of the 2004 IEEE Intelligent Vehicles Symposium, Parma, Italy, 14–17 June 2004; pp. 943–944.
4. Hebert, M.H.; Thorpe, C.; Stentz, A. *Intelligent Unmanned Ground Vehicles*; Springer US: Boston, MA, USA, 1997.
5. Crane, C.D., III; Armstrong, D.G., II; Touchton, R.; Galluzzo, T.; Solanki, S.; Lee, J.; Kent, D.; Ahmed, M.; Montane, R.; Ridgeway, S.; et al. Team CIMAR's NaviGator: An unmanned ground vehicle for the 2005 DARPA grand challenge. *J. Field Robot.* **2006**, *23*, 599–623. [CrossRef]
6. How Google's Self-Driving Car Works—IEEE Spectrum. Available online: http://spectrum.ieee.org/automaton/robotics/artificial-intelligence/how-google-self-driving-car-works (accessed on 12 March 2016).
7. Gietelink, O.; Ploeg, J.; De Schutter, B.; Verhaegen, M. Development of advanced driver assistance systems with vehicle hardware-in-the-loop simulations. *Veh. Syst. Dyn.* **2006**, *44*, 569–590. [CrossRef]
8. 2014 Top Ten Tech Cars—IEEE Spectrum. Available online: http://spectrum.ieee.org/transportation/self-driving/2014-top-ten-tech-cars (accessed on 9 October 2016).
9. Top Tech Cars 2013: Infiniti Q50—IEEE Spectrum. Available online: http://spectrum.ieee.org/transportation/advanced-cars/infiniti-q50 (accessed on 9 October 2016).
10. Wang, H.; Man, Z.; Shen, W.; Cao, Z.; Zheng, J.; Jin, J.; Tuan, D.M. Robust Control for Steer-by-Wire Systems with Partially Known Dynamics. *IEEE Trans. Ind. Inform.* **2014**, *10*, 2003–2015. [CrossRef]
11. Wang, H.; Xu, Z.; Do, M.T.; Zheng, J.; Cao, Z.; Xie, L. Neural-network-based robust control for steer-by-wire systems with uncertain dynamics. *Neural Comput. Appl.* **2015**, *26*, 1575–1586. [CrossRef]
12. Wang, H.; Man, Z.; Shen, W.; Zheng, J. Robust sliding mode control for Steer-by-Wire systems with AC motors in road vehicles. In Proceedings of the 2013 8th IEEE Conference on Industrial Electronics and Applications (ICIEA), Melbourne, VIC, Australia, 19–21 June 2013; pp. 674–679.
13. Wang, H.; Man, Z.; Kong, H.; Shen, W. Terminal sliding mode control for steer-by-wire system in electric vehicles. In Proceedings of the 2012 7th IEEE Conference on Industrial Electronics and Applications (ICIEA), Singapore, 18–20 July 2012; pp. 919–924.
14. Wang, H.; Kong, H.; Man, Z.; Tuan, D.M.; Cao, Z.; Shen, W. Sliding mode control for steer-by-wire systems with AC motors in road vehicles. *IEEE Trans. Ind. Electron.* **2014**, *61*, 1596–1611. [CrossRef]
15. Do, M.T.; Man, Z.; Zhang, C.; Wang, H.; Tay, F.S. Robust Sliding Mode-Based Learning Control for Steer-by-Wire Systems in Modern Vehicles. *IEEE Trans. Veh. Technol.* **2014**, *63*, 580–590. [CrossRef]
16. Iqbal, J.; Shin, K.; Han, C. Neural network based control for steer-by-wire systems vehicles. *Adv. Circuits Syst. Signal Process. Telecommun.* 2014. Available online: http://www.wseas.us/e-library/conferences/2015/Dubai/CSST/CSST-09.pdf (accessed on 16 July 2017).
17. Kim, C.J.; Jang, J.H.; Yu, S.N.; Lee, S.H.; Han, C.S.; Hedrick, J.K. Development of a control algorithm for a tie-rod-actuating steer-by-wire system. *Proc. Inst. Mech. Eng. Part D* **2008**, *222*, 1543–1557. [CrossRef]
18. Park, T.J.; Han, C.S.; Lee, S.H. Development of the electronic control unit for the rack-actuating steer-by-wire using the hardware-in-the-loop simulation system. *Mechatronics* **2005**, *15*, 899–918. [CrossRef]
19. OH, S.W.; CHAE, H.C.; YUN, S.C.; HAN, C.S. The Design of a Controller for the Steer-by-Wire System. *JSME Int. J. Ser. C* **2004**, *47*, 896–907. [CrossRef]
20. Yih, P.; Gerdes, J.J.C. Modification of vehicle handling characteristics via steer-by-wire. *IEEE Trans. Control Syst. Technol.* **2005**, *13*, 965–976. [CrossRef]
21. Yamaguchi, Y.; Murakami, T. Adaptive Control for Virtual Steering Characteristics on Electric Vehicle Using Steer-by-Wire System. *IEEE Trans. Ind. Electron.* **2009**, *56*, 1585–1594. [CrossRef]
22. Polesel, M.; Shyrokau, B.; Tanelli, M.; Savitski, D.; Ivanov, V.; Ferrara, A. Hierarchical control of overactuated vehicles via sliding mode techniques. In Proceedings of the 2014 IEEE 53rd Annual COnference on Decision and Control (CDC), Los Angeles, CA, USA, 15–17 December 2014; pp. 4095–4100.
23. Janbakhsh, A.A.; Bayani Khaknejad, M.; Kazemi, R. Simultaneous vehicle-handling and path-tracking improvement using adaptive dynamic surface control via a steer-by-wire system. *Proc. Inst. Mech. Eng. Part D* **2012**, *227*, 345–360. [CrossRef]
24. Kazemi, R.; Janbakhsh, A.A. Nonlinear adaptive sliding mode control for vehicle handling improvement via steer-by-wire. *Int. J. Automot. Technol.* **2010**, *11*, 345–354. [CrossRef]
25. Sun, Z.; Zheng, J.; Man, Z.; Wang, H. Robust Control of a Vehicle Steer-by-Wire System Using Adaptive Sliding Mode. *IEEE Trans. Ind. Electron.* **2016**, *63*, 2251–2262. [CrossRef]
26. Sun, Z.; Zheng, J.; Wang, H.; Man, Z. Adaptive fast non-singular terminal sliding mode control for a vehicle steer-by-wire system. *IET Control Theory Appl.* **2017**, *11*, 1245–1254. [CrossRef]

27. Wang, H.; Man, Z.; Kong, H.; Zhao, Y.; Yu, M.; Cao, Z.; Zheng, J.; Do, M.T. Design and Implementation of Adaptive Terminal Sliding-Mode Control on a Steer-by-Wire Equipped Road Vehicle. *IEEE Trans. Ind. Electron.* **2016**, *63*, 5774–5785. [CrossRef]

28. Wenzel, T.A.; Burnham, K.J.; Blundell, M.V.; Williams, R.A. Dual extended Kalman filter for vehicle state and parameter estimation. *Veh. Syst. Dyn.* **2006**, *44*, 153–171. [CrossRef]

29. Hong, S.; Lee, C.; Borrelli, F. A novel approach for vehicle inertial parameter identification using a dual Kalman filter. *IEEE Trans. Intell. Ttansp.Syst.* **2015**, *16*, 151–161. [CrossRef]

30. Oh, J.J.; Choi, S.B. Vehicle velocity observer design using 6-D IMU and multiple-observer approach. *IEEE Trans. Intell. Transp. Syst.* **2012**, *13*, 1865–1879. [CrossRef]

31. Imsland, L.; Johansen, T.A.; Fossen, T.I.; Fjær Grip, H.; Kalkkuhl, J.C.; Suissa, A. Vehicle velocity estimation using nonlinear observers. *Automatic* **2006**, *42*, 2091–2103. [CrossRef]

32. Zhao, L.H.; Liu, Z.Y.; Chen, H. Design of a nonlinear observer for vehicle velocity estimation and experiments. *IEEE Trans. Control Syst. Technol.* **2011**, *19*, 664–672. [CrossRef]

33. Guo, H.; Chen, H.; Cao, D.; Jin, W. Design of a reduced-order non-linear observer for vehicle velocities estimation. *IET Control Theory Appl.* **2013**, *7*, 2056–2068. [CrossRef]

34. Du, J.; Liu, Z.; Wang, Y.; Wen, C. An adaptive sliding mode observer for lithium-ion battery state of charge and state of health estimation in electric vehicles. *Control Eng. Pract.* **2016**, *54*, 81–90. [CrossRef]

35. Woodman, O. *An Introduction to Inertial Navigation*; University of Cambridge: Cambridege, UK, 2007.

36. Xu, L.; Yao, B. Adaptive robust control of mechanical systems with non-linear dynamic friction compensation. *Int. J. Control* **2008**, *81*, 167–176. [CrossRef]

37. Slotine, J.; Li, W. *Applied Nonlinear Control*; Prentice Hall: Englewood Cliffs, NJ, USA, 1991.

38. Rajamani, R. *Vehicle Dynamics and Control*; Springer Science & Business Media: New York, NY, USA, 2011.

39. Abe, M. *Vehicle Handling Dynamics: Theory and Application*; Elsevier Science Technology: London, UK, 2015.

40. Benoussaad, M.; Sijobert, B.; Mombaur, K. Robust foot clearance estimation based on the integration of foot-mounted IMU acceleration data. *Sensors* **2016**, *16*. [CrossRef] [PubMed]

41. Simon, D. *Optimal State Estimation: Kalman, H Infinity, and Nonlinear Approaches*; Wiley-Interscience: Hoboken, NJ, USA, 2006.

42. Liu, J.; Wang, X. *Advanced Sliding Mode Control for Mechanical Systems: Design, Analysis and MATLAB Simulation*; Springer Science & Business Media: New York, NY, USA, 2012.

applied
sciences

MDPI

Article

Study on Driving Decision-Making Mechanism of Autonomous Vehicle Based on an Optimized Support Vector Machine Regression

Junyou Zhang *, Yaping Liao, Shufeng Wang * and Jian Han

College of Transportation, Shandong University of Science and Technology, Huangdao District, Qingdao 266590, China; liaoyapingsk@163.com (Y.L.); hanjianzrx@163.com (J.H.)
* Correspondence: junyouzhang@sdust.edu.cn (J.Z.); shufengwang@sdust.edu.cn (S.W.);
  Tel.: +86-139-0532-3314 (J.Z.); +86-186-0532-6013 (S.W.)

Received: 16 November 2017; Accepted: 20 December 2017; Published: 22 December 2017

Featured Application: This work is specifically applied to the driving decision-making system of autonomous vehicles, allowing autonomous vehicles to run safely under complex urban road environment.

Abstract: Driving Decision-making Mechanism (DDM) is identified as the key technology to ensure the driving safety of autonomous vehicle, which is mainly influenced by vehicle states and road conditions. However, previous studies have seldom considered road conditions and their coupled effects on driving decisions. Therefore, road conditions are introduced into DDM in this paper, and are based on a Support Vector Machine Regression (SVR) model, which is optimized by a weighted hybrid kernel function and a Particle Swarm Optimization (PSO) algorithm, this study designs a DDM for autonomous vehicle. Then, the SVR model with RBF (Radial Basis Function) kernel function and BP (Back Propagation) neural network model are tested to validate the accuracy of the optimized SVR model. The results show that the optimized SVR model has the best performance than other two models. Finally, the effects of road conditions on driving decisions are analyzed quantitatively by comparing the reasoning results of DDM with different reference index combinations, and by the sensitivity analysis of DDM with added road conditions. The results demonstrate the significant improvement in the performance of DDM with added road conditions. It also shows that road conditions have the greatest influence on driving decisions at low traffic density, among those, the most influential is road visibility, then followed by adhesion coefficient, road curvature and road slope, while at high traffic density, they have almost no influence on driving decisions.

Keywords: autonomous vehicle; driving decision-making mechanism; road conditions; support vector machine regression; PSO algorithm

---

1. Introduction

With the current rapid economic growth, vehicle ownership is fast increasing, accompanied by more than one million traffic accidents per year worldwide. According to statistics, about 89.8% of accidents are caused by driver's wrong decision-making [1]. So, in order to alleviate traffic accidents, autonomous vehicles have been the world's special attention for its non-driver's participation. Key issues in researching autonomous vehicle include autonomous positioning, environmental awareness, driving decision-making, motion planning, and vehicle control [2]. As an important manifestation of the intelligent level of autonomous vehicles, the driving decision-making has currently become the focus and difficulty for experts in the study of autonomous vehicle [3]. For autonomous vehicle, it needs to rely on driving decision-making mechanism (DDM) to decide accurate driving strategy [4].

Collecting and extracting traffic scene feature by sensors and based on the driving rules, it could not only make accurate driving decisions, but also drive safely in complex traffic environment.

In recent years, many scholars have devoted themselves to the research of DDM for autonomous vehicles. Suh et al. [5] established vehicles' desired steering angle model and longitudinal acceleration model based on vehicle states, and developed a control algorithm for the driving model. Wang et al. [6] established a DDM for car following, free driving, and lane changing, with the decision tree algorithm only considering vehicles' running states on the road. To improve the disadvantage of lacking flexibility existing in the decision tree algorithm, Zheng et al. [7] used traditional artificial neural network to substitute the decision tree algorithm and trained an ANN (Artificial neural networks) driving decision-making model. Noh and An [8] presented a driving decision-making framework for automated driving in highway environment, which considers the interactions between the subject and surrounding vehicles. The previous research works mainly took vehicle states as the reference indexes of DDM, and ignored the influence of road conditions on the driving decision-making.

The empirical studies show that road conditions have a great influence on driving decision-making, including weather-related and road geometry related factors [9,10]. For example, reducing road visibility will change the traffic flow dynamics [11], and changing the geometric layout of the road will easily lead to changes in driving behavior [12]. Hamdar et al. [10] analyzed the impact of road conditions on vehicles' longitudinal operation, and found that extreme environmental conditions could increase the extent to which a vehicle deviated from normal driving behavior. Hoogendoorn et al. [11] conducted a series of driving simulations and found that driving in foggy weather led to the lower speed and acceleration, as well as to consider a larger distance from the lead vehicle. Broughton et al. [13] studied the car following decision-making under three visibility conditions, and the results showed that low visibility could reduce driver's risk identification ability. Wang et al. [14] analyzed the impact of road curvature and slope on car following behavior. They found that when driving on road with different slopes and curvature, the car-following characteristics of vehicles varied greatly, and then established a car following model while considering curve and slope. Olofsson et al. [15] investigated optimal maneuvers for vehicles on different road surfaces, such as asphalt, snow, and ice, and found that there were fundamental differences in the optimal maneuvers depending on tire-road characteristics. Previous studies have shown that driving in abnormal road conditions would increase the incidence of traffic accidents that are caused by incorrect driving behavior. So, road conditions, including road curvature, slope, visibility, and friction coefficient, are important parameters for DDM.

At present, most researches use neural network, decision tree model, and mathematical model to build DDM, but these methods need a large sample size or workload, and their prediction accuracy needs to be improved [5–8]. Support Vector Machine (SVM) is a widely accepted machine learning method with strong generalization ability; it can use nonlinear methods to map the variables to be classified (SVC, support vector machine classification) or regressed (SVR, support vector machine regression) into higher or more infinite dimensional feature spaces. But, in the classification problem, SVR adopts the same principle as SVC, and SVR can predict the value of infinite possible output. In addition, the tolerance margin is set up in SVR to approximate the most accurate classification results [16]. Although SVR is more complex than SVC, it has better flexibility in solving the multi-classification problem. Therefore, in this paper, the SVR algorithm is used to predict the multi-driving decision, and the output threshold range is set for each driving decision. In the research of driving decision, SVR is mainly used to car following behaviors [17,18], driving risk assessment [19], and so on. At present, most researches use SVR to solve the problems by specifying the kernel functions directly [17–19], but sometimes it may be found that the kernel function does not match the target problem. Therefore, in order to make the objective problem automatically choose the optimal kernel function, a new kernel function is proposed to optimize the SVR model.

So, in this paper, by simultaneously referring vehicle states and road conditions, an optimized SVR model is developed to obtain the inherent complexity of driving decisions, including car following, lane changing, and free driving. Specifically, this study makes the following contributions:

(1) A detailed analysis of DDM for autonomous vehicles is conducted, which suggests that the control maneuvers of autonomous vehicle depend on the extracted traffic environment feature, not only including vehicle states, but also road conditions.

(2) A SVR model, optimized by a weighted hybrid kernel function and particle swarm optimization (PSO) algorithm, is developed to establish DDM for autonomous vehicle. In order to validate the effectiveness of the optimized SVR model, the SVR model with a single RBF kernel function and BP neural network (BPNN) model are tested to compare with it.

(3) By comparing the reasoning results of DDM with different reference index combinations, and by the sensitivity analysis, the effect of road conditions on driving decisions is quantitatively evaluated.

2. The Driving Decision-Making Process of Autonomous Vehicle

As shown in Figure 1, with the sensor equipment, the autonomous vehicle can sense and collect traffic information, including vehicle states and road conditions in real time, to input them into the designed data processing program for some data processing to obtain the input variables of DDM.

According to these input variables, the DDM searches the relevant information and matches the accurate driving decision with the learning experiences, and then transmits the decision order to the control system. These learning experiences refer to the driving decision-making rules in DDM that are obtained by learning a lot of real driving experience. Then, the control system will control the actuators (include the steering system, pedals, and automatic gearshift) to carry on with the corresponding operation.

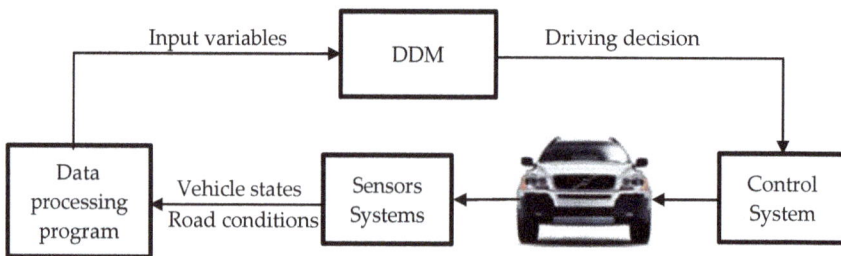

Figure 1. Schematic architecture of the driving decision-making process of autonomous vehicle. DDM: Driving Decision-making Mechanism.

In the whole process of information collection, transmission, and execution, the DDM plays a key role, which is the central system to control the autonomous vehicle. The types of driving decision DDM outputs include free driving, car following, and lane changing. Its input variables are obtained through the preliminary data processing for extracting traffic scenario characteristics as reference indexes and the further data fusion. The method of data fusion adopted in this paper is Principal Component Analysis (PCA). The whole detailed data processing steps in the data processing program are described in Figure 2.

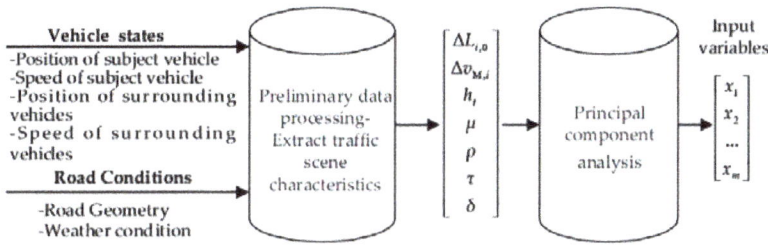

Figure 2. The detailed data processing of data processing program.

The schematic diagram of vehicle states on a road is shown in Figure 3. All of the above obtained reference indexes in Figure 2 are described as follows:

$\Delta L_{i,0}$/(m): The gap difference between L_i and safe distance L_0, and L_i refers to the distance between the subject vehicle M and vehicle i, $i = \{\text{Mbf}, \text{Mlf}, \text{Mlb}\}$;

$\Delta v_{M,i}$/(m/s): The relative speed between vehicle M and vehicle i;

h_t/(s): The time headway of current lane;

μ: Road adhesion coefficient, dimensionless;

ρ/(m^{-1}): Road curvature;

τ: Road slope, percentage; and,

δ/(m): Road visibility.

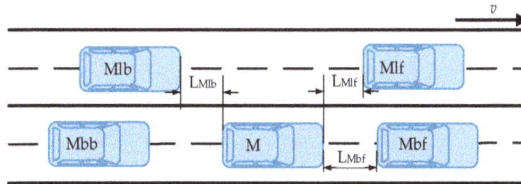

Figure 3. Diagrammatic sketch of vehicle states.

2.1. Support Vector Machine Regression Model

SVR model is a kind of machine learning method based on statistical learning theory, which can improve the generalization ability of learning machine by seeking the minimum structural risk [16,20]. So, SVR model has been widely applied and developed in the fields of pattern recognition, regression analysis, and sequence prediction [18,21].

Let $SV = \{(x_1, y_1), (x_2, y_2), \ldots, (x_m, y_m)\}$ be a set of m training samples, each of samples x_m is the input variable, which is obtained from traffic environment features. y_m is the output driving decision corresponding to x_m. These training samples are fitted by $f(x) = w^T x + b$, and all of the fitted results must be satisfied with error accuracy ε, i.e.,:

$$\left| w^T x_i + b - y_i \right| \leq \varepsilon, i = 1, 2, \ldots, m \tag{1}$$

According to the minimization criteria of structural risk, $f(x)$ should make $\frac{1}{2}\|w\|^2$ minimum. When considering the exiting fitted errors, the relaxation factors are introduced as $\xi_j \geq 0$, $\xi_j^* \geq 0$. The best regression result can be derived from the minimum extreme value of the following function:

$$\Phi(w, \xi_j, \xi_j^*) = \frac{1}{2}\|w\|^2 + C\sum_{i=1}^{m}(\xi_j + \xi_j^*) \tag{2}$$

where C is the penalty factor value, $C > 0$.

Then, adopt the dual principle, and set the Lagrange multiplier α, α^* to establish the Lagrange equation. Through drafting the parameters ω, b, ξ_j, ξ_j^* and making the drafted formulas equal to 0, the regression coefficient ω and constant term b can be obtained:

$$
\begin{cases}
\omega = \sum_{i=1}^{m} (\alpha_i - \alpha_i^*) x_i \\
b = \frac{1}{m} \left\{ \sum_{\substack{0 < \alpha < C \\ 0 < \alpha^* < C}} \left[y_i - \sum_{x_j \in SV} (\alpha_j - \alpha_j^*) \cdot (x_i - x_j) - \varepsilon \right] \right\}
\end{cases}
\tag{3}
$$

After that, the results are substituted into the function $f(x)$ to get the regression function:

$$
f(x) = \sum_{i=1}^{m} (\alpha_i - \alpha_i^*)(x_i \cdot x) + b
\tag{4}
$$

Finally, the original samples are mapped into a high-dimensional feature space with a kernel function $K(x_i, x)$, and calculate the parameters with the same method, as above. The obtained non-linear regression function is:

$$
f(x) = \sum_{i=1}^{m} (\alpha_i - \alpha_i^*) K(x, x_i) + b
\tag{5}
$$

The common kernels are showed as following:

Polynomial kernel function: $K_{Poly}(x, x_i) = (a(x \cdot x_i) + b)^d$

Radial basis function: $K_{Rbf} = exp(-\frac{\|x - x_i\|^2}{\sigma^2})$

Sigmoid kernel function: $K_{Sig} = tan(\tau(x \cdot x_i) - \delta)$

Where the dot denotes the inner-product operation in Euclidean space, d is the degree of polynomial kernel, σ is the constant term determining the width of RBF kernel [18]. With different kernels, it can be structured by different regression surfaces, then different training results may be gotten on driving decision-making. So, it is important to select the proper kernel function and kernel parameters in the SVR model.

2.2. The Optimized Support Vector Machine Regression Model

2.2.1. The Selection of Kernel Function

In the research field of SVR model, the selection of kernel function type is the most popular research problem. The kernel function adopted by most of SVR research is the RBF kernel function. But, for different specific problems, the selected kernel function can reflect some of the characteristics of the problem itself [22]. The kernel function specified by researchers based on experience may not be the best choice for specific problems. So, this requires some ways to choose the optimal kernel function for them. In this paper, in order to avoid complexity and one-sidedness of the selection, and to give full play to the benefits that are brought by various kernel functions for the DDM, a weighted hybrid kernel function is proposed:

$$
K(x, x_i) = \beta_1 [K_{Poly}(x, x_i)]^{e_1} + \beta_2 [K_{Rbf}(x, x_i)]^{e_2} + \beta_3 [K_{Sig}(x, x_i)]^{e_3}
\tag{6}
$$

where $0 \leq \beta_j \leq 1$, $e_j \in R$, $j = 1$, 2, 3, respectively, refer to the weight factor and exponential factor corresponding to each kernel function. Then, combine the exponential factor e_1, e_2 with d and σ respectively, we can simplify this formula:

$$K(x, x_i) = \beta_1 K_{\text{Poly}}(x, x_i) + \beta_2 K_{\text{Rbf}}(x, x_i) + \beta_3 [K_{\text{Sig}}(x, x_i)]^{e_3} \tag{7}$$

The weighting factor needs to be satisfied:

$$\beta_1 + \beta_2 + \beta_3 = 1 \tag{8}$$

When $\beta_j = 0$, it represents that the corresponding kernel function does not play a role in DDM. When $\beta_1 = 1$, $e = 1$ and $\beta_{2,3} = 0$, then the expression of the formula is similar with the primitive type of Polynomial Kernel.

2.2.2. Parameter Optimization

Particle swarm optimization (PSO) algorithm is a new evolutionary and iterative optimization algorithm developed in recent years. PSO algorithm is also started from the random solution and the quality of its solution is evaluated by the fitness. It finds the global optimum following the optimal particles in the solution space [23]. PSO algorithm has a fast convergence rate, and can avoid falling into the local optimum [24,25]. So, in this paper, we adopt PSO algorithm to optimize the undetermined parameters of the SVR model and the weighted hybrid kernel function.

In the PSO algorithm, particles dynamically adjust their positions in the n-dimensional space through their individual and peer flight experience. In n-dimensional space, the number of particles is l, and the position of particle i can be represented as $x_i = [x_{i1}, x_{i2}, \ldots, x_{in}]$, and its flying speed is $v_i = [v_{i1}, v_{i2}, \ldots, v_{in}]$. The best position visited by the particle i so far can be noted as the particle best, i.e., $Pbest_i = [Pbest_{i1}, Pbest_{i2}, \ldots, Pbest_{in}]$, and the best position found by all the particles so far can be noted as the global best, i.e., $Gbest = [Gbest_{i1}, Gbest_{i2}, \ldots, Gbest_{in}]$. At every moment t, the particle will adjust its speed and position by:

$$v_{id}^{t+1} = \mu v_{id}^t + c_1 rand(0, 1)(Pbest_{id} - x_{id}^t) + c_2 rand(0, 1)(Gbest_{id} - x_{id}^t) \tag{9}$$

$$\begin{cases} v_{id} = v_{\max}, & v_{id} > v_{\max} \\ v_{id} = -v_{\max}, & v_{id} < -v_{\max} \end{cases} \tag{10}$$

$$x_{id}^{t+1} = x_{id}^t + v_{id}^{t+1} \tag{11}$$

where $i = 1$, 2, $\ldots$, l, $d = 1$, 2, $\ldots$, n, $v_{\max}$ is the limited maximum flying speed, and $rand(0, 1)$ is the uniform random number on the interval $[0, 1]$, it can increase the searching randomness of particles based on the *Pbest* and the *Gbest*.

Then, the PSO-SVR parameter optimization architecture is established in Figure 4. We set the updated step factor as $\mu = 1$ and the positive acceleration coefficients of particle as $c_1 = c_2 = 2.0$. The limited maximum flying speed $v_{\max}$ is set to 100, and the number of particles l is 50. The number of undetermined parameters is 9, including β, e in selecting kernel function type, parameters of each single kernel function a, b, d, σ, τ, δ, and SVR penalty factor C, it is represented as the dimension of the particle space. The parameter of kernel function σ and penalty factor C are limited in the value range $(-10, 10)$.

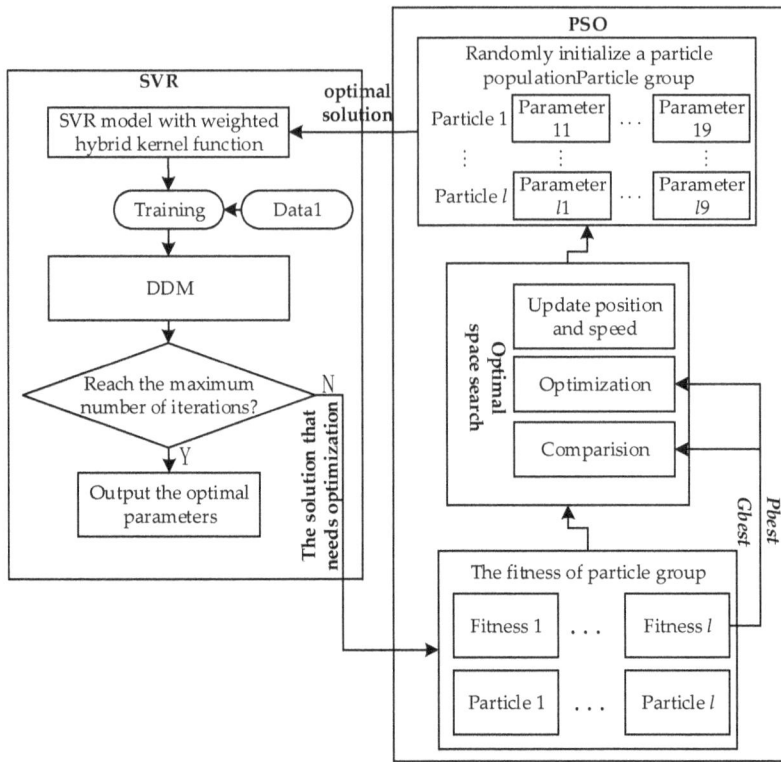

Figure 4. The steps of Particle Swarm Optimization (PSO)—Support Vector Machine Regression (SVR) parameters optimization architecture.

The optimization steps are given as follows:

Step 1   randomly initialize the positions and speeds of all particles;

Step 2   the fitness value of each particle is calculated according to the fitness function of driving decision problem;

Step 3   respectively compare the fitness value of each particle with their own *Pbest* and *Gbest*. If the fitness value is larger than *Pbest*, then update *Pbest* with the fitness value. If the fitness value is larger than *Gbest*, then update *Gbest* with the fitness value;

Step 4   for each update, reset the SVR penalty factor *C* to create a larger research space for particles, avoid falling into the local area of current optimal value;

Step 5   update the position and speed of each particle according to Formulas (8) and (9); and,

Step 6   when the number of iterations reaches the maximum set, stop it and output the optimal parameters. Otherwise, return to Step 2.

In this paper, set the training accuracy as the fitness function in the optimized process. In order to evaluate the predicting effect of model for each driving decision, the average absolute error E_M and relative mean square error E_R are selected as the comprehensive evaluation indexes. The former can reflect the degree of deviation between reasoning and measured values, and the latter is the changing embodiment of the error values, which reflects the output stability of SVR model.

$$E_M(x_i) = \frac{1}{m} \sum_{i=1}^{m} \frac{|y_i - f(x_i)|}{y_i} \times 100\% \tag{12}$$

$$E_R(x_i) = \sqrt{\frac{1}{m}\sum_{i=1}^{m}\left(\frac{|y_i - f(x_i)|}{y_i}\right)^2} \times 100\% \tag{13}$$

3. Experimental Set-Up

A driving experiment needs to be set up to collect relevant data for training the optimized SVR model. Driving simulation is an alternative on-road experiment when the driver desires to use more controllable traffic scenarios to manipulate under certain experimental conditions. By adjusting the light, brightness, motion, audio, etc. in the simulator, it can represent a real traffic scene and an actual vehicle for the driver, which is used to study driving behaviors safely. From the output data, we can obtain the trajectory data of the subject and surrounding vehicles, which are useful to analyze driving decisions.

3.1. Driving Simulator

Driving simulation experiment is performed using the UC-win/Road 12.0 driving simulator platform (12.0 version, Fulamba Software Technology Co., Ltd., Shanghai, China, 2016) at the intelligent transportation experimental center of Transportation College in Shandong University of Science and Technology, which is shown in Figure 5. The hardware is made up of three networked computers and some interfaces, such as the steering system, pedals and the automatic gearshift. The traffic environment is projected onto a large visual screen (Fulamba Software Technology Co., Ltd., Shanghai, China) (this big screen is made up of 3 sub-screens), which can provide a 135° field of view. The resolution of visual scene is 1920 × 1080, the refresh rate of the scene is 20–60 Hz depending on the complexity traffic environment. The simulator can record the position coordinates, speed, acceleration of the subject vehicle, and the surrounding vehicle in real time.

Figure 5. Traffic Simulation Scene of Simulated Driving Test.

3.2. Participants

A total of 31 drivers with different driving experiences are recruited for experiment, including 19 male and 12 female drivers. Before performing driving simulation experiments, a survey for all of the participants is conducted, which is mainly focused on personal driving habits, driving experience, car accident history, physical and psychological status, etc. The average age of the participants is 25.7 years old (std is 3.91 years), ranging from 23 to 37 years. All of the participants have a qualified driver's license, and more than five years of driving experience (std is 4.33 years). None of participant has any visual and psychological problems. Among 31 participants, three participants (two males, one female) had car crashes in the past five years. The participants are trained to be familiar with the driving simulated operation and to complete the driving simulation on all the traffic environments as required.

3.3. Driving Scenario Setting

A two-way with four-lane urban road section is established for this experiment, as shown in Figure 5. Setting different parameters for vehicles, roads, and traffic, we can establish different traffic simulated scenarios. Set all the vehicles running on these scenarios as standard cars, and the traffic density range to 4–32 veh/km (note: 4–16 veh/km is the low density range, 16–28 veh/km is the middle density range, and 28–32 veh/km is the high density range). The traffic flow is running randomly at each density range with a desired speed of 40–50 km/h. The reference values of the road parameters are shown in Table 1, and the initial set of road parameters are standard values, i.e., $(\mu, \rho, \tau, \delta) = (0.75, 0, 0, 1000)$. The data acquisition frequency is 10 Hz.

Table 1. Settings of Road Parameters.

Road Parameters	Setting Values
Adhesion coefficient μ	0.75/0.55/0.28/0.18
Road curvature ρ (m^{-1})	0/3/1.67/1
Road slope τ	0/2%/4%/6%
Road visibility δ/m	1000/500/100/50

3.4. Data Acquisition and Preprocessing

3.4.1. Data Acquisition

The collected data include driving trajectory data of the subject and its surrounding vehicles, their speeds and road environment parameters. According to the following method, the useful driving trajectory data of each driving decision are extracted and classified into the driving decision data set:

(1) Lane changing: The driving trajectory data of 10 s before implementing lane changing are recorded in lane changing data set.
(2) Car following: The driving trajectory data within the 50 m gaps between the subject and its leading vehicle are recorded in car following data set.
(3) Free driving: The driving trajectory data beyond 50 m gaps between the subject and its leading vehicle, and the driving trajectory data output when the subject vehicle with the desired speed are recorded in free driving data set.

After data classification and statistics, a total of 3211 groups of free driving data, 5312 groups of car following data, and 1009 groups of lane changing data are obtained. Each group of driving decision data includes one group of the driving trajectory data, together with their corresponding speeds and the road environment parameters.

3.4.2. Preliminary Data Process

In the preliminary data process, the data contained in all driving decision data sets are calculated to obtain the driving decision samples. From the driving trajectory data, we can obtain $\Delta L_{M_{bf},0}$, $\Delta L_{M_{lf},0}$, $\Delta L_{M_{bl},0}$, h_t and the driving decision (free driving, car following or lane changing), from the speed information, we can obtain $\Delta v_{M,M_{bf}}$, $\Delta v_{M,M_{lf}}$, and $\Delta v_{M,M_{bl}}$, from the road environment parameters, we can obtain the values of μ, ρ, τ, δ. One sample includes one reference index vector $H = \left[\Delta L_{M_{bf},0}, \Delta L_{M_{lf},0}, \Delta L_{M_{bl},0}, \Delta v_{M,M_{bf}}, \Delta v_{M,M_{lf}}, \Delta v_{M,M_{bl}}, h_t, \mu, \rho, \tau, \delta\right]$ and its corresponding driving decision.

3.4.3. The Output and Input Variables of the Optimized SVR Model

(1) The output variables

In this paper, the output variable of the optimized SVR model is a driving decision, may be free driving, car following, or lane changing. We assign the represented values and the output threshold ranges to all of the driving decisions, as seen in Table 2. For example, if an output value of DDM falls within the threshold range $(-1.5, 0.5)$, it represents that the driving decision is free driving.

Table 2. Driving Decision-Making Behaviors.

Driving Decision	Symbol	Represented Value	Threshold Range of the Output Value
Free driving	y_1	-1	$(-1.5, 0.5)$
Car following	y_2	0	$(-0.5, 0.5)$
Lane Changing	y_3	1	$(0.5, 1.5)$

(2) The input variables

Solving practical problems often need to collect a lot of indexes to reflect more information about the research object. If the correlation between these indexes is high, then the information reflected from them will have a certain overlap, which will increase the complexity of processing information. To solve this problem, Principal Component Analysis (PCA) is proposed to analyze data indexes and obtain the needed input variables [26].

PCA is a statistical analysis method. It can transform multiple correlated indexes into a few of uncorrelated indexes. The comprehensive indexes, called the principal components, will keep the original indexes information as much as possible. If there is a p-dimensional random vector $f = (f_1, f_2 \ldots, f_p)'$, using PCA, the p reference indexes can be transformed into a set of uncorrelated principal indexes $x_1, x_2, \ldots, x_p$ as their principal components, as seen in (14).

$$\begin{cases} x_1 = a_{11}f_1 + a_{12}f_2 + \ldots + a_{1p}f_p \\ x_2 = a_{21}f_1 + a_{22}f_2 + \ldots + a_{2p}f_p \\ \ldots\ldots\ldots\ldots\ldots\ldots\ldots\ldots\ldots\ldots\ldots\ldots\ldots \\ x_p = a_{p1}f_1 + a_{p2}f_2 + \ldots + a_{pp}f_p \end{cases} \tag{14}$$

Then, $m(m < p)$ principal components need to be selected from above p principal components to adequately reflect the information represented by f_p. The number of principal components m depends on the cumulative contribution rate of the variance $G(m)$.

$$G(m) = \sum_{i=1}^{m} \lambda_i / \sum_{k=1}^{p} \lambda_k \tag{15}$$

where λ_i is the eigenvalue of x_i.

Usually, when $G(m) > 85\%$, these m principal components can adequately reflect the information of the original p reference indexes.

Then, we use PCA to make the correlation analysis of 11 reference indexes through 200 sets of samples. The analysis process of PCA is shown in Figure 6. The calculated results of PCA for each principal component are shown in Figure 7. According to the cumulative contribution rate of the variance of each principal component, the first five principal components $\mathbf{X} = [x_1, x_2, \ldots, x_5]$ are selected as the input variables of the optimized SVR model.

$$H = \left[\Delta L_{M_{bf},0}, \Delta L_{M_{lf},0}, \Delta L_{M_{bl},0}, \Delta v_{M,M_{bf}}, \Delta v_{M,M_{lf}}, \Delta v_{M,M_{bl}}, h_1, \mu, \rho, \tau, \delta\right]$$

Min-max standardization: all values are limited within (0,1)

Establish a standardized variables covariance matrix

Eigenvalues: $\lambda_i (i = 1,2,...11)$

Feature vector: $U = [u_1, u_2, ...u_{11}]$

Calculate the variance contribution rate:
$$g_i = \lambda_i \bigg/ \sum_{k=1}^{11} \lambda_k$$

Results matrix: $W = HU$

Determine the m vectors of the corresponding principal

$G(m) \geq 85\%$ N

Output new data samples

Y

The number of principal components is m

The optimized SVR model

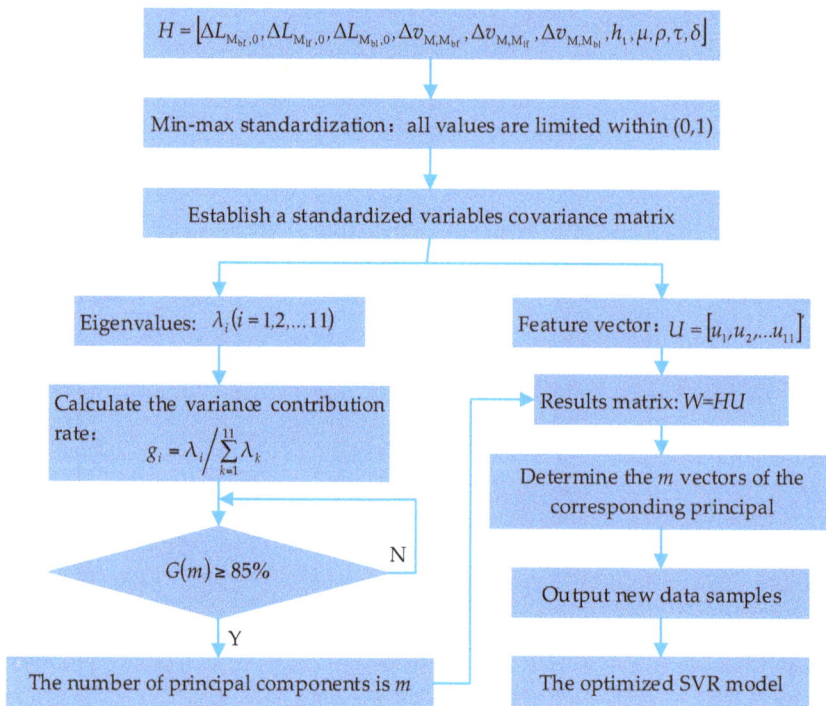

Figure 6. The analysis process of driving decision samples using Principal Component Analysis (PCA).

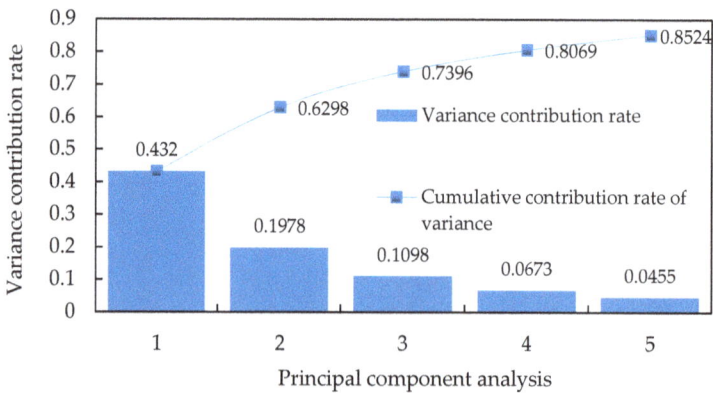

Figure 7. The results of principal component analysis for 11 reference indexes.

4. The Performance of the Optimized SVR Model on Driving Decision-Making

4.1. The Performance of the Weighted Hybrid Kernel Function

In the parameter optimization process of the optimized SVR model, 75% of the driving decision samples are randomly selected for training, and the remaining 25% samples are used for model validation. In order to evaluate the performance of the weighted hybrid kernel function, a SVR model with RBF kernel function is input with the same 75% samples to get its corresponding iteration results. We set to 200 the maximum number of training iterations.

With the PSO algorithm, we can obtain the weighted hybrid kernel function of the optimized SVR model, as shown in formula (16).

$$K(x, x_i) = 3 \times 10^{-3} K_{\text{Poly}}(x, x_i) + 4.21 \times 10^{-1} K_{\text{Rbf}}(x, x_i) + 5.76 \times 10^{-1} [K_{\text{Sig}}(x, x_i)]^{3.13 \times 10^{-1}} \quad (16)$$

The optimal parameters of each basic kernel function incorporated in the weighted hybrid kernel function are shown in the following Table 3. The best penalty factor $C = 5.4142$. In the SVR model with RBF kernel function, the optimal parameters are $\sigma = 1.4142$, $C = 6.0524$. The iterative comparison results of fitted values can be seen from Figure 8.

Table 3. The best parameters of each basic kernel function.

The Basic Kernel Function	Optimal Parameters					
	a	b	d	σ	τ	δ
Polynomial kernel function	12.114	3.741	0.097	-	-	-
Radial basis function	-	-	-	60.565	-	-
Sigmoid kernel function	-	-	-	-	13.255	1.651

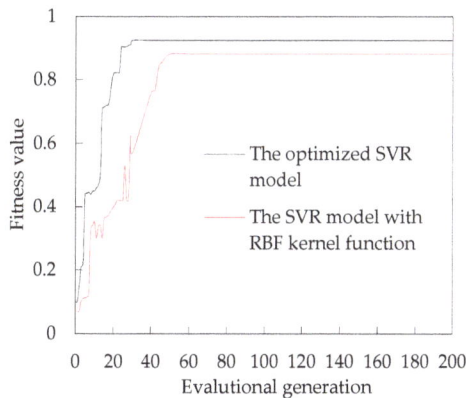

Figure 8. The iterative comparison results of fitted values of two SVR models.

It can be seen that the fitted accuracy of SVR model with weighted hybrid kernel function and RBF kernel function, respectively, are 92.3% after 31 generations and 89.7% after 43 generations. So, when compared with RBF kernel function, the weighted hybrid kernel function shows better performance on driving decision-making.

4.2. The Performance of SVR Model

BP (Back Propagation) neural network (BPNN) is one of the most widely used and successful learning algorithms in current research, and is particularly suitable for solving complex problems with internal mechanisms [27–29]. In order to verify the performance of SVR model, a typical feed-forward BPNN is established to compare with SVR model on the performance of driving decision-making. The BPNN model is established with five layers (an input layer, three hidden layers, and an output layer). Set the Tan-Sigmoid function as the transfer function of BPNN model. The five principal components $X = \{x_1, x_2, \ldots, x_5\}$ obtained above are set as its input layer parameters and the corresponding driving decisions y_k is set as the output layer parameter.

In general, the number range of nodes in the hidden layers depends on the number of nodes in the input and output layer [30]. We use our sample data to check the accuracy performance of BPNNs with different number of nodes in the hidden layers, the final number of nodes in each hidden layer is determined to 7. By the parameter adjustment and the test in MATLAB, the number of iterations is determined to 500, the learning rate is 0.01, and the training goal (mean square error) is 1×10^4. Then, the same 75% samples are input into BPNN model for training to obtain the BPNN-based DDM (BPNN-DDM). In the training process, the weights and bias are adjusted continuously to suit the desired output corresponding to the reference indexes. After 48 iterations, the network converges to the desired error. Then, the remaining 25% samples are input into the trained BPNN-DDM and SVR-DDM with RBF kernel function, the reasoning results of SVR-DDM with weighted hybrid kernel function, SVR-DDM with RBF kernel function and BPNN-DDM can be seen in the Table 4.

Table 4. The reasoning results of three driving decision-making mechanism (DDMs). SVR: Support Vector Machine Regression; RBF: Radial Basis Function; BPNN: Back Propagation neural network.

Type	Driving Decision	Reasoning Accuracy Rate	E_M	E_R	Average Reasoning Time
SVR-DDM with weighted hybrid kernel function	Free driving	93.1%	0.090	0.347	0.004 s
	Car following	94.7%	0.057	0.121	
	Lane changing	89.1%	0.101	0.723	
SVR-DDM with RBF kernel function	Free driving	89.3%	0.127	0.655	0.003 s
	Car following	92.7%	0.091	0.319	
	Lane changing	86.8%	0.142	0.981	
BPNN-DDM	Free driving	89.9%	0.121	0.844	0.009 s
	Car following	91.4%	0.091	0.327	
	Lane changing	87.1%	0.138	1.362	

It can be seen from Table 4 that the SVR-DDM with weighted hybrid kernel function has the best performance in reasoning driving decisions, with the 93.1% accuracy for free driving, 94.7% accuracy for car following, and 89.1% accuracy for lane changing. The reasoning accurate of SVR-DDM with RBF kernel function for three driving decisions is 89.3%, 92.7% and 86.8%, respectively, lower than that of the SVR-DDM with weighted hybrid kernel function, this results are from the optimization of kernel function in SVR Model. When compared with the two SVR-DDMs, the decision reasoning accuracy of BPNN-DDM is lower than SVR-DDM with weighted hybrid kernel function, and has little differences with the SVR-DDM with RBF kernel function. But, the E_R values show that the reasoning stability of the SVR-DDM with RBF kernel function is better than BPNN-DDM. In addition, the three DDMs have the highest accuracy for car following decision, and the lowest accuracy for lane changing. This result may be due to the small number of samples and the complexity of lane changing itself. In summary, the above results support the superior performance of SVR than BPNN in terms of the reasoning accurate, stability, and time, so the SVR model is more suitable for driving decision-making than BPNN model.

4.3. Influence Analysis of Road Conditions on the Reasoning Accuracy of DDM

In order to verify the effects of road conditions on the accuracy of DDM, the reasoning results of three DDMs (include SVR-DDM with weighted hybrid kernel function, SVR-DDM with RBF kernel function and BPNN-DDM) with the following reference index combinations are compared:

1. vehicle states + Road conditions are used as inputs; and,
2. only vehicle states are used as inputs.

Three DDMs with the first reference index combination has already been trained and validated in the previous Table 4.

For the second reference index combination, road conditions information is eliminated from the above 75% training samples and the remaining 25% testing samples. Then, three DDMs without considering road conditions are established using the same training method and tested with the testing samples. The reasoning results of three DDMs without considering road conditions are shown in the following Table 5.

Table 5. The results of three DDMs without considering road conditions.

Type	Driving Decision	Reasoning Accuracy Rate	E_M	E_R	Average Reasoning Time
SVR-DDM with weighted hybrid kernel function	Free driving	82.3%	0.145	1.566	0.003 s
	Car following	85.9%	0.138	1.441	
	Lane changing	78.2%	0.158	1.845	
SVR-DDM with RBF kernel function	Free driving	78.5%	0.157	1.634	0.002 s
	Car following	82.2%	0.155	1.521	
	Lane changing	76.8%	0.173	1.848	
BPNN-DDM	Free driving	78.1%	0.171	1.859	0.006 s
	Car following	80.4%	0.159	1.723	
	Lane changing	75.1%	0.176	1.877	

As illustrated in Tables 4 and 5, after eliminating the information of road conditions from the reference index set, the accuracy of SVR-DDM with weighted hybrid kernel function for free driving, car following and lane changing is reduced from 93.1% to 82.3%, 94.7% to 85.9% and 89.1% to 78.2%, respectively, SVR-DDM with RBF kernel function is reduced from 89.3% to 78.5%, 92.7% to 82.2% and 86.8% to 76.8% respectively, and the BPNN-DDM is reduced from 89.9% to 78.1%, 91.4% to 80.4% and 87.1% to 75.1% respectively. The results support the effectiveness of making driving decision with road conditions. In addition, although the average reasoning time of DDMs with added road conditions is higher than that of DDMs without added road conditions, the reasoning stability of DDMs with added road conditions is much better than that of DDMs without added road conditions. In general, DDM has better performance on reasoning driving decisions with added road conditions, which is further explained that the road condition cannot be ignored in driving decision-making.

4.4. Sensitive Analysis of Road Conditions on Driving Decisions

It can be seen from the above results that road conditions have a great influence on driving decisions. But how does each parameter affect driving decisions? What is the degree of their effects on each driving decision? A solution is provided to quantitatively evaluate their effects with the SVR-DDM with weighted hybrid kernel function (all of the DDMs mentioned in the following analysis refer to the SVR-DDM with weighted hybrid kernel function and with added road conditions).

We quantitatively evaluate the effects of each road parameter on driving decisions by analyzing the sensitivity of DDM to the changes in each road parameter. We take the changes in the road adhesion coefficient μ as an example. Using the driving decision samples under standard road conditions, we first count and calculate the proportions of each driving decision at different traffic density ranges. Then, we make the μ take values at 0.55, 0.28 and 0.18, respectively. The other three road parameters remain standard. Every time that the μ changes, a new set of driving decision samples is obtained and input into the DDM. From the output of DDM, the proportion of each driving decision in different traffic density is calculated. Then, we can get the trend that the proportion of each driving decision varies with the traffic density when μ taken at 0.75, 0.5, 0.25 and 0.18, respectively. In the same way, we can also get the trend that the proportion of each driving decision varies with the traffic density when the other three road parameters take different values, respectively. After this operation and data statistics, the quantitative influence is displayed in Figure 9.

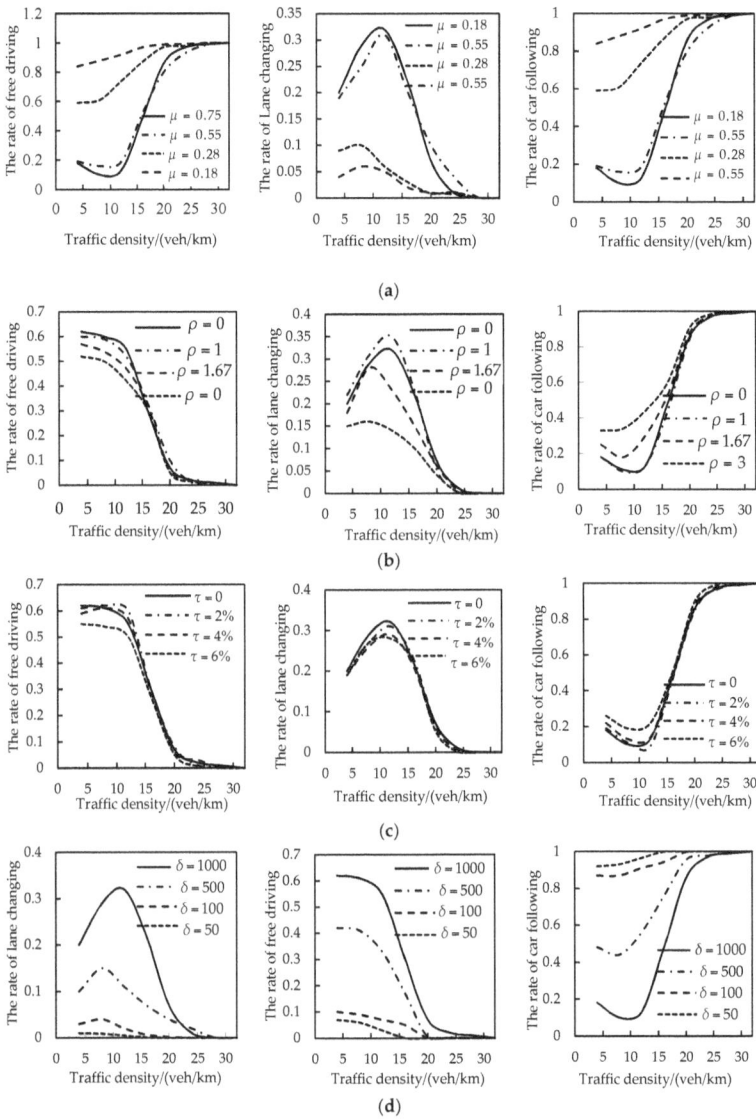

Figure 9. Driving Decision Rate under Different Road Conditions. (The horizontal axis in these diagrams represents the traffic flow density, and the vertical axis represents the rate of each driving decision (between 0 and 1). The solid lines in all diagrams represent the changing trend of the proportion of driving decisions with traffic density under standard road conditions. (**a**) The changing trend of each driving decision rate with traffic density when μ takes different values; (**b**) The changing trend of each driving decision rate with traffic density when ρ takes different values; (**c**) The changing trend of each driving decision rate with traffic density when τ takes different values; (**d**) The changing trend of each driving decision rate with traffic density when δ takes different values. From left to right, each column represents the trend of lane changing rate, free driving rate and car-following rate with traffic density when each road parameter is taken as different values, respectively.).

As shown in Figure 9, it can be seen that the changes of road conditions have the greatest influence on the driving decisions in the low traffic density range (4–16 veh/km) and almost have no influence in the high traffic density range (28–32 veh/km). In the low and middle traffic density range (4–28 km/h), road visibility δ has the greatest effect on driving decision, then followed by adhesion coefficient μ, road curvature ρ, and road slope τ. So, we can conclude: in the low traffic density range, driving decision-making is mainly restricted by the road conditions, in consequence, results are easy to be wrong without considering road conditions. on the other hand, with the high traffic density range, driving decision-making is mainly limited by vehicles states, so even if road conditions are not taken into account, the reasoning results are less affected.

Take the change rate of driving decision in low traffic density range in Figure 9b as an example, when all of the road parameters are taken as the standard values, the average rates of free driving, lane changing, and car following are about 0.469, 0.262, and 0.269, respectively, in the low traffic density range. If δ is changed to 100, the average rates of three driving decisions are changed to about 0.078, 0.034, 0.888, respectively, which means that about 61.9% of samples change their decisions when the road visibility is changed from 1000 m to 100 m. Similarly, if τ is changed to 4%, then the average rates of three driving decisions are changed to 0.515, 0.238, and 0.247, respectively, which means that about 4.6% of samples change their decisions when the road slope is changed from 0 to 4%. The same is true for the analysis of driving decisions corresponding to the changes in other two parameters. Thus, it can be seen that road conditions are important indexes that cannot be ignored in DDM for autonomous vehicle.

5. Conclusions

In this paper, a SVR model was developed to make accurate driving decisions for autonomous vehicle. Our model was optimized by a weighted hybrid kernel function and a PSO algorithm. Road conditions and vehicle states were simultaneously as the reference indexes of DDM. The driving decisions that were made by DDM included free driving, car following, and lane changing. Then, driving simulated experiments with different traffic environments were executed to extract the driving decision samples. The optimized SVR model was trained and validated with the training and testing samples to establish DDM. Our model was compared with: (1) a SVR model with RBF kernel function, and (2) BPNN model. The comparison results showed that the accuracy of our optimized SVR model was the best, with more than 92% accuracy. Besides, the results also showed that our optimized SVR model had a better performance in free driving and car following with 93.1% and 94.7% of accuracy, respectively, than lane changing decision with 89.1% of accuracy.

Finally, we investigated the effect of road conditions on the accuracy of DDM and quantified their effects on each driving decision through the sensitive analysis. The results showed that road conditions almost had almost no influence on driving decisions with high traffic density range, and had the greatest influence with low traffic density range. In the low and middle traffic density, road visibility δ has the greatest effect on the driving decisions, then followed by μ, ρ, and τ. To some extent, the verified results were consistent with the actual driving experience, which indicated the reasonability of the obtained DDM with added road conditions.

Even though the DDM based on the optimized SVR model is able to reason driving decisions, and outperforms other models that are proposed in this paper, there are still some weak points and limits, such as the sample size of lane changing decision is smaller than that of car following and free driving, and that the DDM has not yet been implemented in real road environment, we will improve them in the future. In addition, future research will focus on establishing a DDM used in dangerous driving environments, for example, if a pedestrian or vehicle suddenly present in front of the subject vehicle, then the subject vehicle should make proper driving decision, like steering, braking, or steering and braking.

Acknowledgments: This work was financially supported by the Natural Science Foundation of China under Grant no. 51678320, the Outstanding Young Scientists Research Award Foundation of Shandong Province under Grant no. ZR2016EEB06, the Scientific Research Foundation of Shandong University of Science and Technology for Recruited Talents under Grant no. 2015RCJJ035.

Author Contributions: The corresponding authors Junyou Zhang and Yaping Liao proposed this research and designed the experiments; The corresponding authors Shufeng Wang and Jian Han performed the experiments and analyzed the experimental data; Junyou Zhang was responsible to the analysis of model and drafted the manuscript with the help of Yaping Liao. Shufeng Wang and Jian Han proposed some amendment opinions and helped to improve this paper.

Conflicts of Interest: The authors declare no conflict of interest.

References

1. Lv, W.; Song, W.G.; Liu, X.D.; Ma, J. A Microscopic Lane Changing Process Model for Multilane Traffic. *Phys. A Stat. Mech. Appl.* **2013**, *392*, 1142–1152. [CrossRef]
2. Julia, N.; Brannstrom, M.; Coelingh, E.; Fredriksson, J. Lane Change Maneuvers for Automated Vehicles. *Trans. Intell. Transp. Syst.* **2017**, *18*, 1087–1096.
3. Zheng, J.; Suzuki, K.; Fujita, M. Car-following Behavior with Instantaneous Driver–vehicle Reaction Delay: A Neural-network-based Methodology. *Transp. Res. Part C Emerg. Technol.* **2013**, *36*, 339–351. [CrossRef]
4. Feng, S.M.; Li, J.Y.; Ding, N.; Nie, C. Traffic Paradox on A Road Segment Based on A Cellular Automaton: Impact of Lane-changing Behavior. *Phys. A Stat. Mech. Appl.* **2015**, *428*, 90–102. [CrossRef]
5. Suh, J.S.; Kim, B.J.; Yi, K.S. Design and Evaluation of a Driving Mode Decision Algorithm for Automated Driving Vehicle on a Motorway. *IFAC-PapersOnLine* **2016**, *49*, 115–120. [CrossRef]
6. Wang, X.Y.; Yang, X.Y. Research on Decision Making Mechanism of Driving Behavior Based on Decision Tree. *J. Syst. Simul.* **2008**, *20*, 415–419, 448.
7. Zheng, J.; Suzuki, K.; Fujita, M. Predicting driver's lane-changing decisions using a neural network model. *Simul. Model. Pract. Theory* **2014**, *42*, 73–83. [CrossRef]
8. Noh, S.; An, K. Decision-Making Framework for Automated Driving in Highway Environments. *IEEE Trans. Intell. Transp. Syst.* **2017**, 1–14. [CrossRef]
9. Geng, X.L.; Liang, H.W.; Xu, H.; Yu, B. Influences of Leading-Vehicle Types and Environmental Conditions on Car-Following Behavior. *Int. Fed. Autom. Control* **2016**, *49*, 151–156. [CrossRef]
10. Hamdar, S.H.; Qin, L.Q.; Talebpour, A. Weather And Road Geometry Impact on Longitudinal Driving Behavior: Exploratory Analysis Using An Empirically Supported Acceleration Modeling Framework. *Transp. Res. Part C* **2016**, *67*, 193–213. [CrossRef]
11. Hoogendoorn, R.G.; Hoogendoorn, S.P.; Brookhuis, K.A.; Daamen, W. Simple and multi-anticipative car-following models: Performance and parameter value effects in case of fog. In Proceedings of the Transportation Research Board (TRB) Traffic Flow Theory and Characteristics Committee (AHB45) Summer Meeting, Annecy, France, 27 June 2010; pp. 2–16.
12. McLean, J. Driver speed behavior and rural road alignment design. *Traffic Eng. Control* **1981**, *22*, 208–211.
13. Broughton, K.L.M.; Switzer, F.; Scott, D. Car Following Decisions under Three Visibility Conditions and Two Speeds Tested with a Driving Simulator. *Accid. Anal. Prev.* **2007**, *39*, 106–116. [CrossRef] [PubMed]
14. Wang, H.; Ma, S.-H. Car-following Model and Simulation with Curves and Slopes. *Civ. Eng.* **2005**, *38*, 106–111.
15. Olofsson, B.; Lundahl, K.; Berntorp, K.; Nielsen, L. An Investigation of Optimal Vehicle Maneuvers for Different Road Conditions. *IFAC Proc. Vol.* **2013**, *46*, 66–71. [CrossRef]
16. Shafizadeh-Moghadam, H.; Tayyebi, A.; Ahmadlou, M.; Delavar, M.R.; Hasanlou, M. Integration of genetic algorithm and multiple kernel support vector regression for modeling urban growth. *Comput. Environ. Urban Syst.* **2017**, *65*, 28–40. [CrossRef]
17. Wei, D.L.; Liu, H.C. Analysis of asymmetric driving behavior using a self-learning approach. *Transp. Res. Part B Methodol.* **2013**, *47*, 1–14. [CrossRef]
18. Qiu, X.P.; Liu, Y.L. A car-following model based on support vector machine. *J. Chongqing Jiaotong Univ. (Nat. Sci. Ed.)* **2015**, *34*, 128–132.
19. Yu, R.J.; Abdel-Aty, M. Utilizing support vector machine in real-time crash risk evaluation. *Accid. Anal. Prev.* **2013**, *55*, 252–259. [CrossRef] [PubMed]

20. Hong, W.-C.; Dong, Y.C.; Zheng, F.F.; Lai, C.-Y. Forecasting urban traffic flow by SVR with continuous ACO. *Appl. Math. Model.* **2011**, *35*, 1282–1291. [CrossRef]
21. Zhu, L.L.; Liu, L.; Zhao, X.P.; Yang, D. Study on Vehicle Driving Behavior Recognition Based on Support Vector Machine. *J. Transp. Syst. Eng. Inf.* **2017**, *17*, 91–97.
22. Yang, D.G.; He, C.W.; Li, M.; He, Q.G. Behavioral Recognition of Vehicle Steering and Lane Based on Support Vector Machine. *J. Tsinghua Univ. (Sci. Technol.)* **2015**, *55*, 1093–1097.
23. Gold, C.; Sollich, P. Model selection for support vector machine classification. *Neurocomputing* **2003**, *55*, 221–249. [CrossRef]
24. Rokonuzzaman, M.; Sakai, T. Calibration of the parameters for a hardening-softening constitutive model using genetic algorithms. *Comput. Geotech.* **2010**, *37*, 573–579. [CrossRef]
25. Babazadeh, A.; Poorzahedy, H.; Nikoosokhan, S. Application of particle swarm optimization to transportation network design problem. *J. King Saud Univ. Sci.* **2011**, *23*, 293–300. [CrossRef]
26. Deng, W.; Zhao, H.M.; Yang, X.H.; Xiong, J.X.; Meng, S.; Bo, L. Study on an improved adaptive PSO algorithm for solving muli-objective gate assignment. *Appl. Soft Comput.* **2017**, *59*, 288–302. [CrossRef]
27. Saha, P.; Roy, N.; Mukherjee, D.; Sarkar, A.K. Application of Principal Component Analysis for Outlier Detection in Heterogeneous Traffic Data. *Procedia Comput. Sci.* **2016**, *83*, 107–114. [CrossRef]
28. Chen, Y.; Yi, Z.-C. The BP artificial neural network model on expressway construction phase risk. *Syst. Eng. Procedia* **2012**, *4*, 409–415. [CrossRef]
29. Tang, J.J.; Liu, F.; Zhang, W.H.; Ke, R.M.; Zou, Y.J. Lane-changes prediction based on adaptive fuzzy neural network. *Expert Syst. Appl.* **2018**, *91*, 452–463. [CrossRef]
30. Peng, J.S.; Guo, Y.S.; Fu, R.; Yuan, W.; Wang, C. Multi-parameter prediction of drivers' lane-changing behaviour with neural network model. *Appl. Ergon.* **2015**, *50*, 207–217. [CrossRef] [PubMed]

applied
sciences

MDPI

Article

Cloud Incubator Car: A Reliable Platform for Autonomous Driving

Raúl Borraz *,†, Pedro J. Navarro †, Carlos Fernández † and Pedro María Alcover †

División de Sistemas en Ingeniería Electrónica (DSIE), Universidad Politécnica de Cartagena, Campus Muralla del Mar, s/n, 30202 Cartagena, Spain; pedroj.navarro@upct.es (P.J.N.); carlos.fernandez@upct.es (C.F.); pedro.alcover@upct.es (P.M.A.)
* Correspondence: raul.borraz@upct.es; Tel.: +34-968-32-6546
† These authors contributed equally to this work.

Received: 30 November 2017; Accepted: 11 February 2018; Published: 20 February 2018

Abstract: It appears clear that the future of road transport is going through enormous changes (intelligent transport systems), the main one being the Intelligent Vehicle (IV). Automated driving requires a huge research effort in multiple technological areas: sensing, control, and driving algorithms. We present a comprehensible and reliable platform for autonomous driving technology development as well as for testing purposes, developed in the Intelligent Vehicles Lab at the Technical University of Cartagena. We propose an open and modular architecture capable of easily integrating a wide variety of sensors and actuators which can be used for testing algorithms and control strategies. As a proof of concept, this paper presents a reliable and complete navigation application for a commercial vehicle (Renault Twizy). It comprises a complete perception system (2D LIDAR, 3D HD LIDAR, ToF cameras, Real-Time Kinematic (RTK) unit, Inertial Measurement Unit (IMU)), an automation of the driving elements of the vehicle (throttle, steering, brakes, and gearbox), a control system, and a decision-making system. Furthermore, two flexible and reliable algorithms are presented for carrying out global and local route planning on board autonomous vehicles.

Keywords: autonomous vehicles; intelligent vehicles; control architecture; decision-making system; global and local path planning; Bezier curves

---

1. Introduction

It appears clear that the road transport will be going through enormous changes in the near future (intelligent transport systems), although, presumably, gradual, in which new technologies will have a prevailing role [1]. Thus, in the Horizon 2020 EU work program, smart, green, and integrated transport are placed as the main pillars for transport in the future. Among the program lines the following can be distinguished: "Safe and connected automation in road transport, where specific requirements are imposed on simple automation, limiting its foreseeable scope in the short term." Without a doubt the most disruptive technological challenge is focused on the so called autonomous vehicle and, the next step, autonomous cooperative vehicles [2]. However, some issues need to be addressed: (1) the operation must be completely reliable and safe [3]: this specification is a highly demanding requirement, which usually complies with limiting the application of these vehicles to controlled scenarios, simple applications, or at reduced speeds of operation. Thus, it is understood that autonomous driving is viable in dedicated lanes or areas in situations of congestion or in parking manoeuvres, which is already a reality in many cases. Collision avoidance safety systems are more complex, especially those that take control of the vehicle in emergency manoeuvres; (2) the cost of sensors and control hardware must comply with what the market can assume, so that the mass production of vehicles can be economically affordable [4]; (3) much work is still necessary concerning legal and social issues, i.e., the question of liability where an automated vehicle is involved in an accident causing property damage or personal injury [5].

Regarding the above, it can be said that fully automated driving is not a technological utopia, but its practical implementation on the roads should be carefully analysed [6]. Researchers and manufacturers advocate for automation in successive stages, firstly approaching longitudinal control, then simple side controls, to reach complete automation in scenarios of increasing complexity. In a few words, the change from conventional to assisted driving is now a reality, and the leap is being made towards automated driving in its different stages: partial, high, or fully automated.

Intelligent vehicles (IV) are one of the key elements within intelligent transport systems. When an IV has intelligence—understood as the fact that the control relies on the onboard vehicle processing systems, it can move autonomously. The onboard vehicle intelligence becomes aware of the situation of the vehicle in its environment and performs all the necessary control actions to proceed towards its destination.

Driving automation—considered a low-risk activity for humans—is a matter that needs the continuous evaluation of two main factors: (a) the current vehicle state (position, velocity, acceleration, direction) and (b) the environmental conditions (nearby vehicles, obstacles, pedestrians, etc.). To the extent that these two factors are accurately assessed, appropriate decisions can be taken towards reliable autonomous driving, and we will be closer to a vehicle-centric [6] approach to autonomous driving, where the main goal is the safe movement of the vehicle.

It is not an easy task to define what an IV with autonomous abilities is, although it is possible to describe its capacities, as detailed in Table 1 [7].

Table 1. Capacities of an IV.

Capacity	Means to Achieve That Capacity
Acquisition of information about the environment (at different distances)	Onboard sensors (self-positioning, obstacle detection, driver monitoring)
	Network connection for long-distance information
	Vehicular interconnection
Processing of environment information	Onboard software for critical analysis
	Cloud processing for route planning, traffic control, etc.
Making driving decisions	Cognitive systems
	Agent systems
	Evolutionary algorithms

But, what exactly does autonomous driving mean? Following the policy defined by the NHTSA (National Highway Traffic Safety Administration, USA) [8], the International Society of Automotive Engineers (SAE) has set three categories of advanced driving automation, which imply the automation of both the driving actions (steering, throttle, and break) and the monitoring of the driving environment (see Table 2) [9].

Table 2. Three categories of advanced driving automation.

Category	Main Features	Fault Correction Rely on	Automated Driving Modes Covered
Level 3. Conditional Automation	All dynamic driving tasks are automated. The human driver is expected to act correctly when required.	Human driver	Some driving modes:
			Parking maneuvers
			Low speed traffic jam
Level 4. High Automation	All dynamic driving tasks are automated. The human driver is not expected to act correctly when required.	System	Some driving modes:
			High-speed cruising
Level 5. Full Automation	All dynamic driving tasks are automated for any type of road and environmental conditions usually managed by a human driver.	System	All driving modes
			Urban driving

Platforms for Autonomous Driving Development

The mass production of autonomous vehicles still has to wait because of different concerns, such as reliability, safety, cost, appearance, and social acceptance, to say just a few [10]. Advances in the state-of-the-art of sensing and control software have allowed great improvements in the reliability and safe operation of autonomous vehicles in real-world conditions, in part thanks to a great variety of development platforms [7,11–13]. Table 3 shows the main features of some of the prominent ones.

Table 3. Some platforms for autonomous driving development.

PLATFORM	Vehicle Basis	Automated Systems	Control System	Sensors	Performances	Navigation
BOSS (DARPA)	Chevrolet Thaoe	Steering Brake Throttle	Proprietary (based on multiprocessor system)	1 HD 3D LIDAR 4 Cameras 6 radars 8 2D LIDAR 1 Inertial GPS navigation system.	Obstacle avoidance Lane keeping Crossing detection	Anytime D* Algorithm
JUNIOR (DARPA)	VW Passat wagon	Steering Brake Throttle	Proprietary (based on multiprocessor system)	1 HD 3D LIDAR 4 Cameras 6 radars 2 2D LIDAR 1 Inertial GPS navigation system.	Object detection Pedestrian detection	Moving Frame
BRAiVE (VisLab)	VW Passat	Steering Brake Throttle	Proprietary dSpace	10 cameras 4 lasers 16 laser beams 1 radar 1 Inertial GPS navigation system.	Close loop manoeuvring	NA
AUTOPIA	CITRÖEN C3 Pluriel	Steering Brake Throttle	Proprietary ORBEX (based on fuzzy coprocessor)	2 front cameras 1 RTK-DGPS+IMU	Following a leading vehicle Pedestrian detection	Fuzzy logic

The latest approach to autonomous driving is the so-called cloud-based autonomous driving [14,15]. These clouds provide essential services to support autonomous vehicles. Currently, these services include simulation tests for new algorithm deployment, offline deep learning model training, and High-Definition (HD) map generation. Autonomous driving clouds have infrastructures that include distributed computing, distributed storage, as well as heterogeneous computing. When driving algorithms and strategies are ready, they must be implemented in real control systems for adjusting and testing purposes before getting the car into real traffic conditions. Reliable control systems will be very useful for this purpose. This implies the interest of having a hardware–software development platform for control purposes that can be easily adapted to any car, thus allowing the autonomous driving system developers to concentrate exclusively on the sensory and algorithmic aspects of the prototypes. On the other hand, it is interesting to continue integrating new sensing techniques that allow a better and faster characterization of the environment, thus making possible to develop driving algorithms with a more precise behaviour.

The rest of the manuscript has been divided into three sections. Section 2 details the mechanical, electrical, and electronics modifications carried out to create the CICar platform. Section 3 shows the three components of high-level architecture proposed: a control system, a perception system, and a decision-making system. Furthermore, two novel algorithms are presented and used by the decision-making system for global and local path planning. Finally, Section 4 presents our conclusions.

2. Cloud Incubator Car Platform

As described above, autonomous driving technologies are making the transition from laboratories to the real world, and so must the vehicle platforms used to test and develop them. Research teams develop their own experimental platforms and also develop proprietary control systems.

In the Intelligent Vehicles Lab at the Technical University of Cartagena we have developed a comprehensible and reliable platform for autonomous driving technology development as well as for testing purposes: the CICar (Cloud Incubator Car, Figure 1a). This platform allows a complete experimentation of the aspects related to the integration of third-party sensors (laser range finders, global positioning systems, vision cameras, etc.), control of devices (motors, actuators, etc.), and industrial buses for communication. Software development has been made a comprehensible task by using a well-known hardware–software development platform for control purposes: RIO architecture and LabVIEW [16].

The aforementioned platform is based on an electric commercial vehicle—Renault Twizy—that has been conveniently modified in order to meet the specifications mentioned above. The modifications compound the low label architecture and include: (a) automation of driving elements (throttle, brake, steering, gearbox); (b) installation of external anchors capable of holding a wide variety of sensors; (c) installation of interior racks to accommodate the different control and processing systems (Figure 2b); (d) installation of electrical facilities and communication buses to feed and connect the different sensors and the control and processing hardware.

2.1. Modification of the Steering System

The modification of the steering system is a basic mechanical solution, as shown in Figure 1b, which consists of acting on the steering column or axis. A mechanical gear motor assembly is fixed to the steering column with gears, allowing the axis to turn. The original axis has not been modified, and this allows both steering control modes: automatic and manual. The mechanical implementation of the steering automation complies with the requirements of angular position range and precision, in addition to those of axis movement maximum speed (60 rpm). The steering system actuator is composed of an electronic driver EPOS2 and a motor from MAXON (EC60fl-GP52C-1024IMP-100W from Maxon Motor, Sachseln, Switzerland, acquired in Spain delegation).

2.2. Modification of the Braking System

The modification of the braking system, in the same way as that of the steering system, is not invasive of the original system. It has been implemented using a mechanical assembly with a motor comprising a gear coupled to a cam that presses the brake pedal by means of a short displacement of about 15° with force control (see Figure 1c). The braking system actuator is composed of an electronic driver EPOS2 and a motor from MAXON (EC60fl-GP52C-1024IMP-100W from Maxon Motor, Sachseln, Switzerland, acquired in Spain delegation).

2.3. Throttle and Gearbox Modifications

We have devised an electronic solution for the throttle and gearbox automation. After scanning the control signals, it is possible that the electronic device, which is attached to the throttle pedal, exchanges data with the vehicle's ECU (Electronic Control Unit). In the same way, it is possible to select PDNR (P=Park, D=Driver, N=Neutral, R=Reverse) in the gearbox by means of the control system. We have developed a software module which emulates its operation through an analogue card. It works in parallel with the original system, so both manual and automatic throttle control are offered (see Figure 1d).

Figure 1. Automation of driving elements. (**a**) Renault Twizy; (**b**) steering modification; (**c**) braking system modification; (**d**) throttle modification.

2.4. Sensor Holders

We have installed on the CICar vehicle a varied set of supports capable of holding different types of sensors (cameras VIS, Time of Flight cameras, LIDAR 2D and 3D, IMU, GPS, etc.). For this, quick coupling devices have been used together with electrical connectors and secure communications.

2.5. Electrical and Communications Infrastructure

New electrical facilities have been installed that allow the electrical supply of both the sensors and the control and processing systems. The communication infrastructure is composed of:

(1) A serial Controller Area Network (CAN bus), because of its suitability for high-speed applications using short messages, its robustness, and reliability [17].
(2) A gigabyte Ethernet Network which allows the implementation of TCP and UDP protocols. They are used by the 2D [18] and 3D LIDAR [19], respectively.
(3) A point-to-point set communications, such as RS232 and USB.

Figure 2a shows all the components which compound the low-level architecture.

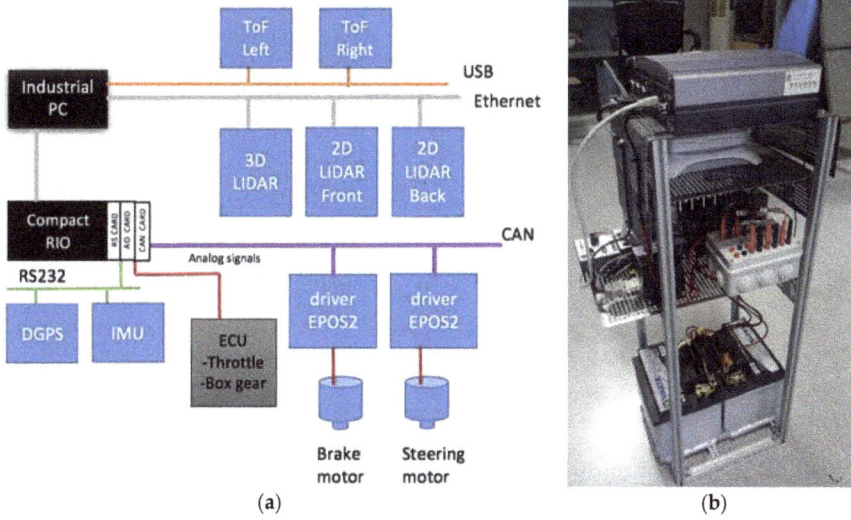

Figure 2. (**a**) Low-level architecture; (**b**) rack to host the control and decision making systems.

3. High-Level CICar Architecture

From a higher abstraction layer point of view, the architecture of the CICar is divided into: (1) a control system, (2) a perception system, and (3) a decision-making system. Figure 3 shows the distribution of the high-level architecture components of the CICar.

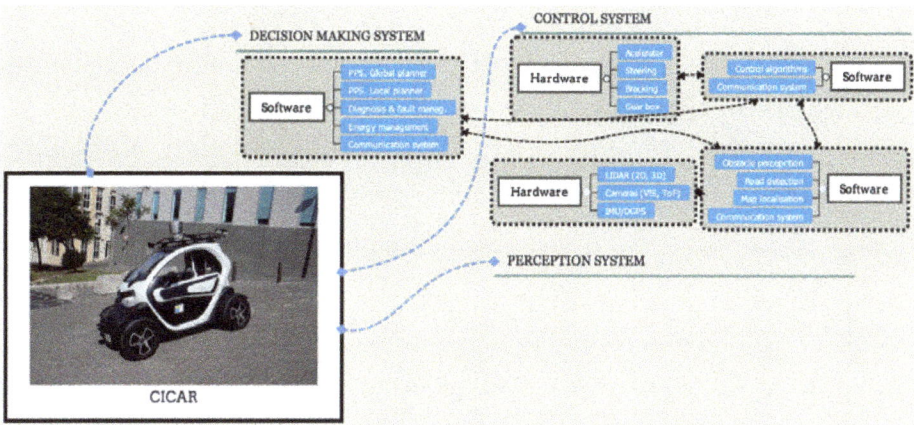

Figure 3. Distribution of the high-level architecture components of the CICar.

3.1. Control System

The goal of the control system is to translate the actions generated by the decision-making system into actuations of the elements which govern the vehicle. The control system has been implemented with a compactRIO 9082 (from National Instruments). The compactRIO processing unit is composed of two processors: an INTEL i7 and a Xilinx FPGA, which confer high performance capacities and flexibility for the testing and implementation of algorithms in the control system.

The compactRIO allows soft and hard real-time processes to be run. The INTEL processor allows less critical processes to be executed, while FPGA allows time-critical processes to be executed with higher clock frequencies. The control system is governed by a Real-Time Operating System (VxWorks from WindRiver). The kernel of the CICar is designed with Finite State Machines implemented in C++ through the QS framework [20].

3.2. Perception System

The perception system is divided into two subsystems: short- (SRS) and long-range (LRS). The perception system forms two concentric rings overlapping around the vehicle (Figure 4). The SRS allows objects to be detected up to 10 m from the front and the rear of the vehicle (2D LIDARs) and around 3 m from the right and left side of the vehicle (ToF cameras). ToF cameras are a seldom used technology in autonomous vehicles but they result to be a very attractive solution for the vehicle for short-range detection. ToFs allow 3D object detection with a high frame rate, are less affected by light conditions and shadows, and have an acceptable cost [21,22]. The objects that enter into the short-range ring suppose an inherent risk and danger for the actions of the CICar. For this reason, the sensors involved in the SRS are commanded by the RT processing unit. On the other hand, the LRS perception is based on a 3D High-Definition LIDAR (HDL64SE from Velodyne) [19]. Its 64 laser beams turn at 800 rpm and can detect objects up to 100 m away with an accuracy of 2 cm. The 3D LIDAR allows trajectories to be established (lane change, obstacle avoidance, speed reduction depending on the traffic conditions, etc.), object tracking, object classification, and behaviour prediction of other drivers. The information from the two subsystems is fused and processed to detect obstacles such as: cars, bikes, pedestrians, traffic lights, and so on. A colour camera is placed on the roof of the vehicle and is used to detect the road, lanes, road lines, and traffic signs.

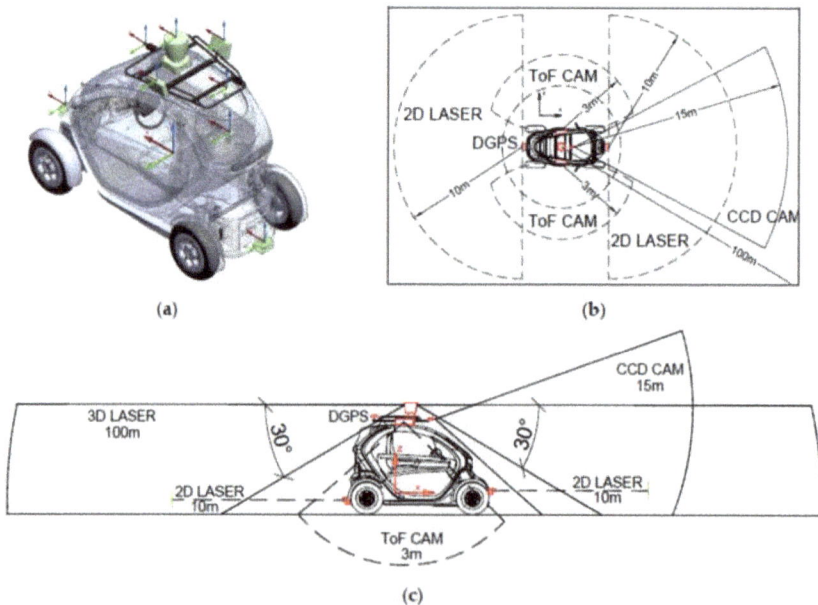

Figure 4. (a) CICar 3D model with reference frames; (b,c) plant and side views of the perception system ranges and components.

The CICar has a localization system based on an RTK unit (Real-Time Kinematic), an IMU (Inertial Measurement Unit), and a relative position encoder installed on the wheel of the vehicle. RTK satellite navigation consists of a technique used to enhance the precision of the position data derived from satellite-based positioning systems (global navigation satellite systems, GNSS) such as GPS, GLONASS, Galileo, and BeiDou. The RTK is composed of a base station and a mobile unit (or rover) and it supplies an accuracy from 20 cm to 1 cm. The localization system cancels out the RTK positioning when the number of satellites detected is less than or equal to four or when the HDOP (Horizontal Dilution of Precision) is greater than five [23,24]. In these cases, the IMU and the positioning encoder installed on the vehicle are used for the localisation task. Table 4 presents a summary of the onboard sensors in the CICar.

Table 4. Sensors features of the perception system.

Sensor	Characteristics and Configurations
SICK LASER scanner 2D TIM551	Operating range: 0.05 m ... 4 m, Aperture angle: 270°
LIDAR VELODYNE HDL64 scanner 3D	120 m range, 1.3 Million Points per Second, 26.9° Vertical FOV
Prosilica GT1290 cam	1.2 Megapixel Ethernet gigabit, RJ45 Ethernet connector, colour
DGPS	GPS aided AHRS
ToF Sentis3D-M420Kit cam	Range Indoor: 7 m Outdoor: 4 m, Horizontal FOV: 90°
IMU Moog Crossbow NAV440CA-202	Pitch and roll accuracy of < 0.4°, Position Accuracy < 0.3 m
EMLID RTK GNSS Receiver	7 mm positioning precision

3.3. Decision-Making System

To operate reliably in the real world, an autonomous vehicle must evaluate and make decisions about the consequences of its possible actions, anticipating the intentions of other traffic participants. The new Intelligent Decision-Making Systems are based on cognitive systems [25], agent systems [26], fuzzy systems [27], neural networks [28], evolutionary algorithms [29], multi-criteria [30] or rule-based methods [31]. The main element of the CICar is a Decision-Making System which incorporates a new local trajectory planner [32] based on Bezier curves [33].

Path-Planning System (PPS)

The main task of the PPS is to establish a global route for short or medium distance from the initial position of the vehicle until the goal. The PPS is composed of a global planner (Figure 5) and a local planner (Figure 6).

Global Planner

Global planners are a key element in autonomous navigation systems. The development of intelligent transport systems and autonomous vehicles has increased the demand for route planners that allow dynamic route generation, capable of adapting to the varying aspects of the road, such as traffic, roadworks, roads that have been cut off, or unexpected events. The initial requirements for the CICar global planner are listed below:

(1)   It must be able to create global routes using digital map sources
(2)   It should allow an easy introduction of traffic restrictions or unplanned events.
(3)   It should be executed in an agile manner and produce results in an acceptable time.

For the map source used for the global route planning, Google maps was selected, the reasons for this being:

(1)   It has an easily accessible Application Programming Interface (API) and is multiplatform.
(2)   The maps can be downloaded with different layers of information.
(3)   It allows georeferencing of the map points.
(4)   It is constantly updated and is available for free.

The first algorithm evaluated for global route generation was the A* Algorithm [34,35]. The following disadvantage was reported with the use of the A* algorithm: the algorithm was not capable of generating routes in a reasonable time using the 512 × 512 resolution binary maps tested. To increase the performance, the size of the binary maps was reduced by a constant factor (2, 3, or 4). The size reduction of the binary map speeded up the calculation time, but, because of the pixelisation effects, the following problems occurred: (1) the destruction of narrow roads, (2) the deterioration of main roads, and, as a result, (3) the routes generated were not smooth and were very abrupt.

As a solution to the problems reported, a new heuristic search algorithm for global route planning, called SCP (Search for Cross Points for obtaining global routes), was developed.

The SCP algorithm is based on a binary map of nxm pixels where the navigable routes are given a value of "1" and the non-navigable areas a value of "0", and an initial point p_0, a goal point p_g, and an initial angle θ_0 are set.

The stages performed by the SCP algorithm to obtain the global route are:

1. A search direction is chosen ('N'-North, 'S'-South, 'E'-East, 'W'-West) according to the position of the angle between the positions p_g and p_0. Equation (1) shows how the search directions are obtained by the algorithm in function of the angle θ_{0g}

$$directions = f(\theta_{0g}) = \begin{cases} N, & -45 \le \theta_{0g} < -135 \\ W, & -135 \le \theta_{0g} < 135 \\ S, & 135 \le \theta_{0g} < 45 \\ E, & 45 \le \theta_{0g} < -45 \end{cases} \qquad (1)$$

2. A clear path is chosen from p_0 to the goal p_g. A route is considered clear if all the points that form the line that joins p_0 and p_g have a value equal to '1'.

    2.1   If the path is not free:

    a. The SCP algorithm searches for the first pair of cross points $[CP_i; CP_{i+1}]$ according to the search direction set as shown in Equation (1). To achieve this, the algorithm goes through the map, starting from p_0 according to an angle θ_1, until it finds a value equal to '0'. The point found is labelled as the first cross point CP_i. Starting from CP_i, and according to the search angle θ_2, the algorithm searches for the next point with a value of '0'. This new point is labelled as the second cross point CP_{i+1}. The angles θ_1, θ_2 are calculated according to the search direction and the following table.

    During the search for the cross points $[CP_i; CP_{i+1}]$, the candidate points $p_i(x_i, y_i)$ are calculated by alternating the values of $\theta = \theta_1$ and $\theta = \theta_2$, based on Equation (2):

$$x_i = x_{i-1} + d\cos(\theta)$$
$$y_i = y_{i-1} + d\sin(\theta) \qquad (2)$$

    with (x_{i-1}, y_{i-1}) being the previous value calculated, and d a constant that allows the CP_i search to advance more quickly.

    b. The Euclidean distance between two consecutive cross points CP_i and CP_{i+1} is calculated.

        i. If the distance is 0, the search direction is changed by evaluating the position of p_g with respect to the last CP_{i+1} that was calculated, p_0 is set to $p_0 = CP_{i+1}$, and the algorithm goes to step 2.

ii.   If the distance is not equal to 0, p_0 is set to $p_0 = CP_{i+1}$, and the algorithm goes to step 2.

2.1   If there is a clear route, the algorithm finishes, and the path is composed of the points $[p_0,pm_1, \dots ,pm_{n-1},pg]$, with $pm_1, \dots ,pm_{n-1}$ being the midpoints obtained from the pairs of points from the cross-point vector $[p_0,CP_1,CP_2, \dots ,CP_n,pg]$.

Figure 5 shows a binary map in the form of a square which will demonstrate the SCP algorithm in operation. Figure 5a shows how the SCP algorithm generates the cross points in the direction 'N', given the information (p_0, p_g, θ_0). In the example with $\theta_0 = 60°$ and Table 5, the angles θ_1 y θ_2 are $-120°$ and $-60°$ respectively.

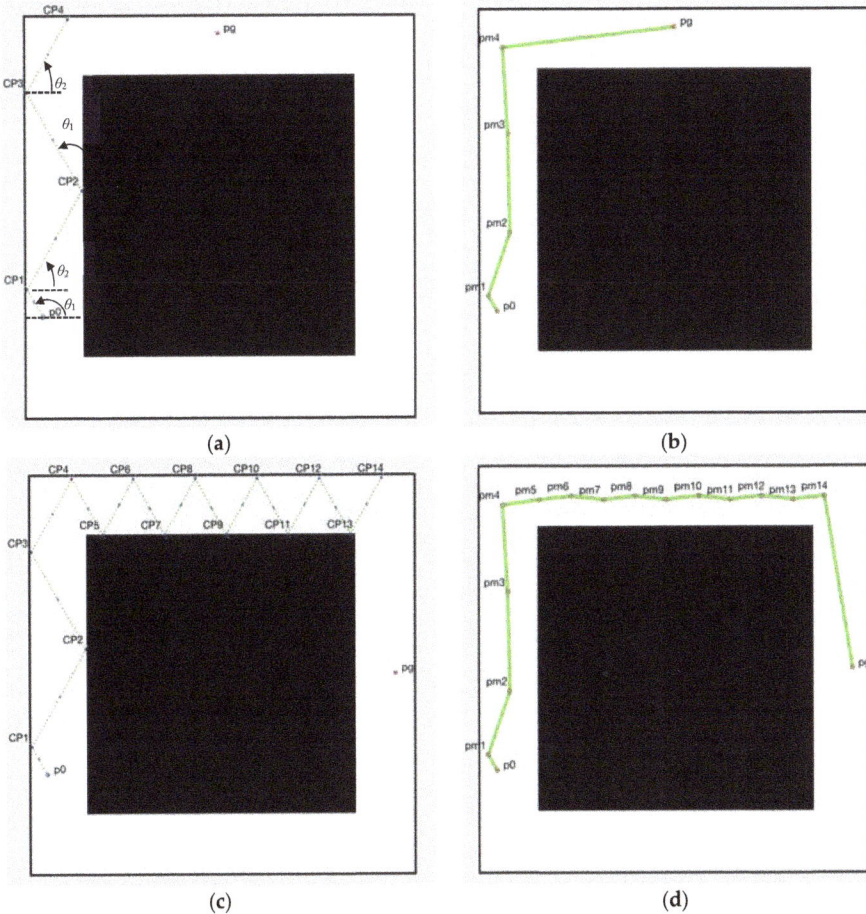

Figure 5. Example of the Search for Cross Points (SCP) working between p_0 and p_g with initial angle $\theta_0 = 60°$. (**a**) Scenario without change of direction; (**b**) Obtained trajectory in scenario without change of direction; (**c**) Scenario with change of direction; (**d**) Obtained trajectory in scenario with change of direction.

Table 5. Calculation of angles θ_1, θ_2, depending on the direction.

Directions	Angle θ_1	Angle θ_2
N	$-180 + \theta_0$	$-\theta_0$
S	θ_0	$180 - \theta_0$
E	θ_0	$-\theta_0$
W	$-180 + \theta_0$	$180 - \theta_0$

As shown in Figure 5a, the cross points CP_i are generated in a zigzag in the direction 'N' until a direct path to the destination is found. Figure 5b shows the route generated between the midpoints of the segments from the set $[p_0CP_1; CP_1CP_2; CP_2CP_3; CP_3CP_4; CP_4p_g]$. In Figure 5c, a more complex scenario with a different p_g is shown. In this case, when the SCP algorithm reaches CP_5, an end of route occurs when it no longer follows the 'N' direction. At this point, the Euclidean distance between CP_4 and CP_5 is 0, and the algorithm changes the search direction from 'N' to 'E', with $\theta_1 = 60°$ and $\theta_2 = -60°$. With the change of direction, it is possible to calculate a new CP_5. Figure 5d shows the route generated between the midpoints of the segments from the set $[p_0CP_1, \ldots, CP_{14}p_g]$.

The SCP algorithm can increase the number of routes calculated with a greater number of initial starting angles. Figure 6a shows a real map of the Technical University of Cartagena where an initial point p_0 and a final point p_g (blue and red crosses) have been selected.

The SCP algorithm has been executed with initial angles θ_0 from 20 to 80 degrees, with 2-degree increments. After its execution, the SCP has calculated three possible trajectories. Figure 6b shows the cross points found, and Figure 6c the trajectories created from the set of midpoints from the three calculated trajectories. Figure 6d shows the optimal path calculated as the shortest between points p_0 and p_g.

The optimal path is partitioned in waypoints ($WP_0, \ldots, WP_i, \ldots, WP_n$). The waypoints will be used by the CICar to establish short missions between them using the local planner module. The CICar location is provided by a DGPS unit on board the vehicle.

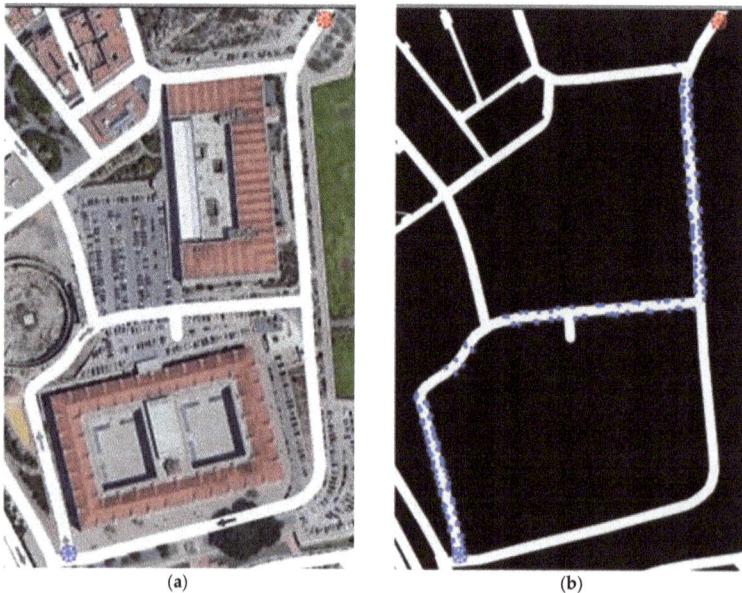

(a)   (b)

Figure 6. *Cont.*

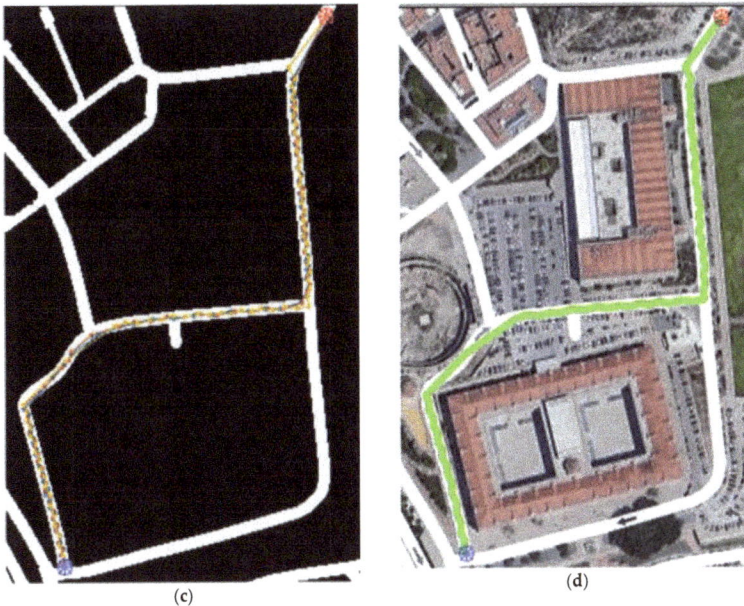

(c)   (d)

Figure 6. SCP applied on a real map. (**a**) Selection of the initial point on the real map; (**b**) cross points found on a binary map; (**c**) possible trajectories on a binary map; (**d**) optimum trajectory on the real map.

Local Planner

The local planner uses a pair of waypoints (WP$_i$, WP$_{i+1}$) supplied by the global planner to generate a set of possible trajectories to reach WP$_{i+1}$ from WP$_i$ (see Figure 7).

For that, the perception system supplies a local map of up to 100 m around the vehicle (Figure 8a), which is completed with three constraint types: (a) the constraints that the vehicle detects on the way by means of the perception system (vehicles, pedestrian, bikes, curbs, and so on), (b) the limitations inherent in the vehicle characteristics (wide, high, kinematic model), c) the constraints derived from the traffic rules (one way, bidirectional way, traffic signs, etc.). To decrease the number of computations and to increase the performance, the perception system supplies a set of keypoints which are a representation of the surroundings of the vehicle. The keypoint set is obtained after applying a filtering process composed of two stages:

1. Distance filter. All XYZ points beyond a maximum distance (d$_{max}$) will be rejected.
2. High filter. All Z points higher than a maximum value (Z$_{max}$) will be discarded.

As shown in Figure 8b, we have simplified the initial XYZ local map to obtain a compact profile of the XYZ representation of the objects in the scene.

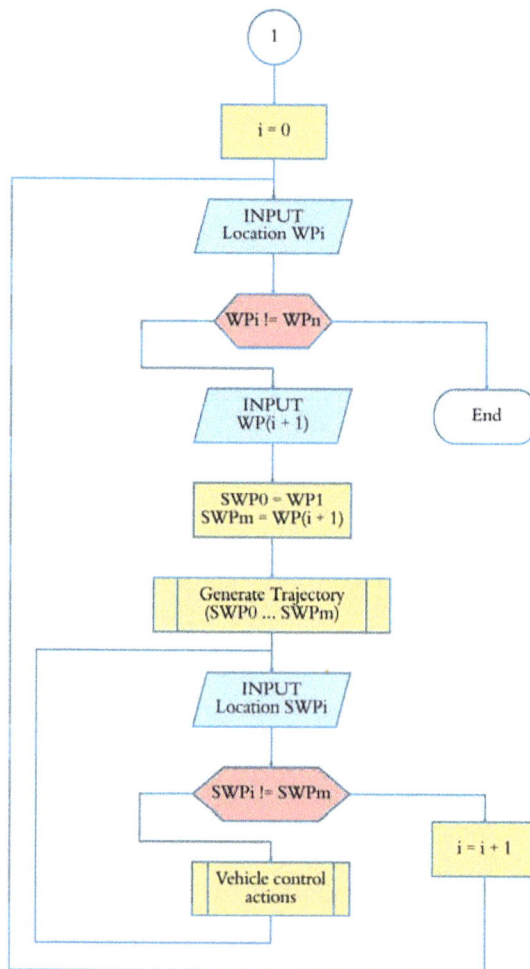

Figure 7. Local planner flowchart.

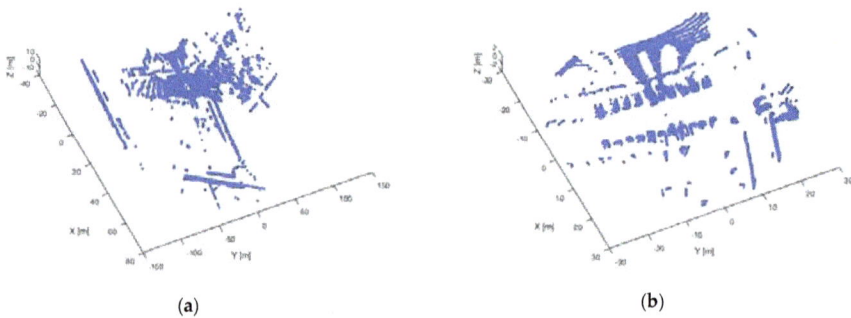

Figure 8. (a) Local map; (b) local filtered map.

After this, the corners of the objects in the scene are computed using the Harris algorithm [36]. The Harris algorithm is applied over an ROI (Region of Interest) in the forward direction of the vehicle. Figure 9 shows the results of corner detection in the ROI of two scenarios extracted from Figure 9b. This group of points calculated by the Harris algorithm are called keypoints (KP$_i$) and will be used in the next stage of the PPS to create a set of trajectories.

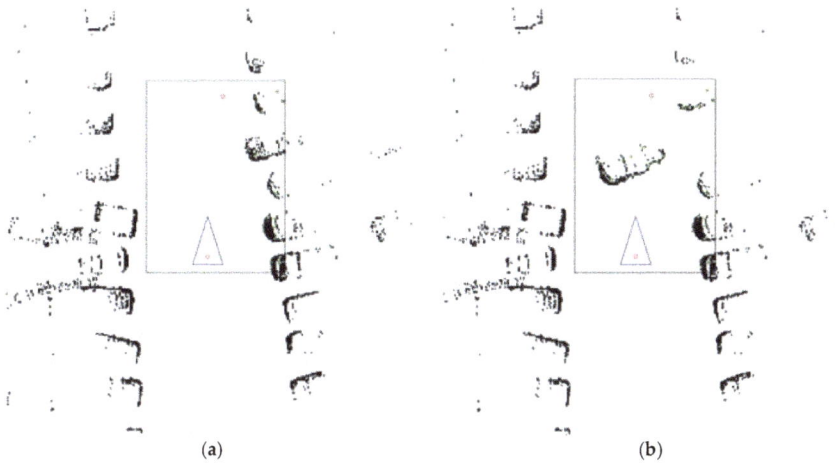

(a)   (b)

Figure 9. Waypoints (WP$_i$, WP$_{i+1}$), ROI, and keypoints in two scenarios (a,b).

In Figure 9, green crosses represent the keypoint set, a triangle with a 2 metre base represents the vehicle, a rectangle shows the ROI where the Harris detector is computed, and two red circles represent the origin and goal locations (WP$_i$, WP$_{i+1}$).

The trajectory generator module receives two waypoints (WP$_i$, WP$_{i+1}$) and a set of keypoints {KP$_0$, ... ,KP$_k$} representative of the environment around the vehicle for computing a set of trajectories between a pair of waypoints. The module uses the Bezier curves to generate possible trajectories between the waypoints, using the keypoints as control points.

A Bezier curve of grade n can be generalized by the Equation (3) [33]:

$$B(t) = \sum_{i=0}^{n} \binom{n}{i} P_i (1-t)^{n-i} t^i, \ t \in [0,1] \tag{3}$$

For example, the second-grade curve (*n* = 2) is (see Equation (4)):

$$B(t) = (1-t)^2 P_0 + 2t(1-t) P_1 + t^2 P_2 , \ t \in [0,1] \tag{4}$$

Being P_0, P_1, and P_2 the control points, and *t* a real value between 0 and 1. A higher step of t allows a higher number of points to be obtained in the Bezier curve generation B(t).

The trajectory generator module implements a set of second-grade Bezier curves displaced along the X-axis control point P1. The development algorithm is shown in Algorithm 1, where *nCurves* indicates the number of trajectory candidates which have been calculated. However, only one set will pass the security distance constraints and will be adopted as the trajectory (Ti[]). For each of them, a new control point P1 is defined, one half to the left of the midpoint between P0 and P2, and the other half to the right. In Figure 9, line 1.2., a control point displacement increment equal to 3.0 is defined. fB is a function that generates a second-grade Bezier curve between P0 and P2. The function fP1 searches for the Pi point with the lowest Euclidean distance to set the Bezier points of B(t).

After a set of trajectories is obtained $T_i = \{B(t)_0,..., B(t)_m\}$, the optimal trajectory can be established under different criteria depending on the conditions in each moment during autonomous driving. In this work, we have implemented three criteria to select the optimal trajectory: (a) the minimum distance travelled, (b) the maximum angle travelled, and (c) the medium trajectory.

Algorithm 1 Trajectory generator algorithm.

Point P0(x0, y0), P1(x1, y1), P2(x2, y2), WPi, WP(i+1)
Point KPi[] % array: set of keypoints
Trajectories (set of points): B(t)
Trajectories: Ti[] % array: set of trajectories
INTEGER: cnt = 0, nCurves % for the WHILE iteration
REAL: distance, safety_dist % safety distance
% Initial assignments:
   INPUT nIter, safety_dist , KPi[]
   P0 = WPi , P2 = WP(i+1)
   P1: ((WPi.x + WP(i+1).x) / 2, (WPi.y + WP(i+1).y) / 2)
   P1.x = P1.x − nCurves * 3.0 / 2
% Process:
BEGIN
1. **WHILE** cnt < nCurves
    1.1.   **IF** P1 is in ROI
       1.1.1.   B(t) = fB(P0, P1, P2)
       1.1.2.   P1 = fP1(KP)
       1.1.3.   cnt = cnt + 1;
       1.1.4.   **IF** distance >= safety_dist AND all B(t) points are in ROI
          1.1.4.1. Ti[] = B(t) is accepted as candidate to trajectory
              END IF % (1.1.4.1)
         END IF % (1.1)
    1.2.   P1.x = P1.x + 3.0
    1.3.   nCurves = nCurves + 1
    END WHILE % (1)
END

The optimal trajectory returns a new group of subwaypoints $(SWP_0, \ldots , SWP_i, \ldots , SWP_n)$, which compound the local route between (WP_i, WP_{i+1}) under local and environment constraints.

To show how the trajectory generator algorithm works, four scenarios have been selected where a car drives back and interrupts the initial trajectory (scenario 1, Figure 10a) between WP_i to WP_{i+1} of the CICar. The scenarios (1 to 4, Figure 10) present an increasing complexity, where the trajectory generator must recalculate the set of trajectories depending of the new keypoints location.

Figure 10 shows how the trajectory generator is capable of resolving a normal traffic situation successfully. The trajectory generator module was configured with 30 iterations and a safety distance of 1 m, and the optimal trajectory criteria were set to the lowest distance travelled. In all the scenarios proposed in Figure 10, the trajectory generator module generated 30 trajectories, but the safety distance constraint only passed: 16 in scenario 1, 11 in scenario 2, 1 in scenario 3, and 2 in scenario 4. In each trajectory set the optimal trajectory was selected according to the lowest distance travelled.

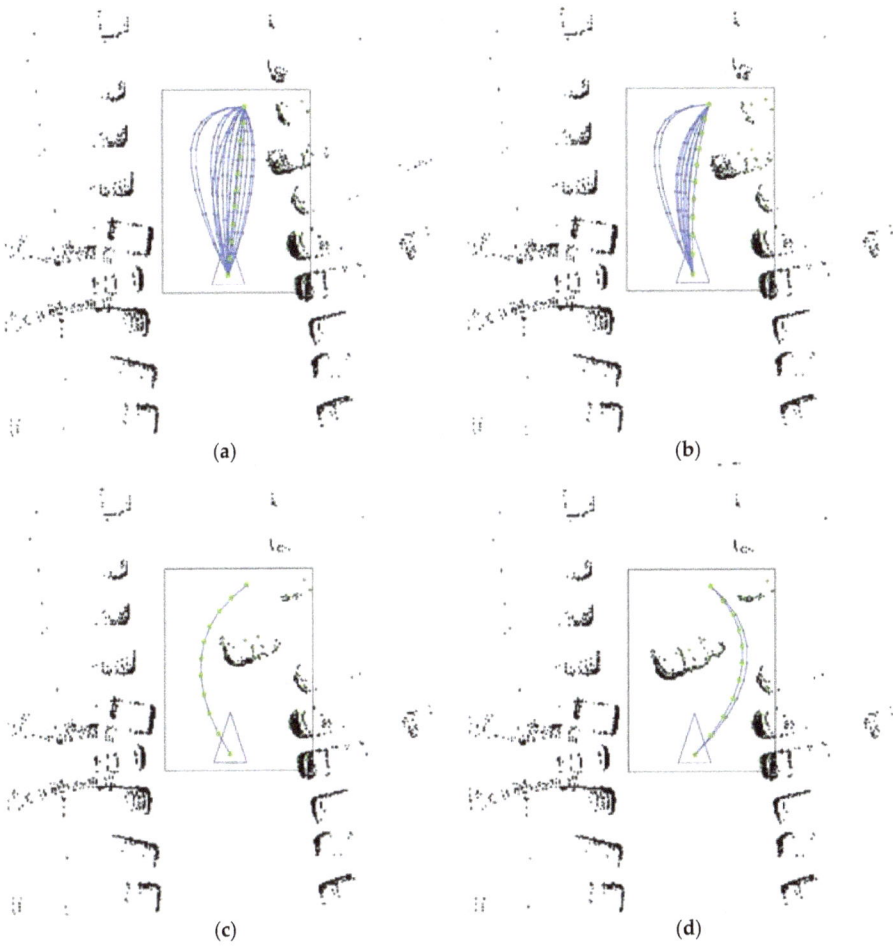

Figure 10. (**a**) Scenario 1; (**b**) scenario 2; (**c**) scenario 3; (**d**) scenario 4 (nCurves = 30, safety distance = 1 m, criteria = minimum distance).

Figure 11 shows the behaviours of the trajectory generator when other criteria and parameters of the algorithm are chosen. In Figure 11a,b, the maximum angle criteria and medium trajectory criteria have been selected from scenario 1, respectively. It is possible to observe that the change in the selection criteria affect the smoothness of the trajectory. In Figure 11c,d, the number of possible trajectories was set to 25. With this change in criteria, the trajectory generator cannot avoid the obstacle and cannot supply any solution in this scenario (see Figure 8c). Nevertheless, a decrement of the safety distance to 0.5 m allows four reliable trajectories to be generated (see Figure 11d).

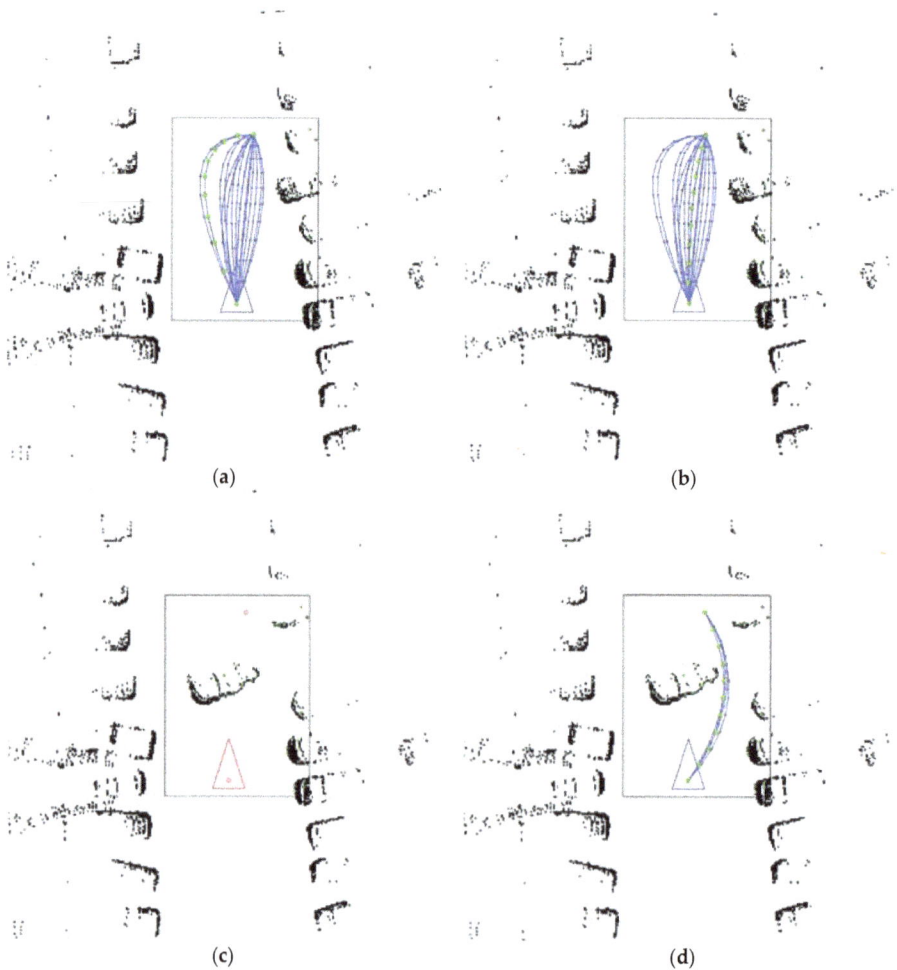

Figure 11. (**a**) Scenario 1 (nCurves = 30, safety distance = 1 m, criteria = maximum angle); (**b**) scenario 1 (nCurves = 30, safety distance = 1 m, criteria = medium trajectory); (**c**) scenario 4 (nCurves = 25, safety distance = 1 m, criteria = minimum distance); (**d**) scenario 4 (nCurves = 25, safety distance = 0.5 m, criteria = minimum distance).

4. Conclusions

The Intelligent Vehicle (IV) will play a central role in the so-called Intelligent Transport Systems. For this reason, a great research effort is devoted to developing algorithms and control strategies for autonomous driving. It is essential, therefore, to have real driving platforms ready to implement and test these algorithms and control strategies.

This paper describes the work done to develop and implement low- and high-level open architectures for an autonomous vehicle: the CICar. The low-level architecture is presented in detail, and comprises a detailed description of all hardware components, sensors, and communication infrastructures on board. Also, the high-level architecture is described, which brings together those systems that provide the vehicle with a certain degree of intelligence.

Appl. Sci. **2018**, *8*, 303

Unlike other platforms presented, which use proprietary control systems, in the Intelligent Vehicles Lab we propose an approach based on well-known hardware–software development platforms for control purposes: RIO architecture and LabVIEW. This means that the current software development approach for self-driving can be easily transferred to control actions on any vehicle equipped with the general purpose control hardware proposed for the CICar.

The sensing infrastructure includes ToF cameras for lateral surveillance. This sensing technology offers advantages compared with the current 2D LIDAR, since a better characterization of the scene is possible, thus offering better capabilities such as: 3D object detection with a high frame rate, lower influence of the light conditions and shadows, and acceptable cost.

As a proof of concept, this paper shows a case study where the implementation of a PPS (path-planning system) on the CICar platform is presented. The implemented PPS uses a map to establish a global route and a local planner to solve the usual traffic issues. The global planer has been resolved by means of the development of a new heuristic algorithm based on a search for cross points (SCP) using binary maps. The SCP algorithm has improved the flexibility, the computing time, and the performance in global route computation. Furthermore, a local planner has also been presented in this work, which implements a trajectory generator based on Bezier curves, which supplies high flexibility during the resolution of unforeseen situations. The trajectory generator allows the setting of: (a) the number of possible trajectories to generate; (b) a safety distance; (c) a criterion to select the optimal trajectory. The CICar autonomous vehicle offers a comprehensible and reliable platform for autonomous driving technology development and testing, which greatly facilitates the development of technology for autonomous driving.

Concerning the PPS, we currently continue to work towards increasing the intelligence to implement it on board the vehicle. We focus our developments on: (a) the improvement of the evaluation criteria for the optimal trajectory; (b) the dynamic modification of the ROI, depending on external driving factors; (c) the incorporation of machine learning technology for the classification of the objects detected by the perception systems.

Acknowledgments: This work was partially supported by ViSelTR (ref. TIN2012-39279), DGT (ref. SPIP2017-02286) and UPCA13-2E-1929 Spanish Government projects, and the "Research Programme for Groups of Scientific Excellence in the Region of Murcia" of the Seneca Foundation (Agency for Science and Technology in the Region of Murcia-19895/GERM/15). We would like to thank Leanne Rebecca Miller for the edition of the manuscript.

Author Contributions: R.B. carried out the design and the mechanical modification of the systems on the vehicle; P.J.N conceived and designed the algorithms involved in the Path-Planning Systems. C.F. and P.A. contributed to improve the mechanical systems and operating system of the vehicle, respectively. R.B, P.J.N, C.P, and P.A. wrote and corrected the draft and approved the final version of the paper.

Conflicts of Interest: The authors declare no conflict of interest.

References

1. Kala, R. *On-Road Intelligent Vehicles: Motion Planning for Intelligent Transportation Systems*; Elsevier Science: Amsterdam, The Netherlands, 2016.
2. Flemming, B.; Gill, V.; Godsmark, P.; Kirk, B. Automated Vehicles: The Coming of the Next Disruptive Technology. In *The Conference Board of Canada*; Conference Board of Canada: Ottawa, ON, Canada, 2015.
3. Corben, B.; Logan, D.; Fanciulli, L.; Farley, R.; Cameron, I. Strengthening road safety strategy development "Towards Zero" 2008–2020 – Western Australia's experience scientific research on road safety management SWOV workshop 16 and 17 November 2009. *Saf. Sci.* **2010**, *48*, 1085–1097. [CrossRef]
4. Engelbrecht, J.; Booysen, M.J.; Bruwer, F.J.; van Rooyen, G.-J. Survey of smartphone-based sensing in vehicles for intelligent transportation system applications. *IET Intell. Transp. Syst.* **2015**, *9*, 924–935. [CrossRef]
5. Thrun, S. Toward robotic cars. *Commun. ACM* **2010**, *53*, 99. [CrossRef]
6. Ibañez-Guzmán, J.; Laugier, C.; Yoder, J.-D.; Thrun, S. Autonomous Driving: Context and State-of-the-Art. In *Handbook of Intelligent Vehicles*; Springer: London, UK, 2012; pp. 1271–1310.

7. Urmson, C.; Anhalt, J.; Bagnell, D.; Baker, C.; Bittner, R.; Clark, M.N.; Dolan, J.; Duggins, D.; Galatali, T.; Geyer, C.; et al. Autonomous Driving in Urban Environments: Boss and the Urban Challenge. *J. F. Robot.* **2009**, *25*, 1–59.

8. NHTSA Federal Automated Vehicles Policy—September 2016 | US Department of Transportation. Available online: https://www.transportation.gov/AV/federal-automated-vehicles-policy-september-2016 (accessed on 29 November 2017).

9. SAE J3016: Taxonomy and Definitions for Terms Related to On-Road Motor Vehicle Automated Driving Systems—SAE International. Available online: http://standards.sae.org/j3016_201401/ (accessed on 29 November 2017).

10. Wei, J.; Snider, J.M.; Kim, J.; Dolan, J.M.; Rajkumar, R.; Litkouhi, B. Towards a viable autonomous driving research platform. In *IEEE Intelligent Vehicles Symposium, Proceedings*; Gold Coast, QLD, Australia, 2013; pp. 763–770.

11. Milanés, V.; Llorca, D.F.; Vinagre, B.M.; González, C.; Sotelo, M.A. Clavileño: Evolution of an autonomous car. In Proceedings of the IEEE Conference on Intelligent Transportation Systems, Proceedings, ITSC, Funchal, Portugal, 19–22 September 2010; pp. 1129–1134.

12. Grisleri, P.; Fedriga, I. The BRAiVE Autonomous Ground Vehicle Platform. *IFAC Proc. Vol.* **2010**, *43*, 497–502. [CrossRef]

13. Levinson, J.; Askeland, J.; Becker, J.; Dolson, J.; Held, D.; Kammel, S.; Kolter, J.Z.; Langer, D.; Pink, O.; Pratt, V.; et al. Towards fully autonomous driving: Systems and algorithms. In *IEEE Intelligent Vehicles Symposium, Proceedings*; Baden-Baden, Germany, 2011; pp. 163–168.

14. Liu, S.; Tang, J.; Wang, C.; Wang, Q.; Gaudiot, J.L. A Unified Cloud Platform for Autonomous Driving. *Computer* **2017**, *50*, 42–49. [CrossRef]

15. Liu, S.; Tang, J.; Wang, C.; Wang, Q.; Gaudiot, J.-L. Implementing a Cloud Platform for Autonomous Driving. *arXiv* **2017**. [CrossRef]

16. NI white papers The LabVIEW RIO Architecture: A Foundation for Innovation. Available online: http://www.ni.com/white-paper/10894/en/ (accessed on 14 February 2018).

17. Hu, J.; Xiong, C. Study on the embedded CAN bus control system in the vehicle. In Proceedings of the 2012 International Conference on Computer Science and Electronics Engineering, ICCSEE 2012, Hangzhou, China, 23–25 March 2012; IEEE; Volume 2, pp. 440–442.

18. Petrovskaya, A.; Thrun, S. Model based vehicle detection and tracking for autonomous urban driving. *Auton. Robots* **2009**, *26*, 123–139. [CrossRef]

19. Navarro, P.; Fernández, C.; Borraz, R.; Alonso, D. A Machine Learning Approach to Pedestrian Detection for Autonomous Vehicles Using High-Definition 3D Range Data. *Sensors* **2016**, *17*, 18. [CrossRef] [PubMed]

20. Samek, M. *Practical UML Statecharts in C/C++: Event-Driven Programming for Embedded Systems*; Newnes/Elsevier: Burlington, MA, USA, 2009.

21. Li, L. *Time-of-Flight Camera—An Introduction*; Technical White Paper; Texas Instruments: Dallas, TX, USA, 2014; p. 10.

22. Lange, R.; Seitz, P. Solid-state time-of-flight range camera. *IEEE J. Quantum Electron.* **2001**, *37*, 390–397. [CrossRef]

23. Piñana-Díaz, C.; Toledo-Moreo, R.; Toledo-Moreo, F.; Skarmeta, A. A Two-Layers Based Approach of an Enhanced-Mapfor Urban Positioning Support. *Sensors* **2012**, *12*, 14508–14524. [CrossRef] [PubMed]

24. Piñana-Diaz, C.; Toledo-Moreo, R.; Bétaille, D.; Gómez-Skarmeta, A.F. GPS multipath detection and exclusion with elevation-enhanced maps. In Proceedings of the IEEE Conference on Intelligent Transportation Systems, Proceedings, ITSC, Washintong, DC, USA, 5–7 October 2011; pp. 19–24.

25. Czubenko, M.; Kowalczuk, Z.; Ordys, A. Autonomous Driver Based on an Intelligent System of Decision-Making. *Cognit. Comput.* **2015**, *7*, 569–581. [CrossRef] [PubMed]

26. Chen, B.; Cheng, H.H. A Review of the Applications of Agent Technology in Traffic and Transportation Systems. *IEEE Trans. Intell. Transp. Syst.* **2010**, *11*, 485–497.

27. Abdullah, R.; Hussain, A.; Warwick, K.; Zayed, A. Autonomous intelligent cruise control using a novel multiple-controller framework incorporating fuzzy-logic-based switching and tuning. *Neurocomputing* **2008**, *71*, 2727–2741. [CrossRef]

28. Belker, T.; Beetz, M.; Cremers, A.B. Learning action models for the improved execution of navigation plans. *Rob. Auton. Syst.* **2002**, *38*, 137–148. [CrossRef]

29. Chakraborty, D.; Vaz, W.; Nandi, A.K. Optimal driving during electric vehicle acceleration using evolutionary algorithms. *Appl. Soft Comput.* **2015**, *34*, 217–235. [CrossRef]
30. Michalos, G.; Fysikopoulos, A.; Makris, S.; Mourtzis, D.; Chryssolouris, G. Multi criteria assembly line design and configuration – An automotive case study. *CIRP J. Manuf. Sci. Technol.* **2015**, *9*, 69–87. [CrossRef]
31. Cunningham, A.G.; Galceran, E.; Eustice, R.M.; Olson, E. MPDM: Multipolicy decision-making in dynamic, uncertain environments for autonomous driving. In Proceedings of the IEEE International Conference on Robotics and Automation, Seattle, WA, USA, 26-30 May 2015; IEEE; Volume 2015-June, pp. 1670–1677.
32. Chu, K.; Lee, M.; Sunwoo, M. Local path planning for off-road autonomous driving with avoidance of static obstacles. *IEEE Trans. Intell. Transp. Syst.* **2012**, *13*, 1599–1616. [CrossRef]
33. Biswas, S.; Lovell, B.C. B-Splines and Its Applications. In *Bézier and Splines in Image Processing and Machine Vision*; Springer: London, UK; pp. 3–31.
34. Hart, P.; Nilsson, N.; Raphael, B. A Formal Basis for the Heuristic Determination of Minimum Cost Paths. *IEEE Trans. Syst. Sci. Cybern.* **1968**, *4*, 100–107. [CrossRef]
35. Dijkstra, E.W. A Note on Two Problems in Connexion with Graphs. In *Numerische Mathematik*; Stichting Mathematisch Centrum: Amsterdam, The Netherlands, 1959; Volume 271, pp. 269–271.
36. Harris, C.; Stephens, M. A combined corner and edge detector. In Proceedings of the Fourth Alvey Vision Conference, Manchester, UK, 31 August–2 September 1988.

applied sciences

MDPI

Article

An Experimental Platform for Autonomous Bus Development

Héctor Montes [1,2,*] ⬛, Carlota Salinas [3], Roemi Fernández [1] ⬛ and Manuel Armada [1] ⬛

1   Centre for Automation and Robotics, (CAR) CSIC-UPM, Ctra. Campo Real, km. 0.200, La Poveda,
    Arganda del Rey, 28500 Madrid, Spain; roemi.fernandez@car.upm-csic.es (R.F.);
    manuel.armada@csic.es (M.A.)
2   Facultad de Ingeniería Eléctrica, Universidad Tecnológica de Panamá, Panama City 0819, Panama
3   Universidad de Alcalá, Ctra. Madrid-Barcelona, km. 33.6, Alcalá de Henares, 28805 Madrid, Spain;
    carlota.salinasmaldo@uah.es
*   Correspondence: hector.montes1@utp.ac.pa or hmontes@gmail.com; Tel.: +34-91-871-1900

Received: 26 September 2017; Accepted: 30 October 2017; Published: 2 November 2017

Abstract: Nowadays, with highly developed instrumentation, sensing and actuation technologies, it is possible to foresee an important advance in the field of autonomous and/or semi-autonomous transportation systems. Intelligent Transport Systems (ITS) have been subjected to very active research for many years, and Bus Rapid Transit (BRT) is one area of major interest. Among the most promising transport infrastructures, the articulated bus is an interesting, low cost, high occupancy capacity and friendly option. In this paper, an experimental platform for research on the automatic control of an articulated bus is presented. The aim of the platform is to allow full experimentation in real conditions for testing technological developments and control algorithms. The experimental platform consists of a mobile component (a commercial articulated bus) fully instrumented and a ground test area composed of asphalt roads inside the Consejo Superior de Investigaciones Científicas (CSIC) premises. This paper focuses also on the development of a human machine interface to ease progress in control system evaluation. Some experimental results are presented in order to show the potential of the proposed platform.

Keywords: Intelligent Transportation Systems (ITS); Bus Rapid Transit (BRT); autonomous driving; autonomous bus; automatic vehicle control; lateral control; longitudinal control; obstacle detection; human machine interface

---

1. Introduction

As it is well known, the California Program on Advanced Technology for the Highway (PATH), a collaboration between the Institute of Transportation Studies at the University of California at Berkeley and the California Department of Transportation (Caltrans), with funding provided by Caltrans and the U.S. Department of Transportation, has done pioneering work on Automatic Vehicle Control (AVC) since 1986 and provided an outstanding stream of new ideas, methods and developments [1]. It is worth mentioning also that at Ohio State University (OSU), a program running from 1964–1980 included studies on headway safety policy, longitudinal control, lateral control and highway system operations [2]. Moreover, the interest in the Intelligent Vehicle Highway System (IVHS) was emphasized by the 1973–1979 projects funded by the Ministry of International Trade and Industry (MITI) of Japan [3]. Early considerations concerning safety implications of vehicle automation using the drive-by-wire concept and the driver workload with respect to the transformation from Cruise Control (CC) to Adaptive Cruise Control (ACC) were presented in [4]. More recently, the problems of adaptive cruise control, stop and go [5], as well as the overtaking maneuvers [6] have been investigated, and solutions based on artificial intelligence have been proposed and experimentally tested by researchers of the AUTOPIA Group at the Centre for Automation and Robotics (CAR) of the Consejo Superior de Investigaciones Científicas (CSIC), Spain. In addition to these works, a vehicle of AUTOPIA (Spanish research project) ran driverless 100 km, in 2012, from El Escorial to Arganda del Rey (Madrid). A leading vehicle, which was driven manually, dynamically generated a high accuracy map to be tracked by the fully-autonomous following car. The journey covered a wide range of driving scenarios, including urban zones, secondary roads and highways, in standard traffic conditions (https://www.car.upm-csic.es/autopia/).

In a motivating survey paper [7] published in 2000, a substantial increment of research in intelligent vehicles and their application areas was acknowledged, where sectors of interest and key supporting technologies were considered. Applications related to passenger cars, heavy trucks, bus and public transport applications, as well as special vehicle applications were analyzed, and relevant functionalities/requirements were identified (collision warning, collision avoidance and driver assistance, driver impairment monitoring intelligent speed adaptation, automated operation, industrial automation, military operations).

Nowadays, especially in urban/suburban areas, there is a clear conflict among collective transportation (high occupancy vehicle) and individual transportation, which results in overcrowding that makes transportation uncomfortable, lengthy, polluting and more expensive than needed. This situation decreases substantially the quality of collective transportation services. Because of this general concern, the development of new concepts and technology for urban/suburban transportation is of major importance. In these circumstances, the interest in Intelligent Transportation Systems (ITS) and Intelligent Vehicles (IV) is increasing because the wide range of possible applications [8].

Among several current developments for urban/suburban transportation, the concepts of Light Rail or Light Rail Transit (LRT) and Bus Rapid Transit (BRT) are some of the most attractive. LRT is a kind of urban public transport using rolling stock quite similar to a tramway, but operating at a higher capacity, while BRT aims to provide a high-quality bus-based transit system with the main advantages of being a flexible and cost-effective urban mobility transit system [9,10]. Both systems operate in segregated paths, and while LRT bear a resemblance to trams, the BRT is essentially characterized by using public transportation vehicles on tires (buses). Furthermore, both systems require a strong support from Intelligent Transportation System (ITS) elements, this aspect being more critical to BRT, which, in the end, has the intrinsic possibility of reaching the state of a truly autonomous system. The Institute of Transportation & Development Policy (ITDP) [11] defines BRT as "a high-quality bus-based transit system that delivers fast, comfortable and cost-effective urban mobility through the provision of segregated right-of-way infrastructure, rapid and frequent operations, and excellence in marketing and customer service". The BRT goal is to combine the capacity and speed of a metro with the flexibility, lower cost and simplicity of a bus system. BRT systems come in many shapes and forms. However, they all aim, to varying degrees, to mimic the high-capacity, high-performance characteristics of urban rail at a much lower price [12]. According to one study, this reduction can be as much as 4–20-times less than Light Rail Transit (LRT) and 10–100-times less than metro rail systems [13]. BRT uses buses on a wide variety of right-of-ways, including mixed traffic, dedicated lanes on surface streets and busways separated from traffic. In fact, BRT has a greater potential for innovation, and it could be considered as the evolution of present collective transportation systems.

Moreover, the added value provided by the latest advances in control, artificial perception and Information and Communication Technologies (ICT) might facilitate the development of new concepts enabling the creation in the near future of more flexible, fast, reliable, economic and efficient BRT systems of wider and safer use in populated areas [14]. In this regard, an interesting analysis has been released [10] reporting the full picture of the successful dissemination of BRT systems in several Latin America cities, providing the additional encouraging side-effect of environmental conditions' enhancement. Some BRT systems incorporated guidance (mechanical, optical and magnetic) to increase speed in narrow corridors, which also implies better safety and the possibility of precision docking [9], and this is why BRT could be considered as an intermediate step from classical urban transport (conventional bus services) to fully-autonomous systems [8]. Within this context, the relevance of autonomous guidance in urban transportation targeting improvement of ITS efficiency has been recognized by several authors [8,15].

The advancement of ITS technology takes advantage of previous and parallel research on Autonomous Guided Vehicles (AGVs) and mobile robots [16]. A theoretical contribution to the automatic steering of a city bus was presented in [17], where two different approaches, linear and nonlinear controllers, were compared. The works related to the kinematic and dynamic models of wheeled mobile robots [18] have been important to understand modelling and therefore to set up the grounds for motion planning and control. For example, in [19], the implementation of intelligent and stable fuzzy Proportional Derivative-Proportional Integral (PD-PI) controllers for steering and speed control of an AGV consisting of an electrically-powered golf car suitably modified for autonomous navigation and control was reported. In [20] active steering under model predictive control was researched. In [21], a small electric mini-bus was used to demonstrate the autonomous driving capability and obstacle avoidance using GPS for localization and LIDAR for environment recognition.

Lateral vehicle control [22] is one major issue that has been studied because of its relevance [23], both for warning about lateral vehicle displacement and possible automated intervening in manual driving [24], as well as for full automation, as was shown in the California PATH program [25] and more recently in the DARPA Urban Challenge 2007. In this competition, Caltech contributed with a highly modified Ford E-350 van, nicknamed Alice [26,27], and the Team AnnieWAY demonstrated the results of the Cognitive Automobiles Project [28], an autonomous vehicle that is capable of driving through urban scenarios. Another very relevant and cited contribution in the field was offered by [29], where a prototype nicknamed Babieca (a modified Citroen Berlingo), equipped with a color camera combined with DGPS and on-board controls, allowed lateral and orientation position of the ego-vehicle with regard to the center of the lane, so steering an autonomous vehicle using vision. For lateral control and guidance, the use of magnetic markers on the floor was thoroughly investigated [30,31], and also, the fusion of magnetic markers with encoders has been considered [32]. Significant works on lateral control of heavy-duty vehicles like tractor semitrailers have been produced by the PATH program, implementing nonlinear and adaptive controllers for lateral control of heavy vehicles and presenting an experimental comparative study [33–35]. Other relevant problems for vehicle automation have been investigated, and, in particular, those related to control systems have received ample consideration: nonlinear control [36,37], observation of lateral vehicle dynamics [38], external disturbance estimation [39], driver assistance systems based on artificial potential fields [40] and fault-tolerant lateral control for heavy vehicles [41].

While lateral control aims to keep the vehicle at the minimum departure distance from the required path, the longitudinal control [42,43] mission is to properly control the vehicle advance along the desired trajectory, and, to achieve this, a number of critical adaptations need to be implemented. The main issues are start/stop, acceleration and deceleration, velocity control, obstacle detection/avoidance [43] and several others related to the safety of the overall operation [44]. From another side, it is important also to consider that for improved accuracy and reliability of autonomous vehicles' lateral and longitudinal controllers should be integrated because there is a coupling between steering and velocity actions, and the cross-effects of each one on the other need to be taken into account [45]. This is especially relevant when increasing the speed of operation, in platoon formation, and in variable road and vehicle conditions (slippage, cornering, tire traction forces, etc.).

An interesting work regarding automatic steering control based on how drivers steer has been performed by using the data of drivers' steering obtained through a number of experiments, showing that drivers in effect execute a naturally robust controller that allows high-gain corrections and is insensitive to variations in vehicle dynamics and speeds [46]. This controller has been implemented and validated on an 18.3 m articulated bus for revenue service [47].

Having examined briefly the major components that could be needed to achieve autonomous vehicle driving, it is worth mentioning that there is a great interest in the development of more complex strategies like those regarding cooperative autonomous driving [48] aiming to obtain intelligent vehicles that could share roads and streets [49–51]. More recently and regarding bus automation, a small electric Cyberbus was used to demonstrate a parametric-based path generation algorithm for roundabouts [52]. Other related works have been concentrated on automated parking [53]. It is also worth mentioning the works using artificial intelligence methods based on fuzzy logic to control autonomous vehicles [54–56] and the latest developments regarding traffic sign detection [57] and an automatic system to detect distraction and drowsiness in drivers [58].

In this paper, an experimental platform for research on the automatic control of an articulated bus is presented [14,15]. The aim of the platform is to allow full experimentation in real conditions for testing technological developments and control algorithms. The experimental platform consists of a mobile unit (a commercial articulated bus, Volvo B10M-ART-RA-IN) fully instrumented and a ground test area composed of asphalt roads inside CSIC's premises. This paper focuses also on the development of an HMI to ease progress in the control system evaluation. Some experimental results are also presented in order to show the potential of the proposed experimental platform.

The paper is structured as follows: Section 2 presents an overview of the experimental platform for the autonomous bus development with details about its instrumentation and the control architecture; Section 3 is devoted to the discussion of the longitudinal and lateral control approaches, as well as showing some experimental results. The paper is completed with the main conclusions, in Section 4.

2. Materials and Methods

2.1. Experimental Platform Overview

The purpose of the experimental platform is to provide a reliable base to carry out research and feasibility studies on automatic control of vehicles of large dimensions. Figure 1 shows the main elements of the experimental platform setup: (a) Volvo B10M-ART-RA-IN articulated bus (18 m in length); (b) private inner road facility located at CSIC's premises in Arganda del Rey (Madrid), with a total length of 385 m. This experimental road has been designed to contain "enough difficult" curvature sections to permit experimentation in demanding situations. Figure 1c shows an analysis of a possible bus trajectory in one section of the path.

Research on autonomous bus control requires deploying multiple sensors and actuators so that they can be used for both monitoring and control. Furthermore, a control system of this kind involves many variables to analyze and monitor during the testing period of the system and also during the various stages of implementation and demonstration. Therefore, it is of considerable importance to organize a robust observation and acquisition information architecture to gather the sensor signals that are set up in the system. The analysis and interpretation of those variables will benefit the design and the comparison of the different control strategies to be implemented for the automation of the articulated bus. Additionally, experimentation with transport systems of large dimensions requires special care for safety, and this requires knowing the state of the vehicle at all times. This information must be known to both the people who go on board the vehicle and to those working at different points of the test facilities. The electronic instrumentation system includes the elements shown in Table 1. In Figure 2, some sensors are shown during the operating tests carried out in the laboratory.

Table 1. Sensors, actuators and computing and control equipment. CSIC, Consejo Superior de Investigaciones Científicas.

Instrument	Function	Model
	Incremental and absolute optical encoders	HEDS 550X; Industrial encoder DH05
Sensors	Proximity sensors	NBB2-12GM50-E0-V1, E2EL cylindrical proximity sensor
	Velocity measurement system (magnetic pick-up)	KATLAX M18 Digital magnetic pick-up sensor

Table 1. *Cont.*

Instrument	Function	Model
	Inertial Measurement Units (IMUs) (MEMS inertial sensor with three accelerometers) and high precision IMU	MEMS inertial sensor LIS3LO2AL IMU440CA
	Catadioptric omnidirectional stereovision system (with CCD RGB cameras)	Ueye UI-1485LE-C/M, resolution 2560×1920 pixels
	LIDAR systems (IP67, statistical error 5 mm, angular resolution 0.25°, range 80 m, view angle 180°)	SICK LMS221, LMS291
	GPS for localization and tracking	TRIMBLE 5700
	Magnetometer systems for magnetic markers detection	Honeywell HMC1501
Actuators	DC motor for steering wheel control	MAXON 24-volt DC motor (150 watts) with a gearbox (74:1) and encoder
	DC motor for brake control	MAXON 24-volt DC motor (70 watts) with gearbox (74:1) and encoder
Computing and control	Telemetry System	PCL-818, 16 I/O channels multifunction board Wireless-G Ethernet Bridge Wireless access point (WAP54G)
	On-board computers	Industrial PCs running QNX RTOS
	DC motors microcontroller-based control boards	CM3 and CM4, proprietary CSIC control boards

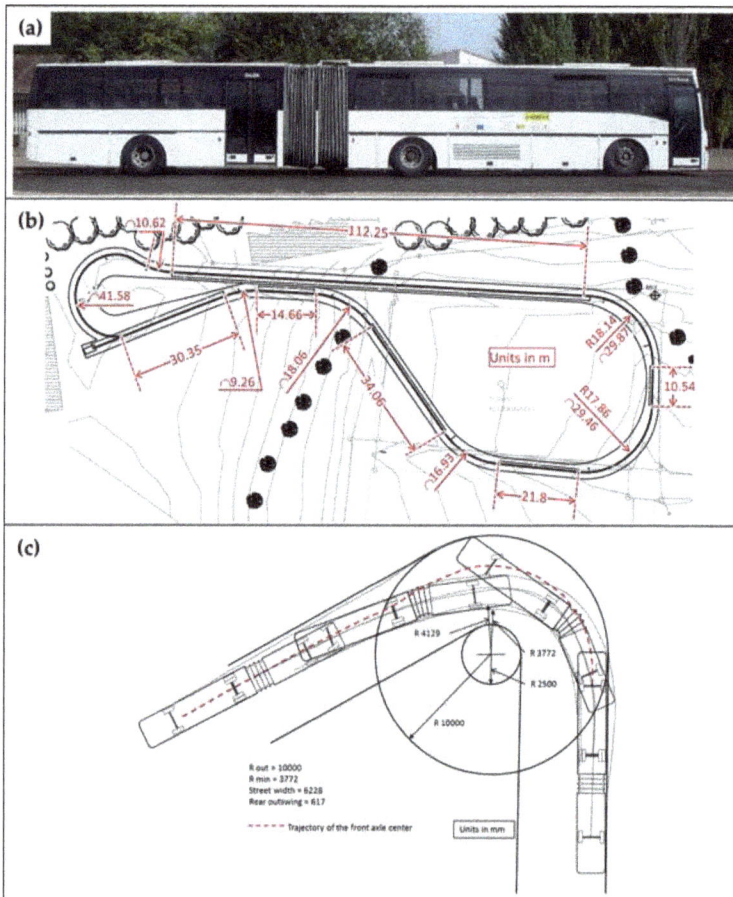

Figure 1. Experimental platform main elements. (**a**) Articulated bus Volvo B10M-ART-RA-IN; (**b**) inner road facility dimensions; (**c**) Bus trajectory analysis in one section of the path.

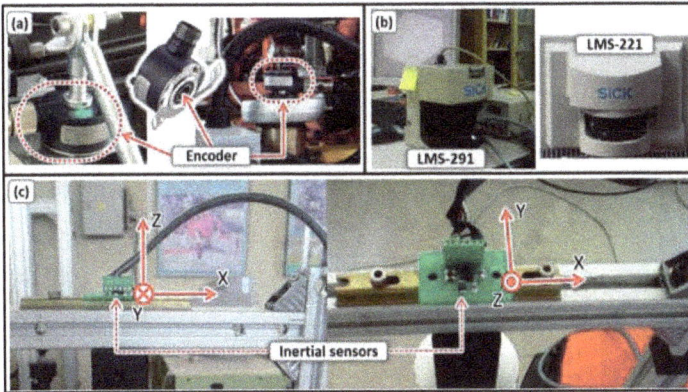

Figure 2. Some sensors used in the proposed experimental platform. (**a**) Industrial-grade absolute optical encoders; (**b**) LIDAR sensors being tested in the lab; (**c**) MEMS inertial sensor testing.

2.2. Velocity Control

The bus velocity control aiming for longitudinal control is performed in the on-board computer by means of two main actions: (1) using an I/O board electronically connected with the bus electronics, where the desired reference advance velocity (throttle) is sent to the bus engine; (2) using the aforementioned I/O board for sending a brake reference signal to the DC motor driver that mechanically controls the brake (pedal) position.

In more detail, and for greater flexibility in the experimental development, the velocity control can be achieved in three ways: (1) the control computer sends the throttle commands to the bus potentiometer that regulates the bus's own throttle control system; (2) one additional potentiometer is installed near the driver's seat and can be used manually to alter the computer command; (3) the driver, if seated, can use the throttle pedal to change the commanded velocity directly.

The implemented brake control system consists of a 24-volt DC motor (70 watts) with a gearbox (74:1) and encoder and a pulley with a steel cable whose free end is connected mechanically to the brake pedal (Figure 3). This system works under digital PID closed-loop control receiving the desired reference velocity from the high level control system and comparing it with the actual velocity (provided by different sensors like encoders installed in the bus wheels or others). This system is very useful in practice, not only to stop the vehicle, but especially to regulate, working in combination with the throttle control, the bus velocity according to the desired values. There is an additional advantage consisting of the automatic regulation of the velocity independently of the road slope (positive or negative). Figure 4 shows the flowchart for the velocity control.

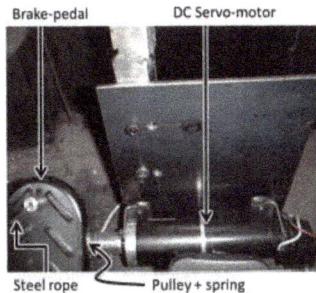

Figure 3. Actuator implementation for the brakes.

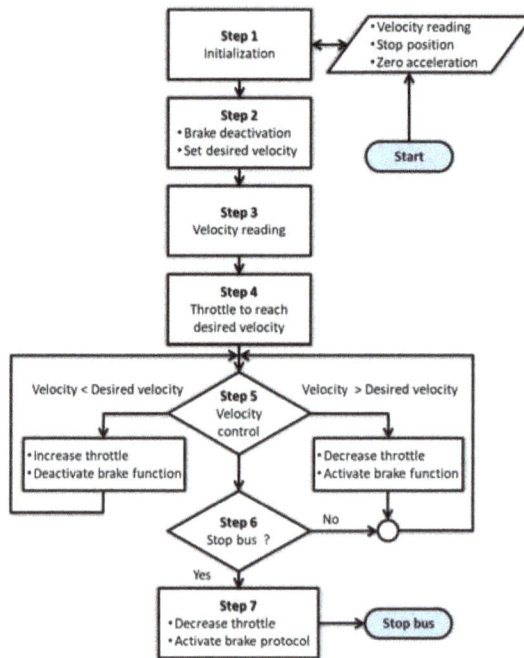

Figure 4. Implemented flowchart for velocity control.

2.3. Steering Control

The steering of the articulated bus aiming for lateral control has been achieved by incorporating a 24-volt DC motor (150 watts) with a gear reduction box (74:1) and an incremental optical encoder (2000 ppr) coupled to the steering wheel axis through a 1:1 toothed gear (Figure 5). Both the DC motor and gearbox were calculated to deliver the required torque to the axis of the bus steering wheel. The optical encoder provides the angular steering wheel value to the PID digital controller in charge of the lateral control. Figure 5 serves to illustrate this assembly.

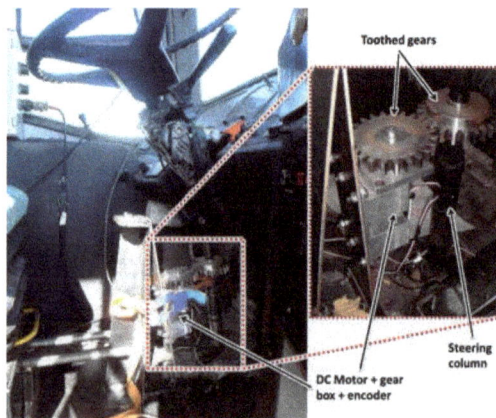

Figure 5. Actuator assembly for driving the steering wheel.

The steering control system has been designed to allow input from different location information sources, the idea being to permit the exploration of using different sensors to assess the accomplishment of the desired trajectory for the bus. Moreover, as will be explained in more detail in Section 2.6, when dealing with the control architecture, it will be possible to employ several location sensors: GPS, magnetometer systems for magnetic markers' detection (hidden magnets buried along the desired bus path), odometer, computer vision and LIDAR. Depending on the chosen sensor, a processing algorithm is needed to transform the supplied data into a reference command signal to the steering control system so that the bus will be able to track the desired trajectory. The experimental setup prepared for this research has been configured to permit the simultaneous use of different, heterogeneous, location sensors, and so, an intelligent combination of sensors (i.e., GPS and computer vision) could be chosen to investigate their complementarity and advantages, normally requiring additional data fusion algorithms.

Regarding the desired trajectory, there are many possible ways to determine it, like manually driving the bus and recording the GPS coordinates, navigation with computer vision or LIDAR and logging data with the odometer or GPS, or just in real time with a magnetometers array looking for (detecting) hidden magnets on the path, among others. In Figure 6, the flowchart for the steering control implemented in this work is shown.

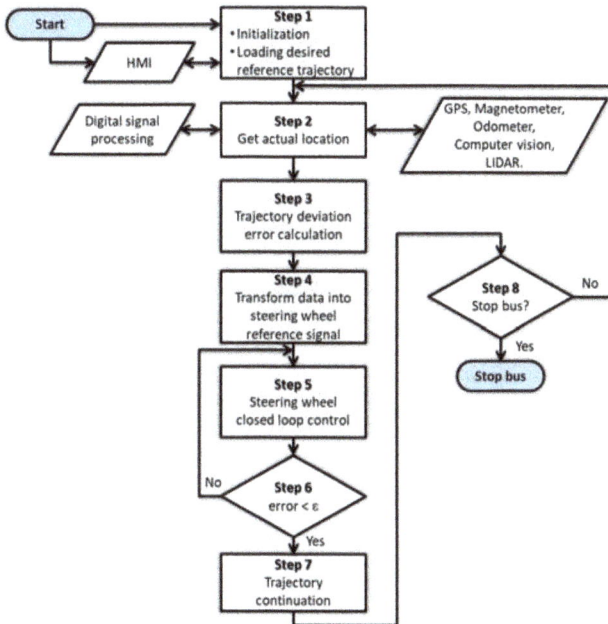

Figure 6. Implemented flowchart for steering control.

2.4. Obstacle Detection

The development of a practical obstacle detection and safety system for vehicles often involves working in complex scenarios. The experimental platform includes two LIDAR systems, the LMS-221 and LMS-291. Both lasers (electronic device based) provide non-contact measurement of the surrounding environment in the form of a cloud of 2D data points. This experimental platform is completed by a catadioptric omnidirectional stereovision system [59,60], which together with the LIDAR make up a reliable obstacle detection system, and it could be useful for other related tasks.

The chosen system is designed to work in harsh conditions (rain, snow, fog, warm or cold weather), mainly using the LMS-221. Some of the applications are: objects' measurement, positioning with respect to other vehicles and obstacle detection (in particular, pedestrian detection). The LIDAR can be used also to calculate the slope of the road, which can help the control system with additional useful information.

Figure 7a shows the LIDAR system installed at the bus front, at a height of 930 mm and 1490 mm above the soil plane. The supporting structure is mechanically adjustable to permit several angles of inclination for these sensors. Figure 7b shows the omnidirectional stereo tracking system, which has been attached to the LIDAR system, with the aim of being used in some experimental trials to evaluate the efficiency of the obstacle detection process. With a range of angular action of 180 degrees and its real-time capabilities, the LIDAR can be used to detect animate or inanimate objects. This system can be used for the bus's own safety and that of its passengers in the case of finding inanimate obstacles close to the bus front and/or for the safety of a moving object crossing the bus path.

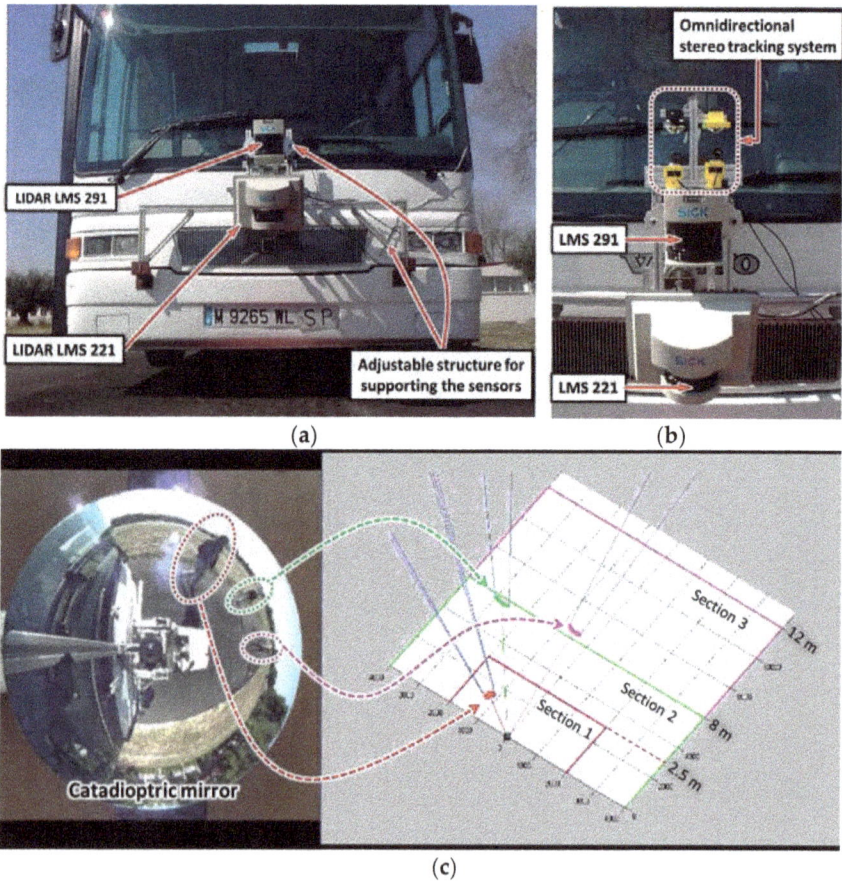

Figure 7. (a) LIDAR sensors placed at the bus front; (b) omnidirectional stereo tracking system; (c) dynamic obstacle detection (more critical close up (Section 1), intermediate (Section 2) and less critical (Section 3)).

In order to test the capabilities of the experimental vehicle, a simplified object classification method is proposed, where the sensory system classifies the sensed scenario into two groups: sections with obstacles and safe path lane areas. The combination of LIDAR systems and cameras is widely used because of its complementary capabilities, and several techniques have been proposed for data fusion [61,62]. In these works, a standard chessboard was utilized to calculate the transformation matrix between sensors. To compute obstacle detection, the data measurements from lasers are split into several subsets/segments by using the jump distance condition. Then, the subsets are classified into one of the two-defined groups by analyzing the geometrical properties of the segments such as the average deviations from the median and the jump distance to each succeeding segment, as described in [63]. The laser segments detected as obstacles are considered as inputs for computing disparities of the regions of interest of the omnidirectional images.

Obstacles detected along the vehicle trajectory are sorted into the three warning sections (see Figure 7c). This classification relies on the distance and position between obstacles and the vehicle. When objects are placed in Section 3 or in Section 2, the vehicle decreases its velocity to 40% or 80%, respectively. In that way, a safety distance of 8–12 m could be achieved. However, when an object is detected within Section 1 (2.5 m), a full stop is carried out. Movement is resumed automatically when the obstacle disappears. Due to the vehicle characteristics, an avoidance maneuver was not considered. Many experimental tests were carried out with success.

2.5. Software and HMI

To speed-up the research of the automatic control system for the articulated bus and to monitor its performance during the experiments, it is convenient to incorporate a Human Machine Interface (HMI). In this section the main characteristics of this HMI will be described. The environmental constraints are considered a key point for designing this HMI. First, the mobile system is a long vehicle, with multiple variables, working in tough conditions under a real-time operation system. Because it must cover large displacements, the operator's safety (and comfort whenever possible) is a priority. The HMI must display comprehensible and significant information, and the real-time requirements must be satisfied.

The main purpose of the HMI is to establish friendly, remote and intuitive communications between the vehicle and the people who operate it, and it also includes an adaptable and open configuration. Thus, the HMI contains the following properties:

- Multiple sensors and variables can be visualized; two and three dimensions are considered for some sensors.
- Real-time visualization of the information.
- Multiple and simultaneous clients connection can be accepted.
- Adaptable according to the requirements/configuration of every client.
- Data storage for every testing session.
- The old session can be visualized: offline mode.
- The TCP/IP communication protocol has been used to communicate between the vehicle and the users (researchers).

The HMI client-server architecture is based on a server running on-board the vehicle, operated on the RTOS (real-time operating system) named QNX® [64] and a client that can run on any PC and that has been developed in MATLAB® [65]. A simultaneous connection of several clients, from $i = 1 \ldots n$, is possible, where each client is denoted as the i-client. Figure 8 shows the general diagram of the remote system communication between the server and the n-clients.

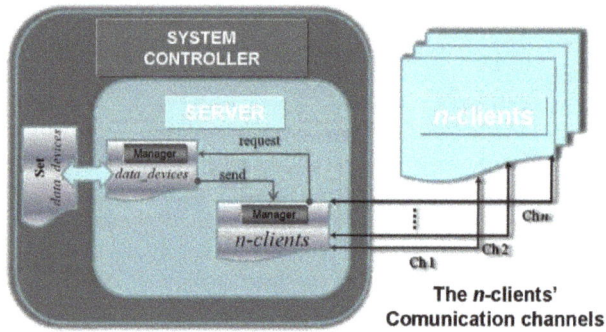

Figure 8. General diagram of the remote system communication.

The server consists of a "Clients Manager", whose purpose is to create and control the communication channels with every connected *i*-client. It also has to handle the clients' requests considering the clients' initial configuration. In order to communicate between the server and the main controller, an interface has been created called "data_devices Manager", which is in charge of acquiring the information of the active devices and variables of the system.

The *i*-client consists of a graphical interface and a "Configuration_Profile Manager". The graphical environment is intended to display the multiple variables of the vehicle; the variables are classified and positioned in four groups or modules (see Figure 9). Module #1 displays the variables strongly associated with the steering control: the steering wheel position (angle) and the trajectory deviation angle. Module #2 shows the variables related to vehicle movement, i.e., the velocity and the interpretation of the brake and accelerator pedals. Module #3 is used for the depiction of the information acquired by the range sensor (LIDAR), whether it is displaying the data in 2D or 3D. Module #4 shows the bus trajectory. The Configuration_Profile Manager is in charge of supervising the execution of the *i*-client, according to the user's requirements.

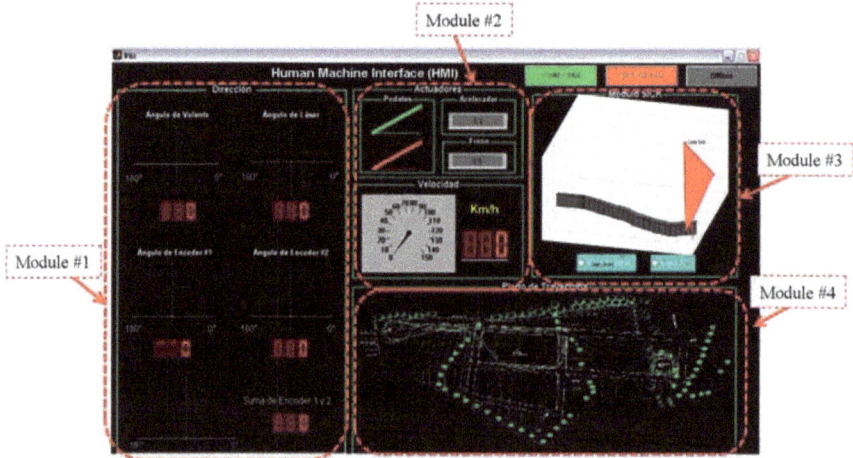

Figure 9. HMI graphical environment.

The remote communication begins by means of any *i*-client. The first action is the *i*-client identification; then, it sends its initial configuration profile, previously established, where the user chooses the variables of the system to be monitored. Once the configuration has been confirmed, the server starts the data sending procedure. This process will be performed until one of the linked parts breaks the connection. Due to the communication system being designed to operate with multiple and simultaneous clients/users, the server performance is not affected when an *i*-client is disconnected or a new *i*-client is connected. The Clients Manager creates an independent channel for every *i*-client, and when the server is disconnected, it sends a shutting down signal to all connected clients. The communication experiments were carried out at the CSIC premises where the network and the data traffic were monitored by the IT services. An optimal wireless connection for the on-board and the client PCs was created with the combination of an access point and range extender devices. In that way, the maximum bandwidth of the network was available. Although a detailed scaling analysis was not performed, because it was not needed for the experiments, there was a testing of the communication channels and the multiple-thread safety situation, and the maximum number of clients connected at the same time was ten. With ten connected clients, no significant problem regarding the server and the channel resources was detected. Figure 10 shows the HMI on board the bus during a test on the CSIC inner road.

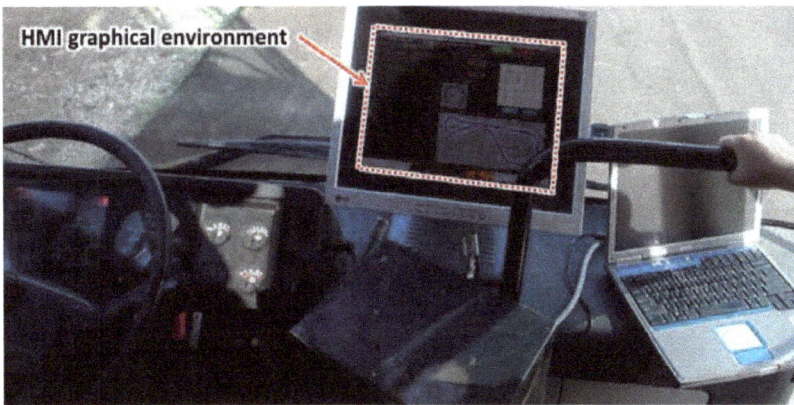

Figure 10. HMI graphical environment on board the bus during a test on the CSIC inner road.

2.6. System Architecture

In order to achieve great flexibility in the experimental platform, the reference trajectory for the lateral control might be provided by different systems that could be employed as a single source of reference points or to investigate a suitable combination of sensors and sensor fusion algorithms: GPS, computer vision (B&W, RGB, stereo, omnidirectional), LIDAR, TOF, magnetometers detecting hidden magnets and odometers whose data are provided by encoders. Some of these sensors could be used also for longitudinal control (i.e., to detect obstacles like pedestrians with computer vision or LIDAR, to locate and identify traffic signals, to determine distance and velocity approach to other vehicles, etc.).

The source data for the bus's desired trajectory are processed by the control system, and specific filtering is applied depending on the selected source. The overall system architecture is illustrated in Figure 11, and a general view of the installed instrumentation inside the bus is shown in Figure 12.

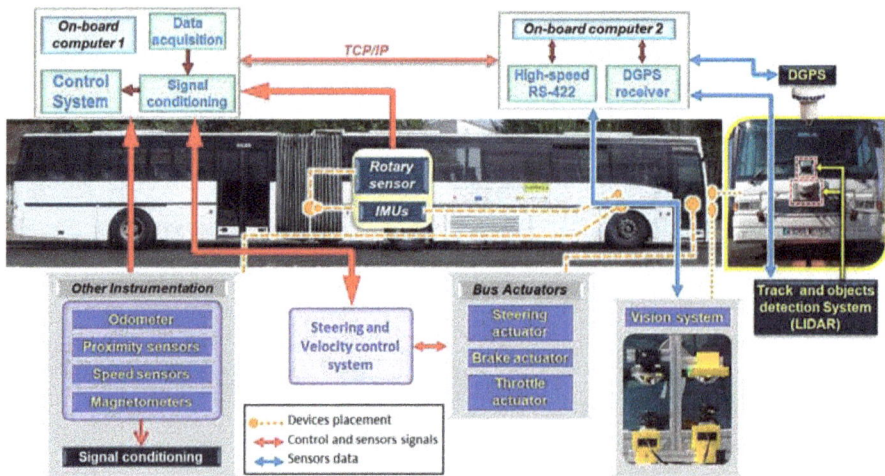

Figure 11. The architecture of the experimental platform.

Figure 12. On-board instrumentation and control system of the experimental platform.

3. Results and Discussion

The automatic control of an articulated bus, presents, at a high level, some features similar to other automatic vehicles, mainly regarding instrumentation and communications [66–69]. However, from the point of view of practical implementation and low-level control, there are some challenging differences. One major difference is inertia, which means greater power is involved and, so, more demanding control and safety measures must be applied, those being of special relevance when coping with down slopes. As a consequence, braking strategies and accelerations (longitudinal, lateral) are now of greater relevance with respect to those of cars or small buses as is the case of the European project CityMobil2 [70]. A second key difference with automatic cars or small buses [66] is the very long length of an articulated bus (about 18 m), which, adding the position of the steering wheel ahead of the front wheels (the bus driver is seated ahead of the front wheels), makes turning maneuvers very cumbersome (in fact, during our experiments, one of the most "striking" sensations was when approaching a curve and the steering wheel did not turn for a while, then it started to turn when the front wheels where required to follow the curve). Furthermore, because of the bus length and the presence of its central articulation, the trajectories of the front, middle and rear wheels present more differences than in the case of a shorter vehicle [66], where practically the rear wheels are able to track the front wheels' trajectory [27]. This intrinsic articulated bus characteristic requires special attention when calculating the required steering angles for the desired bus trajectory.

The longitudinal control of the articulated bus is carried out so that the vehicle will be able to move ahead of the test lane at a desired speed. The speed used in most experiments was relatively slow due to safety reasons and to the intricacy of the trajectory (please refer to Figure 1) intended more for checking control performance than to go very fast. The average speed in the experiments was ranging from 10 km/h to 25 km/h. However, in the straight section of the path, in some controlled experiments, the speed went up to 60 km/h.

Figure 13 shows the longitudinal and lateral control scheme implemented in the bus Volvo B10M ART-RA-IN. Regarding longitudinal control, the system acts on the PotBox of the bus, which has an action range of 0 to 5 DC volts, rather than acting on the throttle, because the control over acceleration/speed of the bus is better in this case and more controllable. The acceleration is limited (saturated) in order to protect the PotBox and not to exceed the desired speed over limits that cannot be controlled, according to the experimental tests.

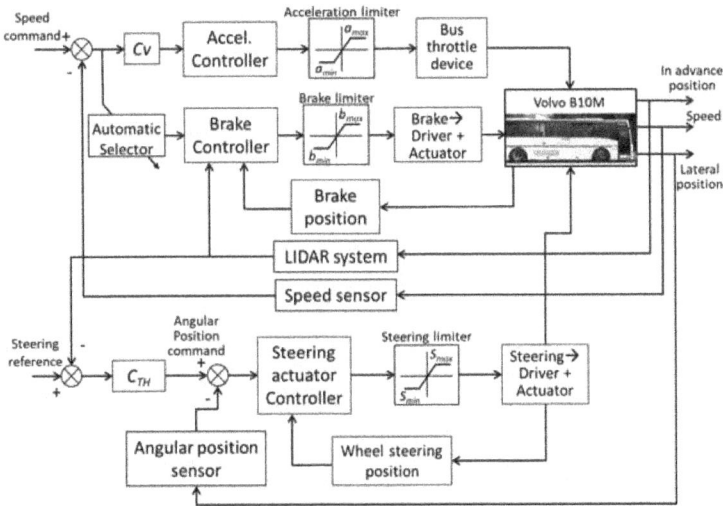

Figure 13. Longitudinal and lateral control scheme.

When the bus exceeds the desired speed, then the acceleration controller stops and selects the brake controller. The brake controller has minimum and maximum limits, which determine the workspace for this controller. Within these limits, the brake controller generates a saw-tooth signal to act on the brake actuator, which consists of a DC servomotor (see Figure 3). This makes the brake subsystem perform stopping of the bus as a gentle deceleration. When the speed decreases to a pre-set threshold, the brake subsystem stops acting and starts the accelerator subsystem. The bus speed is measured by a speed sensor whose resolution is greater than that offered by the original bus tachometer.

Regarding lateral control, an electric actuator composed of a DC motor, gearbox and encoder has been installed on the steering column (see Figure 5). The actuator receives commands from the lateral control system, so that the vehicle will track the desired trajectory. It takes into account the steering limit, which in this case is ±900°. At the same time, it makes the axle of the front wheel range between ±45°. The feedback to the lateral control system is obtained by means of actual (lateral) position information in the form of the angular position (obtained from the different sensing possibilities: GPS, magnetometers, encoders, etc.), which is compared with the angular position command.

The results of longitudinal control for one of the experiments can be noticed in Figure 14, where a comparison among the speed of the vehicle, the position of the brake and the acceleration along the longitudinal axis (bus forward motion) obtained through an IMU installed on the front axle is presented. In this experiment and during most of the trajectory, the commanded speed was about 10 km/h, because the test lane contains small radii of curvature and slopes. In the last trajectory segment of the test lane (straight line), the speed was increased up to 20 km/h. It can be noticed in Figure 14 that 25 s after the start, the brake system applies a brake signal to check if the system is working properly. Then, the main application of the brake controller is to regulate the speed commanded mainly during the downward slope section. This takes place between the times 125 and 300 s, approximately. It can be noticed in Figure 14, for this period of time, that the brake signal (and correspondingly, the acceleration signal) shows an oscillation due to the chosen algorithm for the brake system, where a command signal in the shape of a saw-tooth is entered into the brake controller, which acts as a Proportional-Integral (PI) action and provides a comfortable sensation of very near constant speed, as should be expected.

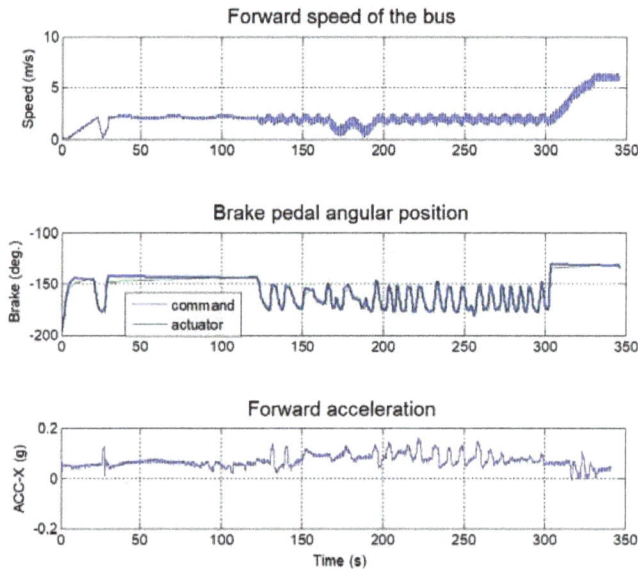

Figure 14. Comparison among bus speed, brake position and x-acceleration during the tracking trajectory. Brake control is applied 25 s after the start and from the times 125 to 300 s to cope with a downslope in the circuit.

Figure 15 shows the experimental results obtained by applying the lateral control of the bus for tracking the required trajectory. This figure shows the steering command, the lateral acceleration of the bus and the angular rate on the z-axis (azimuth). The angular rate sensor follows the curves of the trajectory commanded by the bus steering. The azimuthal speeds have been measured at two different points; the first, on the bus front axle, and the second, on the rotatory junction of the bus, in order to know both characteristics of its behavior. One of them is to know the position of the last section with respect to the first section of the bus. The lateral acceleration signal can be used in a feedback loop lateral control to improve the tracking of the trajectory. Using this signal in the control system, small lateral movements will be diminished.

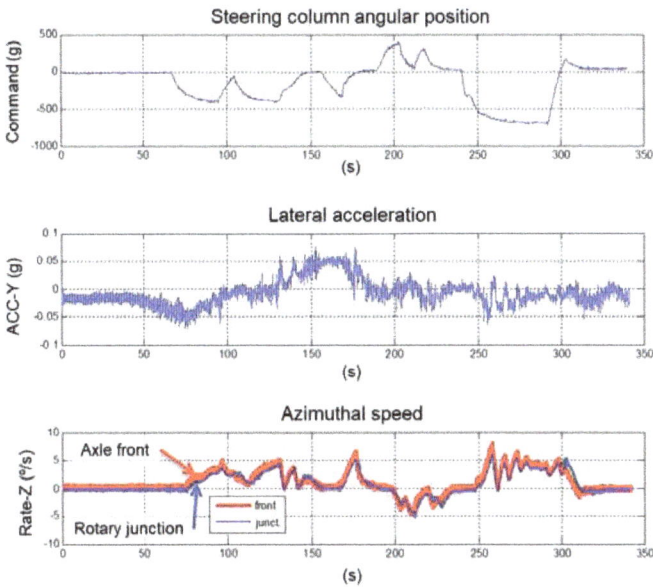

Figure 15. Comparison among steering command, lateral acceleration and angular rate in the z-axis during the tracking of the trajectory.

The lateral error during the trajectory tracking was relatively small and within the safety margins established in the control strategy. Figure 16 shows that the maximum peaks of the lateral error were approximately 0.35 m, these peaks being located in the smallest curvature radii sections, where the bus front is more separated from the theoretical trajectory to precisely cope with the curvature. During most of the trajectory, the lateral error was about or less than 0.10 m, which is acceptable for a bus of these dimensions.

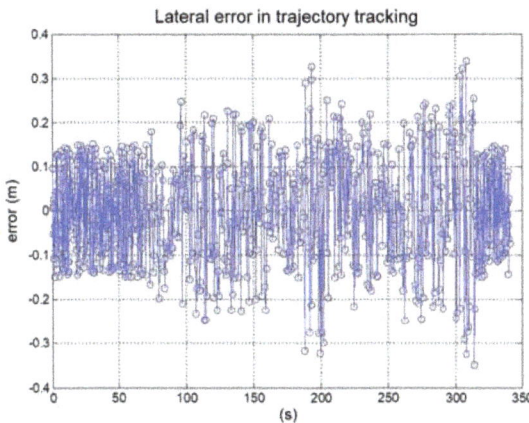

Figure 16. Lateral error during the tracking of the test lane.

Many experiments have been performed within the CSIC experimental test area. Figure 17 illustrates the automatic control of the articulated bus following a required trajectory. The steering wheel angular rotation experimental data are shown as the correlation with the real trajectory. In this way, many longitudinal and lateral control algorithms were compared using this experimental platform, where reliable external references, sensors and electronic instrumentation have been demonstrated to be very useful. Figure 18 shows the steering wheel in automatic mode in several positions (video snapshots) when the bus is passing by one curved section of the test lane, during one of the experimental proofs.

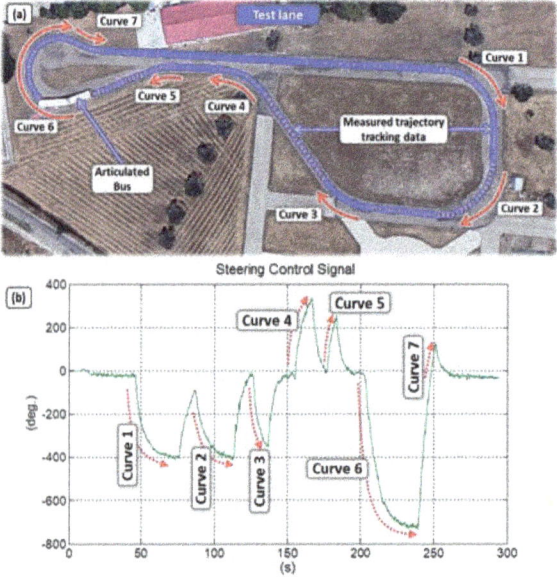

Figure 17. (**a**) Test lane and trajectory at the CSIC; (**b**) steering wheel angular value during an experiment conducted on the test lane.

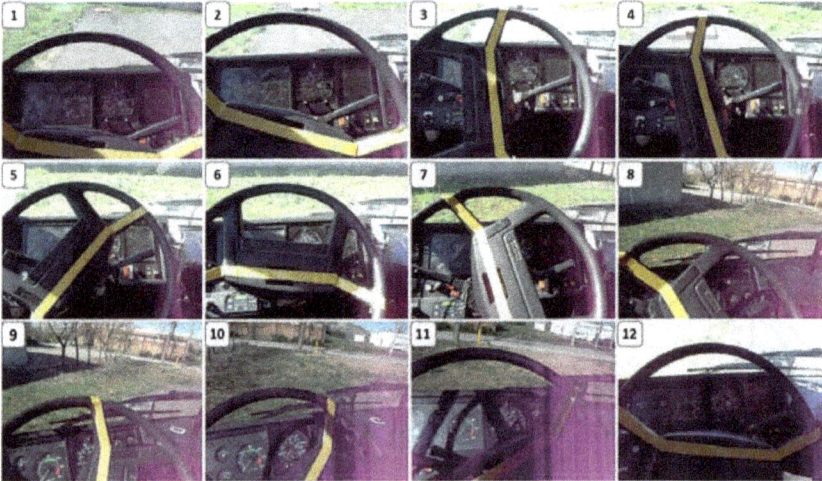

Figure 18. Details of the steering wheel during one curved section of the experimental tests.

4. Conclusions

In this paper, an experimental platform for research on the automatic control of an articulated bus aiming to allow full experimentation in real conditions for testing technological developments and control algorithms in the field of Intelligent Transportation Systems (ITS) has been introduced. A short review of the state of the art in ITS served as a motivation for the undertaken work, and it was followed by the presentation of the experimental platform, which consisted of a mobile component (a commercial articulated bus fully instrumented) and of a ground test area composed of asphalt roads inside the CSIC premises in Madrid. A robust information acquisition architecture composed of a heterogeneous and complementary set of sensors was implemented and integrated on the platform in order to provide real-time monitoring and control of the multiple variables involved in the autonomous operation of the articulated bus. Details regarding instrumentation and the implementation of bus velocity and steering control, as well as obstacle detection and a human machine interface developed to ease progress in the control system evaluation were provided.

Experimental testing of longitudinal and lateral control approaches were also presented and illustrated with the results of many tests performed within the experimental area. Experimental results exhibit that the proposed velocity control, which is based on a PID closed-loop controller, is capable of providing automatic regulation of the velocity independently of the road slope and the demanding layout of the bus track. On the other hand, the steering control was shown to keep the lateral error below 0.10 m during the trajectory tracking experiments, which is quite acceptable for a bus of these dimensions. The safety system was confirmed to be effective in detecting dynamic obstacles and sorting them into three predefined warning sections. This strategy allows the articulated bus either to stop or to reduce its velocity depending on the distance between the obstacle and the vehicle. Finally, the HMI proved to be a friendly and intuitive means of communication between the human operator and the controlled articulated bus. The adaptable and open configuration of the HMI enables the real-time control of the system, as well as the real-time visualization of the multiple variables of interest acquired by the sensors.

In future works, we will implement and evaluate different controllers for this experimental platform in order to compare them in order to reach new conclusions.

Acknowledgments: The authors acknowledge partial funding of this research under: IMADE (Instituto Madrileño de Desarrollo) PIE/62/2008 (Comunidad de Madrid), RoboCity2030 (Phase III) S-2013/MIT-2748, Comunidad de Madrid, and ROBSEN (Robotica y Sensores para los Retos Sociales) PI-201650E050, CSIC. Héctor Montes acknowledges support from Universidad Tecnológica de Panamá. Roemi Fernández acknowledges the financial support from the Ministry of Economy, Industry and Competitiveness under the Ramón y Cajal Programme.

Author Contributions: The work presented in this article was carried out in cooperation with all authors. Hector Montes designed and wrote the manuscript in collaboration with Manuel Armada. Hector Montes, Carlota Salinas, Roemi Fernández and Manuel Armada conceived of and designed the experiments. Héctor Montes and Carlota Salinas performed the experiments for autonomous control of the bus and for data acquisition. Carlota Salinas, Roemi Fernández and Héctor Montes processed and analyzed the data and drew the main conclusions. Héctor Montes implemented the architecture of the experimental platform. Carlota Salinas implemented the HMI. All authors contributed to the review of the manuscript.

Conflicts of Interest: The authors declare no conflict of interest.

References

1. Shladover, S.E.; Desoer, C.A.; Hedrick, J.K.; Tomizuka, M.; Walrand, J.; Zhang, W.-B.; McMahon, D.H.; Peng, H.; Sheikholeslam, S.; McKeown, N. Automatic vehicle control developments in the path program. *IEEE Trans. Veh. Technol.* **1991**, *40*, 114–130. [CrossRef]
2. Fenton, R.; Mayhan, R. Automated highway studies at the Ohio State University-an overview. *IEEE Trans. Veh. Technol.* **1991**, *40*, 100–113. [CrossRef]
3. Collier, W.C.; Weiland, R.J. Smart cars-smart highways. *IEEE Spectr.* **1994**, *31*, 27–33. [CrossRef]
4. Stanton, N.A.; Young, M.; McCaulder, B. Drive-by-Wire: The case of driver workload and reclaiming control with adaptive cruise control. *Saf. Sci.* **1997**, *27*, 149–159. [CrossRef]

5. Naranjo, J.E.; González, C.; García, R.; de Pedro, T. Cooperative Throttle and Brake Fuzzy Control for ACC+Stop&Go Maneuvers. *IEEE Trans. Veh. Technol.* **2007**, *56*, 1623–1630. [CrossRef]

6. Naranjo, J.E.; González, C.; García, R.; de Pedro, T. Lane-Change Fuzzy Control in Autonomous Vehicles for the Overtaking Maneuver. *IEEE Trans. Intell. Transp. Syst.* **2008**, *9*, 438–450. [CrossRef]

7. Bishop, R. A survey of intelligent vehicle applications worldwide. In Proceedings of the IEEE Intelligent Vehicles Symposium, Dearborn, MI, USA, 3–5 October 2000; pp. 25–30.

8. Yoshioka, L.R.; Marte, C.L.; Micoski, M.; Costa, R.D.; Fontana, C.; Sakurai, C.A.; Cardoso, J.R. Bus Corridor Operational Improvement with Intelligent Transportation System based on Autonomous Guidance and Precision Docking. *Int. J. Syst. Appl. Eng. Dev.* **2014**, *8*, 116–123.

9. Hidalgo, D.; Muñoz, J.C. A review of technological improvements in bus rapid transit (BRT) and buses with high level of service (BHLS). *Public Transp.* **2014**, *6*, 185–213. [CrossRef]

10. Mejía-Dugand, S.; Hjelm, O.; Baas, L.; Ríos, R.A. Lessons from the spread of bus rapid transit in Latin America. *J. Clean. Prod.* **2013**, *50*, 82–90. [CrossRef]

11. Zeng, H. China Transportation Briefing: Bus Rapid Transit in China—On the Way. TheCityFix. Available online: http://thecityfix.com/blog/bus-rapid-transit-brt-china-transportation-briefing-series-guangzhou-beijing-heshuang-zeng/ (accessed on 26 October2017).

12. Cervero, R.; Dai, D. BRT TOD: Leveraging transit oriented development with bus rapid transit investments. *Transp. Policy* **2014**, *36*, 127–138. [CrossRef]

13. Wright, L.; Hook, W. Introduction. In *Bus Rapid Transit Planning Guide*, 3rd ed.; Institute of Transportation & Development: New York, NY, USA, 2007; pp. 10–33.

14. Montes, H.; Salinas, C.; Sarria, J.; Armada, M. An Experimental Platform for Research on Automatic Control of Articulated Bus. In Proceedings of the IARP Workshop on Service Robotics and Nanorobotics, Beijing, China, 28–29 October 2009.

15. Salinas, C.; Montes, H.; Armada, M. A perception system for accurate automatic control of an articulated bus. In Proceedings of the 13th International Conference of Climbing and Walking Robots, CLAWAR 2010, Nagoya, Japan, 31 August–3 September 2010; pp. 1021–1028.

16. Le-Anh, T.; De Koster, M.B.M. A review of design and control of automated guided vehicle systems. *Eur. J. Oper. Res.* **2006**, *171*, 1–23. [CrossRef]

17. Ackermann, J.; Guldner, J.; Siegel, W.; Steinhauser, R.; Utkin, V.I. Linear and Nonlinear Controller Design for Robust Automatic Steering. *IEEE Trans. Control Syst. Technol.* **1995**, *3*, 132–143. [CrossRef]

18. Campion, G.; Bastin, G.; D'Andrea-Novel, B. Structural Properties and Classification of Kinematic and Dynamic Models of Wheeled Mobile Robots. *IEEE Trans. Robot. Autom.* **1996**, *12*, 47–62. [CrossRef]

19. Kodagoda, K.R.S.; Wijesoma, W.S.; Teoh, E.K. Fuzzy Speed and Steering Control of an AGV. *IEEE Trans. Control Syst. Technol.* **2002**, *10*, 112–120. [CrossRef]

20. Falcone, P.; Borrelli, F.; Asgari, J.; Tseng, H.E.; Hrovat, D. Predictive Active Steering Control for Autonomous Vehicle Systems. *IEEE Trans. Control Syst. Technol.* **2007**, *15*, 566–580. [CrossRef]

21. Fernández, C.; Domínguez, R.; Fernández-Llorca, D.; Alonso, J.; Sotelo, M.A. Autonomous navigation and obstacle avoidance of a micro-bus. *Int. J. Adv. Robot. Syst.* **2013**, *10*, 1–9. [CrossRef]

22. Fenton, R.; Melocik, G.; Olson, K. On the steering of automated vehicles: Theory and experiment. *IEEE Trans. Autom. Control* **1976**, *21*, 306–315. [CrossRef]

23. Netto, M.S.; Chaib, S.; Mammar, S. Lateral adaptive control for vehicle lane keeping. In Proceedings of the American Control Conference, Boston, MA, USA, 30 June–2 July 2004; pp. 2693–2698.

24. Mammar, S.; Glaser, S.; Netto, M. Vehicle Lateral Dynamics Estimation using Unknown Input Proportional-Integral Observers. In Proceedings of the American Control Conference, Minneapolis, MN, USA, 14–16 June 2006; pp. 14–16.

25. Chen, C.; Guldner, J.; Kanellakopoulos, I.; Tomizuka, M. Nonlinear Damping in Vehicle Lateral Control: Theory and Experiment. In Proceedings of the American Control Conference, Philadelphia, PA, USA, 26 June 1998; pp. 2243–2247.

26. Linderoth, M.; Soltesz, K.; Murray, R.M. Nonlinear Lateral Control Strategy for Nonholonomic Vehicles. In Proceedings of the American Control Conference, Seattle, WA, USA, 11–13 June 2008; pp. 3219–3224.

27. DARPA Urban Challenge. Available online: http://archive.darpa.mil/grandchallenge/ (accessed on 6 September 2017).

28. Kammel, S.; Ziegler, J.; Pitzer, B.; Werling, M.; Gindele, T.; Jagzent, D.; Schröder, J.; Thuy, M.; Goebl, M.; von Hundelshausen, F.; et al. Team AnnieWAY's Autonomous System for the 2007 DARPA Urban Challenge. *J. Field Robot.* **2008**, *25*, 615–639. [CrossRef]

29. Sotelo, M.A. Lateral control strategy for autonomous steering of Ackerman-like vehicles. *Robot. Auton. Syst.* **2003**, *45*, 223–233. [CrossRef]

30. Hessburg, T.; Peng, H.; Tomizuka, M.; Zhang, W.B.; Kamei, E. An Experimental Study on Lateral Control of a Vehicle. In Proceedings of the American Control Conference, Boston, MA, USA, 26–28 June 1991; pp. 3084–3089.

31. Tan, H.S.; Bougler, B. Vehicle Lateral Warning, Guidance and Control Based on Magnetic Markers: PATH Report of AHSRA Smart Cruise 21 Proving Tests. In *California Partners for Advanced Transit and Highways (PATH)*; UC Berkeley: California Partners for Advanced Transportation Technology; UC Berkeley: Berkeley, CA, USA, 2001; pp. 1–71. Available online: http://escholarship.org/uc/item/3jb3r4p5 (accessed on 7 September 2017).

32. Xu, H.G.; Wang, C.X.; Yang, R.Q.; Yang, M. Extended Kalman Filter Based Magnetic Guidance for Intelligent Vehicles. In Proceedings of the IEEE Intelligent Vehicles Symposium, Tokyo, Japan, 13–15 June 2006; pp. 169–175.

33. Chen, C.; Tomizuka, M. Lateral Control of Tractor-Semitrailers for Automated Highway Systems. In *California Partners for Advanced Transit and Highways (PATH)*; UC Berkeley: California Partners for Advanced Transportation Technology; UC Berkeley: Berkeley, CA, USA, 1996; pp. 1–29. Available online: http://escholarship.org/uc/item/5235j21k (accessed on 7 September 2017).

34. Hingwe, P.; Wang, J.Y.; Tai, M.; Tomizuka, M. Lateral Control of Heavy Duty Vehicles for Automated Highway System: Experimental Study on a Tractor Semi-trailer. In *California Partners for Advanced Transit and Highways (PATH)*; UC Berkeley: California Partners for Advanced Transportation Technology; UC Berkeley: Berkeley, CA, USA, 2000; pp. 1–47. Available online: http://escholarship.org/uc/item/9jj235kx (accessed on 7 September 2017).

35. Tai, M.; Tomizuka, M. Robust Lateral Control of Heavy Duty Vehicles: Final Report. In *California Partners for Advanced Transit and Highways (PATH)*; UC Berkeley: California Partners for Advanced Transportation Technology; UC Berkeley: Berkeley, CA, USA, 2003; pp. 1–80. Available online: http://escholarship.org/uc/item/8j2692w0 (accessed on 7 September 2017).

36. Chen, C. Backstepping Design of Nonlinear Systems and Its Application to Vehicle Lateral Control in Automated Highway Systems. Ph.D. Thesis, University of California Berkeley, Berkeley, CA, USA, 1996.

37. Fernández, R.; Aracil, R.; Armada, M. Control de tracción en robots móviles con ruedas. *Rev. Iberoam. Autom. Inf. Ind.* **2012**, *9*, 393–405. [CrossRef]

38. Kiencke, U.; Daiß, A. Observation of lateral vehicle dynamics. *Control Eng. Pract.* **1997**, *5*, 1145–1150. [CrossRef]

39. Lin, C-F.; Ulsoy, A.G.; LeBlanc, D.J. Vehicle Dynamics and External Disturbance Estimation for Vehicle Path Prediction. *IEEE Trans. Control Syst. Technol.* **2000**, *8*, 508–518. [CrossRef]

40. Gerdes, J.C.; Rossetter, E.J. A unified approach to driver assistance systems based on artificial potential fields. In Proceedings of the ASME International Mechanical Engineering Congress and Exposition, Nashville, TN, USA, 14–19 November 1999; pp. 431–438.

41. Talbot, C.M.; Papadimitriou, I.; Tomizuka, M. Fault Tolerant Autonomous Lateral Control for Heavy Vehicles. In *California Partners for Advanced Transit and Highways (PATH)*; UC Berkeley: California Partners for Advanced Transportation Technology; UC Berkeley: Berkeley, CA, USA, 2004; pp. 1–69. Available online: http://escholarship.org/uc/item/7xd2r0cc (accessed on 7 September 2017).

42. Mcmahon, D.H.; Hedrick, J.K. Longitudinal Model Development for Automated Roadway Vehicles. In *California Partners for Advanced Transit and Highways (PATH)*; UC Berkeley: California Partners for Advanced Transportation Technology; UC Berkeley: Berkeley, CA, USA, 1989; pp. 1–69. Available online: http://escholarship.org/uc/item/4746j7jj (accessed on 7 September 2017).

43. Tan, Y.; Kanellakopoulos, I. Longitudinal Control of Commercial Heavy Vehicles: Experimental Implementation. In *California Partners for Advanced Transit and Highways (PATH)*; UC Berkeley: California Partners for Advanced Transportation Technology; UC Berkeley: Berkeley, CA, USA, 2002; pp. 1–27. Available online: https://trid.trb.org/view.aspx?id=726231 (accessed on 7 September 2017).

44. Kim, S.; Song, B.; Song, H. Integrated Fault Detection and Diagnosis System for Longitudinal Control of an Autonomous All-Terrain Vehicle (ATV). *Int. J. Autom. Technol.* **2009**, *10*, 505–512. [CrossRef]

45. Lim, E.H.M.; Hedrick, J.K. Lateral and Longitudinal Vehicle Control Coupling for Automated Vehicle Operation. In Proceedings of the American Control Conference, San Diego, CA, USA, 2–4 June 1999; pp. 3676–3680.

46. Tan, H.S.; Huang, J. Design of a high-performance automatic steering controller for bus revenue service based on how drivers steer. *IEEE Trans. Robot.* **2014**, *30*, 1137–1147. [CrossRef]

47. Huang, J.; Tan, H.S. Control System Design of an Automated Bus in Revenue Service. *IEEE Trans. Intell. Transp. Syst.* **2016**, *17*, 2868–2878. [CrossRef]

48. Lefèvre, S.; Carvalho, A.; Borrelli, F. A learning-based framework for velocity control in autonomous driving. *IEEE Trans. Autom. Sci. Eng.* **2016**, *13*, 32–42. [CrossRef]

49. Justino, J.C.; da Silva, L.A.; Rocha, A.; Cardoso Filho, B.D.J. Aspects of the operation of regular ultra fast charging e-Bus in high grade BRT routes. In Proceedings of the IECON 40th Annual Conference of the IEEE Industrial Electronics Society, Dallas, TX, USA, 29 October–1 November 2014; pp. 3101–3107.

50. Baber, J.; Kolodko, J.; Noël, T.; Parent, M.; Vlacic, L. Cooperative Autonomous Driving. Intelligent Vehicles Sharing City Roads. *IEEE Robot. Autom. Mag.* **2005**, *12*, 44–49. [CrossRef]

51. Jiménez, F.; Clavijo, M.; Naranjo, J.E.; Gómez, O. Improving the Lane Reference Detection for Autonomous Road Vehicle Control. *J. Sens.* **2016**, *2016*. [CrossRef]

52. González, D.; Pérez, J.; Milanés, V. Parametric-based path generation for automated vehicles at roundabouts. *Exp. Syst. Appl.* **2017**, *71*, 332–341. [CrossRef]

53. Szadeczky-Kardoss, E.; Kiss, B. Path planning and tracking control for an automatic parking assist system. In *European Robotics Symposium 2008*; Springer Tracts in Advanced Robotics; Bruyninckx, H., Přeučil, L., Kulich, M., Eds.; Springer: Berlin, Germany, 2008; Volume 44, pp. 175–184. ISBN 978-3-540-78315-2.

54. Naranjo, J.E.; González, C.; Reviejo, J.; Garcia, R.; de Pedro, T. Adaptive fuzzy control for inter-vehicle gap keeping. *IEEE Trans. Intell. Transp. Syst.* **2003**, *4*, 132–142. [CrossRef]

55. Naranjo, J.E.; González, C.; García, R.; de Pedro, T. ACC + Stop&go maneuvers with throttle and brake fuzzy control. *IEEE Trans. Intell. Transp. Syst.* **2006**, *7*, 213–225. [CrossRef]

56. Naranjo, J.E.; Sotelo, M.A.; González, C.; García, R.; de Pedro, T. Using Fuzzy Logic in Automated Vehicle Control. *IEEE Intell. Syst.* **2007**, *22*, 36–45. [CrossRef]

57. Villalon-Sepulveda, G.; Torres-Torriti, M.; Flores-Calero, M. Traffic sign detection system for locating road intersections and braking advance. *Rev. Iberoam. Autom. Inf. Ind.* **2017**, *14*, 152–162. [CrossRef]

58. Fernandez, A.; Usamentiaga, R.; Casado, R. Automatic System to Detect Both Distraction and Drowsiness in Drivers Using Robust Visual Features. *Rev. Iberoam. Autom. Inf. Ind.* **2017**, *14*, 307–328. [CrossRef]

59. Salinas, C.; Montes, H.; Fernandez, G.; Gonzalez de Santos, P.; Armada, M. Catadioptric Panoramic Stereovision for Humanoid Robots. *Robotica* **2012**, *30*, 799–811. [CrossRef]

60. Fernández, R.; Salinas, C.; Montes, H.; Armada, M. Omnidirectional stereo tracking system for humanitarian demining training. In Proceedings of the 8th International Symposium "Humanitarian Demining 2011", Šibenik, Croacia, 26–28 April 2011; pp. 113–116.

61. Salinas, C. A Non-Feature Based Method for Automatic Image Registration Relying on Depth Dependent Planar Projective Transformations. Ph.D. Thesis, Universidad Complutense de Madrid, Madrid, Spain, 2015.

62. Li, Y.; Ruichek, Y.; Cappelle, C. 3D triangulation based extrinsic calibration between a stereo vision system and a LIDAR. In Proceedings of the 14th International IEEE Conference on Intelligent Transportation Systems (ITSC), Washington, DC, USA, 5–7 October 2011; pp. 797–802.

63. Mozos, O.M. *Semantic Place Labeling with Mobile Robots*; Springer Tracts in Advanced Robotics (STAR); Springer: Berlin, Germany, 2010; ISBN 978-3-642-11209-6.

64. QNX® Software System Ltd. QNX 6.4. 2008. Available online: http://www.qnx.com/download/ (accessed on 15 September 2017).

65. The MathWorks. Matlab®. 2013. Available online: https://es.mathworks.com/ (accessed on 15 September 2017).

66. FORTUNE. Available online: http://fortune.com/2016/12/13/google-self-driving-car-waymo-alphabet/ (accessed on 13 October 2017).

67. TESLA. Available online: https://www.tesla.com/ (accessed on 13 October 2017).

68. OLLI. Available online: http://meetolli.auto (accessed on 13 October 2017).

69. OTTO. Available online: https://www.ottomotors.com/ (accessed on 13 October 2017).
70. Alessandrini, A.; Cattivera, A.; Holguin, C.; Stam, D. CityMobil2: Challenges and Opportunities of Fully Automated Mobility. In *Road Vehicle Automation*; Meyer, G., Beiker, S., Eds.; Springer: Cham, Switzerland, 2014; pp. 169–184. ISBN 978-3-319-05989-1.

MDPI

St. Alban-Anlage 66

4052 Basel

Switzerland

Tel. +41 61 683 77 34

Fax +41 61 302 89 18

www.mdpi.com

Applied Sciences Editorial Office

E-mail: applsci@mdpi.com

www.mdpi.com/journal/applsci

www.ingramcontent.com/pod-product-compliance
Lightning Source LLC
Chambersburg PA
CBHW051844210326

41597CB00033B/5769